CLASSICAL AND QUANTUM GRAVITY

PROCEEDINGS OF THE
FIRST IBERIAN MEETING ON GRAVITY

CLASSICAL AND QUANTUM GRAVITY

Évora, Portugal 21 – 26 September 1992

Editors

M C Bento
O Bertolami
J M Mourão
R F Picken

Instituto Superior Técnico, Portugal

World Scientific
Singapore • New Jersey • London • Hong Kong

Published by

World Scientific Publishing Co Pte Ltd
5 Toh Tuck Link, Singapore 596224
USA office: 27 Warren Street, Suite 401-402, Hackensack, NJ 07601
UK office: 57 Shelton Street, Covent Garden, London WC2H 9HE

British Library Cataloguing-in-Publication Data
A catalogue record for this book is available from the British Library.

CLASSICAL AND QUANTUM GRAVITY

ISBN-13 978-981-02-1369-5
ISBN-10 981-02-1369-7

PREFACE

"Então, a Península Ibérica moveu-se

um pouco mais, um metro, dois metros,

a experimentar a força."

in

A Jangada de Pedra

José Saramago, 1986

The First Iberian Meeting on Gravity, which took place in Évora, Portugal, from September 21 to September 26, 1992, was the first of a series of meetings planned to be held every two years alternately in Portugal and Spain.

The meeting, which had as its venue the premises of Évora University (formerly a Divinity School of Jesuits), provided a unique opportunity for Iberian students and researchers in the field of gravity and related topics to interact and discuss the most important recent developments in the subject. This was achieved through a series of lectures delivered by internationally renowned researchers on topics which ranged from experimental gravity to quantum gravity, as well as a set of short communications presented by participants in the afternoon sessions.

The First Iberian Meeting on Gravity could not have taken place if it were not for the generous support provided to the organizing committee from many quarters. First of all, we owe a great debt of gratitude to the following sponsors: Junta Nacional de Investigação Científica e Tecnológica, Instituto Nacional de Investigação Científica, Instituto Superior Técnico, Grupo Teórico de Altas Energias, Câmara Municipal de Évora, Região de Turismo de Évora, Fundação Luso-Americana de Desenvolvimento, British Airways and Universidade de Évora. We were particularly gratified that the financial support received enabled us to assist some participants, in particular from the Community of Independent States and Latin America, who would not otherwise have been able to attend.

We would also like to thank Fundação Calouste Gulbenkian for contributing to the cost of publishing this volume and World Scientific Publishing Company for their patience and kindness throughout the process of publication of these proceedings.

Furthermore we would like to express our gratitude to our Spanish counterparts for accepting the challenge of organizing the Iberian Meetings on Gravity with us.

Mrs. Maria de Fátima Maia Loureiro deserves very great praise for her dedicated activity at the Meeting's various organisational fronts, and our very special thanks are due to Alfredo Barbosa Henriques, Ana Mourão, Paulo Moniz and Paulo Sá for their invaluable help, without which the First Iberian Meeting could not have taken place.

Last but very far from least we would like to thank all those who contributed to this volume, and, of course, all participants, for coming to Évora and helping to make the First Iberian Meeting such a success. We look forward to seeing you again at the next Meeting!

The Editors

Lisbon, 31st January 1993

CONTENTS

Classical Cosmology

Quantum Gravity and Quantum Cosmology

INVITED SPEAKERS

Quantum Geometry

Abhay Ashtekar
Physics Department, Syracuse University, Syracuse, NY 13244-1130, U.S.A.

Abstract

Using a non-perturbative approach to quantum general relativity, the nature of quantum geometry is probed. It is found that: i) While local operators such as the metric at a point may not be well-defined, there do exist *non-local* operators, such as the area of a given 2-surface, which can be regulated in a way that respects the diffeomorphism invariance of the theory and which are finite *without* renormalization; ii) There exist quantum states which approximate a given metric at large scales, but which exhibit a discrete structure of a specific type at the Planck scale. The non-perturbative framework also suggests that differential geometry which underlies classical general relativity will have to be transcended in the Planck regime and provides hints as to what could replace it.

1. Introduction

It is by now generally accepted that perturbative approaches to quantum gravity fail because they assume that space-time geometry can be approximated by a smooth continuum at all scales. What is needed are non-perturbative approaches which can *predict* –rather than assume– the true nature of the micro-structure of this geometry is. In such an approach, background fields such as a classical metric or a connection cannot play a fundamental role; quantum theory must be formulated in a diffeomorphism invariant fashion. An important task in these programs is then to introduce techniques needed to describe geometry and to "explain" from first principles how smooth geometries can arise on macroscopic scales.

Over the past five years, two avenues have been pursued to test if quantum general relativity can exist non-perturbatively. The first is based on numerical simulations [1], while the second is based on canonical quantization[2-7]. In this article, we will use the second method. While the canonical approach itself was introduced by Dirac in the late fifties, the recent work departs from the early treatments in three respects:

i) Dirac's program for quantization of constrained systems is suitably modified to handle certain peculiarities of general relativity [2]. Some of these arise because the theory has no background fields and hence no obvious symmetry group to select the vacuum state. (One can not use the diffeomorphism group here because the quantum constraints require that *every* physical state is diffeomorphism invariant.) Other peculiarities arise because of the nature of the canonical variables that are being used.

ii) The canonical variables consist of a (complex-valued) $SU(2)$ connection A_a^i and its conjugate electric field $\widetilde{E}_i^a$; general relativity is recast as a theory of connections rather than of metrics [5]. This shift in emphasis brings about several technical simplifications. In particular, all the equations of the theory are low order polynomials in the new canonical pair. Furthermore, it enables one to employ in quantum gravity various techniques from gauge theories. The suggestion is that while metrics, distances and light cones are the key concepts in the classical theory, it is the connections, parallel transport of spinors around closed loops and knot classes to which these loops belong that are the primary ingredients in the quantum theory.

iii) Quantization is carried out in a *loop representation* where states arise as suitable functions on the space of loops on a ("spatial") 3-manifold [6]. The passage from connections to loops can be carried out using holonomies as "integral kernels." The loop representation seems to be well-tailored to solve quantum constraint equations. The *general* solution to the diffeomorphism constraint can be written down: solutions depend only on the *knot classes* to which loops belong. The Hamiltonian constraint then restricts the support of these wave functions. Certain knot invariants can now be used to solve *all* quantum constraints [8]. (For a recent review, see, e.g., [7].)

In this chapter, however, I will set aside the interesting developments related to quantum dynamics –i.e., solutions to quantum constraints– and focus instead just on the picture of quantum geometry that arises already at a kinematical level from the use of the loop variables. All the main results reported here were obtained in collaboration with Rovelli and Smolin [9].

To explore geometry non-perturbatively, we must first introduce operators that carry the metric information and regulate them in such a way that the final operators do *not* depend on any background structure introduced in the regularization. Our first result is that such operators do exist and that they are finite without renormalization. Using these operators one can ask if there exist loop states which approximate smooth geometries at large scales. One often takes for granted that the answers to such questions would be obviously "yes". However, in genuinely non-perturbative treatments, this is by no means clear a priori; one may be working in the sector of the theory which does *not* admit the correct or unambiguous classical limit. For example, the sector may correspond to a confined phase which has no classical analog or the limit may yield the wrong number even for the macroscopic dimension of space-time. Our second result is that not only is the answer to the above question in the affirmative but, furthermore, these states exhibit a discrete structure of a specific type at the Planck scale. A discreteness at this scale was anticipated on general grounds since the 30's. Indeed, there exist a number of quantum gravity programs that *begin* by postulating discrete structures at the Planck scale and then attempt to recover from it the known macroscopic physics. The key difference is that, in our approach, discreteness is *arrived at* by combining general relativity with quantum mechanics using loop variables.

2. Preliminaries

In the classical, Hamiltonian description, the configuration variable A_a^i is a complex, $SU(2)$-connection its conjugate momentum, $\tilde{E}_i^a$ –the mathematical analog of the electric field in Yang-Mills theory– is a triad with density weight one [5].[1] The idea is to introduce on this phase space certain (manifestly $SU(2)$-gauge invariant) functions labelled by loops which are closed under the Poisson bracket, construct from them an (abstract) algebra of quantum operators and represent this algebra on the space of functions on the loop space. With this machinery at hand, we will be able to introduce operators representing 3-geometry by suitably smearing and regulating the loop operators.

The simplest of the classical loop variables are just the traces of holonomies, i.e. the Wilson loop functions : Given a closed loop γ on the 3-manifold Σ, we set

$$T[\gamma] = \tfrac{1}{2}\,\mathrm{tr}\,\mathcal{P}\exp G \oint_\gamma A_a dl^a, \tag{1}$$

where G is Newton's constant and where the 2-dimensional representation of the gauge group is used to evaluate traces. (The factor of G is essential because it is GA_a^i that has the usual physical dimensions –inverse length– of a connection.) These are "configuration variables" since they do not depend on $\tilde{E}_i^a$. Variables with momentum dependence are constructed by inserting $\tilde{E}_i^a$ at various points on the loop before

1 Throughout we will let $a, b, \ldots$ denote the spatial indices and $i, j, \ldots$, the $SU(2)$ internal indices. A tilde will denote a field with density weight 1. Thus, the momenta are vector densities of weight 1 with values in the Lie algebra of $SU(2)$. Note that this is also true in gauge theories. However, there we have a background metric whose determinant is implicitly used to "de-densitize" the electric fields. In absence of a background metric, we must let the density weight remain.

taking the trace. Thus, for example, the loop variable quadratic in momenta is given by:

$$\tilde{\tilde{T}}^{aa'}[\gamma](y,y') = \frac{1}{2}\,\mathrm{tr}\left[(\mathcal{P}\exp G\int_{y'}^{y} A_b d\gamma^b)\,\tilde{E}^a(y')\,(\mathcal{P}\exp G\int_{y}^{y'} A_c d\gamma^c)\,\tilde{E}^{a'}(y)\right], \quad (2)$$

where y and y' are any two points on the loop γ. Note that in the limit when the loop γ shrinks to a single point x, $\tilde{\tilde{T}}^{aa'}[\gamma]$ tends to $-4\tilde{E}^a_i(x)\,\tilde{E}^{a'i}(x)$. Now, while $\tilde{E}^a_i$ arises in the Hamiltonian framework as the "electric field" conjugate to the connection, in terms of space-time geometry, it represents a triad of density weight 1. That is, the spatial metric q_{ab} is related to $\tilde{E}^a_i$ via: (det q)$q^{ab} \equiv \tilde{\tilde{q}}^{ab} = \tilde{E}^a_i\tilde{E}^{bi}$. Hence, on the classical phase space, where all fields are assumed to be smooth, one can recover the metric from the loop variable $\tilde{\tilde{T}}^{aa'}$. We will see below that other geometrical q uantities such as areas of surfaces can also be so recovered. In the classical theory, this procedure is rather indirect and awkward. In quantum theory, however, it will provide a natural regularization scheme for the corresponding operators.

To quantize the system, we first construct an abstract algebra of loop operators such as $\hat{T}[\gamma]$ and $\hat{T}^{aa'}[\gamma;y,y']$ in the usual fashion. To find representations of this algebra, let us begin by specifying the vector space V of states. V will be the space of functions $\psi(\gamma)$ of loops satisfying the following two conditions:

i) Continuity: If a sequence of loops γ_i converges to a loop γ pointwise on Σ, then the complex numbers $\psi(\gamma_i)$ converge to $\psi(\gamma)$.

ii) If $\sum c_i T[\alpha_i](A) = 0$ for all $A \in \mathcal{A}/\mathcal{G}$, the space of connections modulo gauge transformations, then $\sum \psi c_i[\alpha_i] = 0$, where c_i are constants.

The first condition can be stated directly on the space of loops without any reference to the connection. The second condition, however, knows about connections, whence also about the gauge group and the effective configuration space $\mathcal{A}/\mathcal{G}$. It is an open question whether this condition can also be recast purely in the language of loops, without reference to connections, *in a concise fashion.* Finally, it is very likely that in a more precise treatment one would need to impose additional "regularity conditions" on the loop functions: examples of topological field theories as well as of linearized gravity [10] and Maxwell theory [11] make this amply clear. However, I will refrain from imposing any conditions that will not be needed directly in the calculations that follow, thereby maintaining a flexibility that may be useful subsequently.

Let us now specify the action of operators on these loop states. We will set[2]:

$$(\hat{T}[\gamma] \circ \psi)(\alpha) = \tfrac{1}{2}(\psi(\alpha\sharp\gamma) - \psi(\alpha\sharp\gamma^{-1})) \tag{3}$$

where, $\alpha\sharp\gamma$ is the "eye-glass loop" obtained by joining α and γ by any line segment, the same segment being used in both terms on the right side. Thus, the operator acts by breaking, re-routing and joining the loops in the argument of the wave function. It is convenient to use a bra-ket notation to express this action more succinctly. Setting $\psi(\gamma) = \langle \gamma \mid \psi \rangle$, we have:

$$\langle \alpha \mid \circ \hat{T}[\gamma] = \tfrac{1}{2}(\langle \alpha\sharp\gamma \mid - \langle \alpha\sharp\gamma^{-1} \mid) \tag{4}$$

The content of these two equations is precisely the same. In what follows, we will often use the notationally more compact form (4) which avoids the use of the dummy variable ψ in each term and enables us to visualize various operators more directly in terms of what they do to loops. In this notation, the second order momentum operator has the following action. If the loop β does not intersect γ at (at least) one of the points $\gamma(s)$ or $\gamma(s')$ where the electric field "sits," then $\langle \beta \mid \circ \hat{T}^{aa'}[\gamma; s, s']$ vanishes identically while if an intesection does occur, then we have:

$$\begin{aligned}\langle \beta \mid \circ \hat{T}^{aa'}[\gamma; s, s'] :=& (\hbar G)^2 \left(\oint_\beta dt \delta^3(\beta(t), \gamma(s)) \dot{\beta}^a(t) \cdot \oint_\beta dt' \delta^3(\beta(t'), \gamma(s')) \dot{\beta}^{a'}(t') \right) \\ & \times \Big(\langle \gamma_s^{s'} \beta_{t'}^{t} (\gamma_{s'}^{s})^{-1} (\beta_t^{t'})^{-1} \mid + \langle \gamma_s^{s'} (\beta_t^{t'})^{-1} (\gamma_{s'}^{s})^{-1} \beta_{t'}^{t} \mid \\ & + \langle \gamma_s^{s'} \beta_{t'}^{t} \cup \gamma_{s'}^{s} \beta_t^{t'} \mid + \langle \gamma_s^{s'} (\beta_t^{t'})^{-1} \cup \gamma_{s'}^{s} \beta_{t'}^{t} \mid \Big). \end{aligned} \tag{5}$$

Here, we have made the reroutings in the four terms explicit by specifying the compositions of various segments involved and where in the last two terms we have used the notation of footnote 2 for disjoint union of loops. These explicit expressions of

---

2 Rovelli and Smolin [6] considered wave functions $\psi(\{\beta\}) := (\psi_0, \psi_1(\beta), \psi_2(\beta_1 \cup \beta_2), \ldots \psi_n(\beta_1 \cup \ldots \cup \beta_n) \ldots)$ of multi-loops and defined the action of operators $\hat{T}[\gamma]$ as follows: $(\hat{T}[\gamma] \circ \psi)(\{\beta\}) := \psi(\gamma \cup \{\beta\})$. At first, it was thought that the existence of such states would let one introduce "raising" and "lowering" operators and endow the space of states with a natural Fock like structure, states with support on single loops playing the role of the "first excited" or, "one particle" states. This interpretation turned out to be incorrect because multi-loop states can be identified with linear combinations of single-loop states in 3 as well as 4 dimensional gravity (and also in the Fock space of spin-2 gravitons). Thus, at a fundamental, conceptual level, we only need to consider functions of single loops. In calculations, however, it is often convenient to consider the combination $\frac{1}{2}(\psi(\gamma\sharp\beta) + \psi(\gamma\sharp\beta^{-1}))$ as a 2-loop state and denote it by $\psi(\gamma \cup \beta)$. We will use this notation later.

the loop operators provide us with a concrete representation of the loop algebra in which the commutator of two loop operators is just $i\hbar$ times the quantum analog of the Poisson bracket between the corresponding loop functions, modulo terms of the oder $\hbar^2$ and higher which arise if one or both operators are quadratic (or higher order) in momenta. (For details, see the original paper by Rovelli and Smolin [6].)

From the viewpoint of our general quantization program, one is free to choose any V one likes and represent the quantum algebra in any manner one pleases so long as the result is a proper representation of the algebra $\mathcal{A}$; there is no a priori need to "justify" these choices. However, one's results may depend sensitively on these choices —and we will see that, in a certain sense, ours do. Therefore, one would like a degree of control and, ideally, a uniqueness theorem which says that these are the only possibilities subject to certain well-motivated physical and mathematical restrictions. At the current stage of the program, we are quite far from such a theorem. However, we can motivate the choices as follows. First, the choices we made mirror the structure we found in the well-understood loop representation of the 2+1 theory (see, e.g., chapter 17 in [1]). Second, in the connection representation, it is natural to represent A_a^i by a multiplication operator and E_i^a by a functional derivative. Now, if one performs a heuristic transform to the loop representation:

$$\psi(\beta) = \int_{\mathcal{A}/\mathcal{G}} d\mu\, \overline{T[\beta](A)}\, \Psi(A) \tag{6}$$

assuming that the measure has certain properties, one is led to the representation of loop operators given above. (For details on a precise form of this loop transform, see [12].) Thus, in a certain sense, the choices made above are the "simplest" ones consistent with the algebraic structure of the loop operators.

This completes the summary of the basic framework.

3. Regularization

I can now discuss the first of the two main results of this article: regularization of geometrical operators in a way that respects the underlying diffeomorphism invariance of the theory.

Let us begin with the spatial metric. As noted in section 2, in the present framework, the spatial metric (of density weight two) is a "composite" field given by $\tilde{\tilde{q}}ab(x) = \tilde{E}^{ai}(x)\tilde{E}_i^b(x)$. In the quantum theory, therefore, this operator must be regulated. The obvious possibility is point splitting. One might set $\tilde{\tilde{q}}^{ab}(x) =$

$\lim_{y\to x} \tilde{E}^{ai}(x)\tilde{E}^b_i(y)$. However, the procedure violates gauge invariance since the internal indices at two *different* points have been contracted. As we saw in section 2, a gauge invariant prescription is to use the Rovelli-Smolin loop variable $\tilde{\tilde{T}}^{aa'}[\gamma](y,y')$ defined in the classical theory by Eq. (2): in the limit γ shrinks to zero, $\tilde{\tilde{T}}^{aa'}[\gamma](y,y')$ tends to $-4\tilde{\tilde{q}}^{aa'}$. We also saw that, in quantum theory, one can define the action of the operator $\hat{T}^{aa'}[\gamma](y,y')$ directly on the loop states via Eq. (5). Its action is rather simple: if a loop β does not intersect γ at y or y', the operator simply annihilates the bra $\langle\beta\mid$ while if an intersection does occur, it breaks and re-routes the loop β, each routing being assigned a specific weight. One may therefore try to define a quantum operator $\hat{q}^{aa'}$ as a limit of $\hat{T}^{aa'}[\gamma]$ as γ shrinks to zero.

The resulting operator does exist after suitable regularization *and* renormalization. However, because of the density weights involved, the operator necessarily carries memory of the background metric used in regularization. Before going into details of the regularization scheme let me sketch a rough argument to see the origin of this problem intuitively. The operator in question is analogous to the product $\delta^3(x)\cdot\delta^3(x)$ of distributions at the same point. To regulate it, we may introduce a background metric. The final result is a distribution of the form $\tilde{N}(x)\delta^3(x)$ where, because $\delta^3(x)$ is a density of weight one, the renormalization parameter $\tilde{N}$ is now a density of weight one, proportional to the determinant of the background metric. Since the final answer carries a memory of the particular metric used to regulate the operator, we have violated the diffeomorphism invariance. Although there is no definitive proof, there do exist arguments which suggest that any *local* operator carrying the information about geometry will face the same problem.

There do exist, however, *non-local* operators which can be regulated in a way that respects diffeomorphism invariance.

As the first example, consider the function $q(\omega)$ –representing the smeared 3-metric– on the classical phase space, defined by

$$Q(\omega) := \int d^3x\,(\tilde{\tilde{q}}^{ab}\omega_a\omega_b)^{\frac{1}{2}}, \tag{7}$$

where ω_a is any smooth 3-form of compact support. (Note that the integral is well-defined without the need of a background volume element because $\tilde{\tilde{q}}^{ab}$ is a density of weight two. Also, in spite of the notation, $Q(\omega)$ is not obtained by smearing a *distribution* with a test field; there is a square-root involved.) To define the corresponding operator, we can proceed as follows. Let us choose on Σ test fields $f_\epsilon(x,y)$ which are densities of weight one in x, which satisfy:

$$\lim_{\epsilon\to 0}\int_\Sigma d^3x f_\epsilon(x,y)g(x) = g(y) \tag{8}$$

for all smooth functions of compact support $g(x)$. If Σ is topologically $\mathbb{R}^3$, for example, we can construct these test fields as follows:

$$f_\epsilon(x,y) = \frac{\sqrt{h(x)}}{\pi^{\frac{3}{2}}\epsilon^3} \exp -\frac{|\vec{x}-\vec{y}|^2}{2\epsilon^2}, \tag{9}$$

where $\vec{x}$ are the cartesian coordinates labeling the point x and $h(x)$ is a "background" scalar density of weight 2. Next, let us define

$$\tilde{\tilde{q}}_\epsilon^{aa'}(x) = -\frac{1}{4}\int_\Sigma d^3y \int_\Sigma d^3y' f_\epsilon(x,y) f_\epsilon(x,y') \tilde{\tilde{T}}^{aa'}(y,y'). \tag{10}$$

As ϵ tends to zero, the right side tends to $\tilde{\tilde{q}}^{ab}$ because the test fields force both the points y and y' to approach x, and hence the loop passing through y, y', used in the definition of $\tilde{\tilde{T}}^{aa'}(y,y')$, to zero. It is now tempting to try to define a local metric operator $\hat{q}^{aa'}$ corresponding to $\tilde{\tilde{q}}^{aa'}$ by replacing $\tilde{\tilde{T}}^{aa'}(y,y')$ in Eq. (10) by its quantum analog and then taking the limit. One finds that the limit does exist provided we first renormalize $\hat{q}_\epsilon^{aa'}$ by an appropriate power of ϵ. However, the answer depends on the background structure (such as the density $h(x)$) used to construct the test fields $f_\epsilon(x,y)$. If, on the other hand, one tries to construct the quantum analog of the *non-local* classical variable $Q(\omega)$, this problem disappears. To see this, let us first express $Q(\omega)$ using Eq. (10) as:

$$Q(\omega) = \lim_{\epsilon\to 0} \int_\Sigma d^3x (\tilde{\tilde{q}}_\epsilon^{aa'} \omega_a \omega_{a'})^{\frac{1}{2}}. \tag{11}$$

The required quantum operator $\hat{Q}(\omega)$ on the loop states can now be obtained by replacing $\tilde{\tilde{T}}^{aa'}(y,y')$ by the operator $\hat{T}^{aa'}(y,y')$ defined in Eq. (5). A careful calculation shows that: i) the resulting operator *is* well-defined on loop states; ii) no renormalization is necessary, i.e., the limit is automatically *finite*; and, iii) the final answer carries no imprint of the background structure (such as the density $h(x)$ or, more generally, the specific choice of the test fields $f_\epsilon(x,y)$) used in regularization. To write out its explicit expression, let me restrict myself to smooth loops γ without any self-intersection. Then, the action is given simply by:

$$\langle\gamma \mid \circ\hat{Q}(\omega) = l_P^2 \oint_\gamma ds |\dot{\gamma}^a \omega_a| \cdot \langle\gamma\mid, \tag{12}$$

where $l_P = \sqrt{G\hbar}$ is the Planck length, s, a parameter along the loop and $\dot{\gamma}^a$ the tangent vector to the loop. (Recall that the G in l_P comes from the fact that GA_a^i

has the usual dimensions of a connection while $\hbar$ comes from the fact that $\hat{E}_i^a$ is $\hbar$ times a functional derivative.) Thus, on non-intersecting loops, the operator acts simply by multiplication: the loop representation is well-suited to find states in which the 3-geometry –rather than its time evolution– is sharp.

The second class of operators corresponds to the area of 2-surfaces. Note first that, given a smooth 2-surface S in Σ, its area $\mathcal{A}_S$ is a function on the classical phase space. We first express it using the classical loop variables. Let us divide the surface S into a large number N of area elements $S_I, I = 1, 2...N$, and set $\mathcal{A}_I^{\rm appr}$ to be

$$\mathcal{A}_I^{\rm appr} = -\frac{1}{4}\left[\int_{S_I} d^2S^{bc}(x)\eta_{abc}\int_{S_I} d^2S^{b'c'}(x')\eta_{a'b'c'}\,\tilde{\tilde{T}}^{aa'}(x,x')\right]^{\frac{1}{2}}, \tag{13}$$

where η_{abc} is, as usual, the (metric independent) Levi-Civita density of weight -1. Since $\tilde{\tilde{T}}^{aa'}$ approximates $-4(\det q)q^{ab}$ for smooth metrics, $\mathcal{A}_I^{\rm appr}$ approximates the area function (on the phase space) defined by the surface elements S_I, the approximation becoming better as S_I –and hence loops with hands at x and x' used in the definition of $\tilde{\tilde{T}}^{aa'}$– shrink. Therefore, the total area $\mathcal{A}_S$ associated with S is given by

$$\mathcal{A}_S = \lim_{N\to\infty}\sum_{I=1}^{N}\mathcal{A}_I^{\rm appr}. \tag{14}$$

To obtain the quantum operator $\hat{\mathcal{A}}_S$, we simply replace $\tilde{\tilde{T}}^{aa'}$ in Eq. (13) by the quantum loop operator $\hat{T}^{aa'}$. This somewhat indirect procedure is necessary because, as indicated above, there is no well-defined operator-valued distribution that represents the metric or its area element *at a point*. Again, the operator $\hat{\mathcal{A}}_S$ turns out to be finite. Its action, evaluated on a nonintersecting loop γ for simplicity, is given by:

$$\langle\gamma|\circ\hat{\mathcal{A}}_S = \frac{l_P^2}{2}\, I(S,\gamma)\cdot\langle\gamma|, \tag{15}$$

where $I(S,\gamma)$ is simply the *unoriented* intersection number between the 2-surface S and the loop α. (One obtains the *unoriented* intersection number here and the absolute sign in the integrand of Eq. (12) because of the square-root operation involved in the definition of these operators.) Thus, in essence, a loop γ contributes half a Planck unit of area to any surface it intersects.

The fact that the area operator also acts simply by multiplication on non-intersecting loops lends further support to the idea that the loop representation is well-suited to "diagonalize" operators corresponding to 3-geometry. Indeed, we can immediately construct a large set of simultaneous eigenbras of the smeared metric and the area operators. There is one, $\langle\gamma|$, associated to every nonintersecting loop γ. Note that the corresponding eigenvalues of area are *quantized* in integral multiples of $l_P^2/2$. There are also eigenstates associated with intersecting loops which, however, I will not go into to since the discussion quickly becomes rather involved technically.

4. Weaves

With operators representing 3-geometry on hand, we can now turn to the construction of weaves. Recall that the goal here is to introduce loop states which approximate a given 3-metric, say, h_{ab} on Σ on scales large[3] compared to l_p. The basic idea is to weave the classical metric out of quantum loops by spacing them so that on an average only one line crosses every surface element whose area, *as measured by the given* h_{ab} is one Planck unit. Such loop states will be called *weaves*. Note that these states are not uniquely picked out since our requirement is rather weak. Indeed, given a weave approximating a given classical metric, one can obtain others, approximating the same classical metric. Let us begin with a concrete example of such a state which will approximate a *flat* metric h_{ab}. To construct this state, we proceed as follows. Using this metric, let us introduce a random distribution of points on $\Sigma = R^3$ with density n (so that in any given volume V there are $nV(1+\mathcal{O}(1/\sqrt{nV}))$ points). Center a circle of radius $a = (1/n)^{\frac{1}{3}}$ at each of these points, with a random orientation. We assume that $a << L$, so that there is a large number of (non-intersecting but, generically, *linked*) loops in a macroscopic volume L^3. Denote the collection of the circles by Δ. As noted in footnote 2, due to $SL(2,\mathbb{C})$ trace identities, multi-loops are equivalent to single loops, whence there is a well-defined bra $\langle\Delta|$. I would like to claim that this is a weave state with the required properties.

Let us first consider the observable $\hat{Q}[\omega]$. To see if $\langle\Delta|$ reproduces the geometry determined by the classical metric Lh_{ab} on a scale $L >> l_p$, let us introduce a 1-form ω_a which is *slowly varying on the scale* L and compare the value $Q[\omega](h)$ of the classical $Q[\omega]$ evaluated at the metric h_{ab}, with the action of the quantum operator $\hat{Q}[\omega]$ on $\langle\Delta|$. A detailed calculation yields:

$$\langle\Delta| \circ \hat{Q}[\omega] = \left[\frac{\pi}{2}\,(\frac{l_P}{a})^2\, Q[w](h) + \mathcal{O}(\frac{a}{L})\right] \cdot \langle\Delta|. \tag{16}$$

Thus, $\langle\Delta|$ is an eigenstate of $\hat{Q}[\omega]$ and the corresponding eigenvalue is closely related

3 Note, incidently, that the large scale limit is equivalent to the semi-classical limit since in source-free, non-perturbative quantum general relativity, $\hbar$ and G always occur in the combination $\hbar G = l_p^2$.

to $Q[\omega](h)$. However, even to the leading order, the two are unequal *unless* the average distance, a, between the centers of loops *equals* $\sqrt{\pi/2H}\, l_p$. More precisely, Eq. (16) can be interpreted as follows. Let us write the leading coefficient on the right side of this equation as $I(1/4)(2\pi a/l_p)(nl_p^3)$. Since this has to be unity for the weave to reproduce the classical value (to leading order), we see that Δ should contain, on an average, one fourth Planck length of curve per Planck volume, where lengths and volumes are measured using h_{ab}.

The situation is the same for the area operators $\hat{A}_S$. Let S be a 2-surface whose extrinsic curvature varies slowly on a scale $L >> l_P$. One can evaluate the action of the area operator on $\langle\Delta|$ and compare the eigenvalue obtained with the value of the area assigned to S by the given flat metric h_{ab}. Again, the eigenvalue can be re-expressed as a sum of two terms, the leading term which has the desired form, except for an overall coefficient which depends on the mean separation a of loops constituting Δ, and a correction term which is of the order of $\mathcal{O}(\frac{a}{L})$. We require that the coefficient be so adjusted that the leading term agrees with the classical result. This occurs, again, precisely when $a = \sqrt{\pi/2}l_p$. It is interesting to note that the details of the calculations which enable one to express the eigenvalues in terms of the mean separation are rather different for the two observables. In spite of this, the final constraint on the mean separation is *precisely* the same.

Finally, I would like to make two technical remarks on the structure of weave states. First, it is not difficult to extend the above construction to obtain weave states for curved metrics q_{ab} which are slowly varying with respect to a flat metric h_{ab}. Given such a metric, one can find a slowly varying tensor field $t_a{}^b$, such that the metric q_{ab} can be expressed as $t_a{}^c t_b{}^d h_{cd}$. Then, given a weave of the type $\langle\Delta|$ considered above approximating h_{ab}, we can "deform" each circle in the multi-loop Δ using $t_a{}^b$ to obtain a new weave $\langle\Delta|_t$ which approximates q_{ab} in the same sense as $\langle\Delta|$ approximates h_{ab}. The second remark is that since the weaves are eigenbras of the operators that capture the 3-geometry, none of them is a candidate for representing the vacuum state. In the linearized Theory, the vacuum is a coherent state. It is neither an eigenstate of the linearized metric operator, nor of the linearized extrinsic curvature (or, connection). Rather, it is peaked, with *minimum* uncertainty spreads for both operators, at their zero values. The candidates for vacuum state in the full theory would have a similar characteristic. Some insight into this issue has come from a recent analysis of the relation between the exact and the linearized theory in the loop representation particularly due to Iwasaki and Rovelli [13].

Let us explore the meaning and implications of all these results.

1) As is emphasized in the general context of decoherence, to obtain classical behavior from quantum theory, one needs two things: i) appropriate coarse graining, and, ii) special states. In our procedure, the slowly varying test fields

ω_a enable us to perform the appropriate coarse graining while weaves –with the precisely tuned mean separation a– are the special states. There is, however, something rather startling: The restriction on the mean separation a –i.e., on the *short distance* behavior of the multi-loop Δ– came from the requirement that $\langle\Delta|$ should approximate the classical metric h_{ab} on *large scales* L!

2) In the limit $a \to \infty$, the eigenvalues of the two operators on $\langle\Delta|$ go to zero. This is not too surprising. Roughly, in a state represented by any loop γ, one expects the quantum geometry to be excited just at the points of the loops. If the loops are *very* far away from each other as measured by the fiducial h_{ab}, there would be macroscopic regions devoid of excitations whence the quantum geometry would seem as if it corresponds to a zero metric.

3) The result of the opposite limit, however, *is* surprising. One might have naively expected that the best approximation to the classical metric would occur in the continuum limit in which the separation a between loops goes to zero. However, the explicit calculation outlined above shows that this is not at all the case: as a tends to zero, the leading terms in the eigenvalues of $\hat{Q}[\omega]$ and $\mathcal{A}_S$ actually diverge[4]! It is, however, easy to see the underlying reason. Intuitively, the factors of the Planck length in Eqs. (12) and (15) force each loop in the weave to contribute a Planck unit to the eigenvalue of the two geometrical observables. In the continuum limit, the number of loops in any fixed volume (relative to the fiducial h_{ab}) grows unboundedly and the eigenvalue diverges.

4) It is important to note the structure of the argument. In non-perturbative quantum gravity, there is no background space-time. Hence, terms such as "slowly varying" or "microscopic" or "macroscopic" have, a priori, no physical meaning. One must do some extra work, introduce some extra structure to make them meaningful. The required structure should come from the very questions one wants to ask. Here, the questions had to do with approximating a classical geometry. Therefore, we could *begin* with classical metric h_{ab}. We used it repeatedly in the construction: to introduce the length scale L, to speak of "slowly varying" fields ω_a and surfaces S, and, to construct the weave itself. The final result is then a consistency argument: If we construct the weave

4 One's first impulse from lattice gauge theories may be to say that the limit is divergent simply because we are not rescaling, i.e., renormalizing the operator appropriately. Note, however, that, in contrast to the calculations one performs in lattice theories, here, we *already* have a well defined operator in the continuum. We are only probing the properties of its eigenvectors and eigenvalues, whence there is nothing to renormalize. Even in non-relativistic quantum mechanics the spectrum of respectable operators are typically unbounded whence the Hilbert space admits sequences of eigenstates with the property that the corresponding sequence of eigenvalues diverges in the limit.

according to the given prescription, then we find that it approximates h_{ab} on macroscopic scales L provided we choose the mean separation a to be $\sqrt{\pi/2}l_p$, where all lengths are measured relative to the same h_{ab}.

5) Note that there is a considerable non-uniqueness in the construction. First of all, as we noted already, a given 3-geometry can lead to distinct weave states; our construction only serves to make the existence of such states explicit. For example, there is no reason to fix the radius r of the individual loops to be a. For the calculation to work, we only need to ensure that the loops are large enough so that they are generically linked and small enough so that the values of the slowly varying fields on each loop can be regarded as constants plus error terms which we can afford to keep in the final expression. Thus, it is easy to obtain a 2-parameter family of weave states, parametrized by r and a. The condition that the leading order terms reproduce the classical values determined by h_{ab} then gives a relation between r, a and l_P which again implies discreteness. Clearly, one can further enlarge this freedom considerably: There are a lot of eigenbras of the the smeared-metric and the area operators whose eigenvalues approximate the classical values determined by h_{ab} up to terms of the order $\mathcal{O}(\frac{l_p}{L})$ since this approximation ignores Planck scale quantum fluctuations. Thus, this non-uniqueness is not very surprising. There is, however, a possibility of a more subtle non-uniqueness in the opposite direction: A given weave may approximate two *different* 3-metrics! This could happen precisely because the notion of "slowly varying" is tied to the metric h_{ab} we are trying to approximate. Suppose, two metrics h_{ab} and h'_{ab} which are *not* slowly varying with respect to one another, lead to quite distinct classes of slowly varying test fields ω_a and ω'_a. Then, it could happen that there is a single weave $\langle\Delta|$ which has the property that, to leading order, the eigenvalues of $\hat{Q}[\omega]$ are equal to $Q[\omega](h)$ while those of $\hat{Q}[\omega']$ are equal to $Q[\omega](h')$. While it is likely that such an ambiguity exists, whether it in fact does is not quite clear. In a broader context, I suspect that such ambiguities will arise in the classical interpretations of quantum states in *any* framework simply because of the freedom in the choice of coarse graining that is needed to pass to this limit. In most current discussions, "obvious" choices are generally made and ambiguities may well be simply overlooked.

6) Finally, I would like to emphasize that, at a conceptual level, the important point is that the eigenvalues of $\hat{Q}[\omega]$ and $\hat{A}[S]$ can be discrete; quantized in multiples of half Planck units. Given a specific eigenstate, one can examine the micro-structure of the geometry it defines. The precise characteristics of the discreteness in that structure will vary from state to state. Since many of these weaves may define the same macroscopic geometry, it is also clear that we

cannot, at least at this stage, associate a specific discrete structure to a given *classical* geometry. All detailed claims refer only to eigenstates of the geometric operators.

5. Discussion

Since these results of the last two sections are both unexpected and interesting, it is important to probe their origin. We see no analogous results in familiar theories. For example, the eigenvalues of the fluxes of electric or magnetic fields are not quantized in QED nor do the linearized analogs of our geometric operators admit discrete eigenvalues in spin-2 gravity. Why then did we find qualitatively different results? The technical answer is simply that the familiar results refer to the *Fock representation* for photons and gravitons while we are using a completely different representation here. Thus, the results *are* tied to our specific choice of the representation. Why do we not use Fock or Fock-like states? It is not because we insist on working with loops rather than space-time fields such as connections. Indeed, one *can* translate the Fock representation of gravitons and photons to the loop picture. (See, e.g., [10,11].) And then, as in the Fock space, the discrete structures of the type we found in sections 3 and 4 simply disappear. However, to construct these loop representations, one must use a flat background metric and essentially every step in the construction violates diffeomorphism invariance. Indeed, there is simply no way to construct "familiar, Fock-like" representations without spoiling the diffeomorphism invariance. Thus, the results we found are, in a sense, a direct consequence of our desire to carry out a genuinely non-perturbative quantization without introducing any background structure.

As remarked earlier, however, we do not have a uniqueness theorem singling out the representation we are using. The construction of a proper representation theory has only just begun. Therefore, one cannot rule out the possibility of the existence of other, sufficiently rich representations which also maintain diffeomorphism invariance but in which geometric operators considered above have only continuous spectra. The one which we are using has resulted from the "simplest" choices of states and operators along the way and we do not yet have a good control of what would have happened if different choices had been made. In a sense the status is no different from that of the procedure one generally adopts in interacting quantum field theory in Minkowski space, where again there is no a priori justification for using the Fock representation. (Moreover, in 4 dimensions, there are good mathematical reasons for *not* using it!) However, in that case, over the years one has acquired a great deal of theoretical experience and, more importantly, there exist a vast number of experimental tests which support the strategy as a good working hypothesis. In non-perturbative quantum gravity, on the other hand, one lacks both and hence the issue of "control" and "uniqueness" become much more important. (Incidently, although this point is rarely raised on other approaches to quantum gravity, such as quantum geometrodynamics or string theory, it is equally valid there.) However, I should also emphasize that the choices made in the construction of the loop representation *are* well motivated. Furthermore, within the chosen representation, the analysis has now reached a high level of precision and is on a sound mathematical footing by standards conventionally used in theoretical (as opposed to mathematical) physics. Finally, although important technical problems remain, a general, systematic approach which may justify the choices we have made *is* now in sight [12, 14].

My overall viewpoint therefore is that one should simultaneously proceed along two lines: i) Since the results are intuitively appealing, one should take them as an indication that we are on the right track and push this particular representation as far as possible. This path is being actively pursued by a large number of researchers; ii) One should try to better understand this representation by, e.g., changing it in small steps, with the hope of arriving at a uniqueness result. Some recent work of Gambini and his collaborators [15] provides a framework which is appropriate for such an investigation.

Acknowledgments: I am grateful to Chris Isham, Carlo Rovelli and Lee Smolin for innumerable discussions and for collaboration. This work was supported in part by the NSF grants PHY90-12099, PHY90-16733 and INT88-15209 and by research funds provided by Syracuse University.

[1] B.S. DeWitt, E. Myers, R. Harrington and A. Kapulkin, Nucl. Phys.B. (Proc. Suppl.) **20**, 744 (1991); M.E. Agishtein and A. A. Migdal, Mod. Phys. Lett. **7**, 1039 (1992).

[2] A. Ashtekar, *Non-perturbative canonical quantum gravity* (Notes prepared in collaboration with R.Tate) (World Scientific, Singapore, 1991).

[3] C. Rovelli, Class. & Quant. Grav. **8**, 1613 (1991).

[4] L. Smolin, To appear in The Proceedings of the 1991 GIFT International Seminar on Theoretical Physics (World Scientific, Singapore, in press); Syracuse pre-print SU-GP-92/2-2,

[5] A. Ashtekar, Phys. Rev. Lett. **57**, 2244 (1986), Phys. Rev.**D36**, 1587 (1987).

[6] C. Rovelli and L. Smolin, Phys. Rev. Lett. **61**, 1155 (1988); Nucl. Phys. **B133**, 80 (1990).

[7] A. Ashtekar; To appear in The Proceedings of the 1992 Les Houches School on Gravitation and Quantization, B. Julia ed (North Holland, Amsterdam, in press); Syracuse pre-print SU-GP-92/11-2.

[8] B. Brügmann, R. Gambini and J. Pullin, Phys. Rev. Lett. **68**, 431 (1992); to appear in The Proceedings of the XXth International Conference on Differential Geometric Methods in Physics, S.Cota and A. Rocha eds. (World Scientific, Singapore, in press); Syracuse pre-print SU-GP-91/8-4.

[9] A. Ashtekar, C. Rovelli and L. Smolin, Phys. Rev. Lett. **69**, 237 (1992).

[10] A. Ashtekar, C. Rovelli and L. Smolin, Phys. Rev. **D44**, 1740 (1991).

[11] A. Ashtekar and C. Rovelli, Class. Quan. Grav. **9**, 1121 (1992).

[12] A. Ashtekar and C. J. Isham, Class. Quan. Grav. **9** 1433 (1992).

[13] J. Iwasaki and C. Rovelli, Int J. Mod. Phys. **D** (in press); J. Zegwaard, Nucl. Phys. **B378**, 288 (1992).

[14] A. Ashtekar and J. Lewandowski (in preparation); A. Rendall, Syracuse pre-print.

[15] Di Bortolo, R. Gambini and J. Griego, Comm. Math. Phys. (in press).

The Cosmological Kibble Mechanism in the Laboratory: String Formation in Liquid Crystals*

Mark J. Bowick[1]

Physics Department
Syracuse University
Syracuse, NY 13244-1130, USA

Abstract

I report on the observation of the production of strings (disclination lines and loops) via the Kibble mechanism of domain (bubble) formation in the isotropic to nematic phase transition of a sample of uniaxial nematic liquid crystal. The probablity of string formation per bubble is measured to be 0.33 ± 0.01. This is in good agreement with the theoretical value $1/\pi$ expected in two dimensions for the order parameter space $S^2/\mathbf{Z}_2$ of a simple uniaxial nematic liquid crystal.

* Invited talk delivered at the workshop "String Quantum Gravity and Physics at the Planck Energy Scale," Erice, June 21–28, 1992 and at the 1st Iberian Meeting on Gravity, Évora, Portugal, September 21-26, 1992.

[1]E-mail: Bowick@suhep.bitnet *or* Bowick@suhep.phy.syr.edu.

The appearance of topological defects in phase transitions associated with spontaneous symmetry breaking is common in quantum field theories, particle physics models, condensed matter physics and cosmology. Given a phase transition at a scale μ involving the reduction of explicit symmetry from a symmetry group G to a symmetry group H

$$G \to H, \tag{1}$$

the manifold of equivalent vacua or ground-states is the coset space $\mathcal{M} = G/H$. Topological defects arise if $\mathcal{M}$ has non-trivial topology. In 3 dimensions the possibilities are classified the homotopy groups $\pi_0(\mathcal{M})$ through $\pi_3(\mathcal{M})$. Non-vanishing $\pi_0(\mathcal{M})$ occurs if $\mathcal{M}$ is not connected and leads to two-dimensional defects called *domain walls*. Non-vanishing $\pi_1(\mathcal{M})$ occurs if $\mathcal{M}$ is not simply connected and leads to one-dimensional *string* or *line* defects. Non-vanishing $\pi_2(\mathcal{M})$ leads to zero-dimensional *point* defects such as monopoles. Finally non-vanishing $\pi_3(\mathcal{M})$ gives rise to defects which have been generically called *texture*. In the quantum field theory case an example is the Skyrmeon.

Nematic Liquid Crystals (NLCs) form a class of matter with a rich variety of topological defects. While the classification and static properties of these defects have been thoroughly studied [1–5] the mechanism of their formation and their dynamics is relatively unexplored in the laboratory. Recently the evolution in three dimensions of the dense tangle of string defects (disclinations) produced by a pressure-induced isotropic to nematic phase transition of a sample of NLC ("K15") was studied experimentally [6,7]. It was shown that the string length density (total length in string per volume) decays linearly in time and that the coarsening of the string density obeys the scaling hypothesis $\rho(t) \approx \frac{1}{\xi(t)^2}$ where the length scale $\xi(t)$ is assumed to be both the mean radius of curvature of strings and the mean separation between strings and grows in time as t^ν with $\nu = 0.51 \pm 0.02$. It was also shown that a string loop of radius $R(t)$ shrinks in time as $R(t) \propto (t_0 - t)^\alpha$, with the exponent α being 0.5 ± 0.03. This agrees with the exponent $\frac{1}{2}$ obtained from an analysis of the nematodynamic equations of a disclination line moving through a nematic medium with a constant velocity v [1,7].

An important motivation for this recent work was to verify several of the string interactions postulated in the study of cosmic strings and fundamental strings. In this talk I report on some new work described in [8]. We have observed the *formation* of string defects in the supercooling of the NLC $K15$. Preceeding the production of strings we see the formation of uncorrelated domains (bubbles) as described theoretically by Kibble [9] some years ago in his pioneering analysis of the production of defects (domain walls, cosmic strings and monopoles) in the cooling of the early Universe. In our experiment disclinations are analogous to *global* cosmic strings rather than *local* cosmic strings as there are no analogues of gauge fields in simple uniaxial NLCs. The idea of testing cosmological theories of string formation in condensed matter systems was in fact proposed by Zurek [10], who suggested a cryogenic experiment to detect superfluid helium flow arising from the random formation of uncorrelated domains produced by a rapid pressure quench in an annular geometry.

NLCs are *thermotropic* materials typically consisting of rod-like molecules roughly $20\AA$ in length and $5\AA$ in width. They often have two benzene rings associated with different functions and short flexible exterior chains. They exhibit a high-temperature *isotropic* phase characterized by disorder with respect to translational and rotational symmetries and a low-temperature *nematic* phase in which there is long-range orientational order but translational disorder. The nematic phase is thus liquid with respect to translations and crystalline with respect to orientations. Configurations of NLCs are specified by a director field $\mathbf{n}(\vec{x})$. This is a three-dimensional directionless unit vector giving at each point $\vec{x}$ the local orientation of the rod molecules. An order parameter for the first-order isotropic to nematic phase transition is $\langle\frac{3cos^2\theta-1}{2}\rangle$ where θ is the angle of the director field to some arbitrary fixed axis. This order parameter vanishes in the isotropic phase and is non-zero in the nematic phase. For $K15$ the isotropic to nematic transition occurs at $T_{NI} = 35.3°C$ at atmospheric pressure. The easily accessible temperatures (typically $0° - 200°C$) of nematic phase transitions is one reason these systems are simple to work with in the laboratory (especially compared to superfluid helium).

The effective energy density describing NLCs is similar to that of an $O(3)$ non-linear sigma model except that the NLC system is invariant only under simultaneous spatial and internal rotations rather than independent spatial and internal rotations [6]. This means that the effective energy density for NLCs consists of three terms of different symmetry associated with splay, twist and bend deformations respectively [5], as well as a surface term. Corresponding to each term is a potentially different elastic (Frank) constant. The non-linear sigma model field theory is recovered only in the isotropic limit in which all three elastic constants are equal (the one-constant approximation).

NLCs exhibit three classes of topological defects. The space of ground-state nematic configurations $\mathcal{M}$ (ground-state manifold) is given by $SO(3)/D_\infty$, where $SO(3)$ is the symmetry group of the disordered (isotropic) phase and D_∞ is the unbroken subgroup consisting of rotations about the molecular axis and 180° rotations about axes perpendicular to the molecular axis. This space is isomorphic to $S^2/\mathbf{Z}_2$, the two-sphere with antipodal points identified, which in turn is isomorphic to the projective plane $\mathbf{RP}^2$. The existence of topological defects is determined by the homotopy groups of the ground-state manifold. The non-vanishing homotopy groups of $\mathcal{M}$ are $\pi_1(\mathcal{M}) = \mathbf{Z}_2$, $\pi_2(\mathcal{M}) = \mathbf{Z}$ and $\pi_3(\mathcal{M}) = \mathbf{Z}$. Thus there are $\mathbf{Z}_2$ disclinations (string or line defects), integrally-charged monopoles (point defects) and integrally charged *texture* defects (whose field theory analogues are Skyrmeons) associated with non-trivial elements of the first, second and third homotopy groups respectively [2]. We will be concerned here only with the π_1 defects. Note that non-trivial strings arise from the lack of polarity of liquid-crystal molecules which gives rise to the director field identification $\mathbf{n} \equiv -\mathbf{n}$. This is encoded in the $\mathbf{Z}_2$ isotropy group of $\mathcal{M} = S^2/\mathbf{Z}_2$. A path in $\mathcal{M}$ which winds from $\mathbf{n}$ to $-\mathbf{n}$ as one goes around a closed loop in space will not be continuously deformable to a point and hence will represent a non-trivial disclination (string) defect.

To see more clearly the director field configurations associated with disclinations we now give a very brief description of them. The basic string defect along the z-axis is characterized by the planar configuration of the director field [5]

$$n_x = \cos\phi, \quad n_y = \sin\phi \quad \text{and} \quad n_z = 0, \tag{2}$$

where we have taken the director to lie in the $x-y$ plane. For an isotropic NLC (all elastic constants equal) the equations of motion minimizing the free energy are

$$\nabla^2\phi = 0. \tag{3}$$

Solutions independent of the radial coordinate $r = \sqrt{x^2+y^2}$ are given by

$$\phi = s\alpha + c, \tag{4}$$

where $\alpha = \tan^{-1}(y/x)$ and c is a constant between 0 and π. In the core of the string (of typical radius 100Å) the director changes discontinuously and the most reasonable assumption is that the system is in the isotropic phase [4].. The parameter s labels the strength of the disclination. As the core of the string is encircled the director rotates by an angle $2\pi s$. The basic disclinations are the type-$\pm\frac{1}{2}$ disclinations. They can be continuously rotated into each other and so are topologically equivalent. Integral strength disclinations are continuously deformable to trivial configurations without a singular core by escape into the third dimension [2] and hence are topologically trivial although their dynamics is still of considerable interest. The energy density per unit length of a disclination is proportional to the square of its strength and hence the strength $\pm\frac{1}{2}$ disclinations dominate the statistical mechanics of these systems.

The formation of topological defects in a phase transition as described by Kibble [9] in the context of the early Universe, arises due to the existence of uncorrelated regions of space (domains). As these domains come into contact with each other, the variation of the order parameter field from one domain to another may be such that topological defects get trapped at the junctions of these domains. The probability of defect formation can be calculated by assuming that the order parameter field varies randomly from one domain to another. Let us illustrate this by considering a simpler case of 2 spatial dimensions when the ground state manifold is a circle S^1. The value of the order parameter field in the ground state is then given by an angle θ between 0 and 2π. The elementary strings will arise when θ winds by 2π around a closed path in space. Consider now the situation when three domains meet at a point and further that θ takes random values (say θ_1, θ_2 and θ_3) in these three domains. The value of θ in between any two domains will be governed by energetic considerations and will be such that it amounts to smallest variation of θ from one domain to another. With this, the string formation at the junction will correspond to the case when one of the angles (say θ_3) is such that $\theta_1 + \pi < \theta_3 < \theta_2 + \pi$. [Assuming that $\theta_2 + \pi > \theta_1 + \pi$, otherwise the order of θ_1 and θ_2 should be reversed. All angles here are taken mod(2π).] The maximum and minimum values of this angular span available for θ_3 for string formation are π and 0 with an average of $\pi/2$. As θ_3 can assume any value between 0 and 2π, the probability that a string will form at the junction is 1/4. The calculation of probability of formation of other defects (and for other ground state manifolds) can be carried out along similar lines. For NLCs the ground state manifold is $\mathbf{RP}^2$ and the probability of string formation in 2 spatial dimensions can be calculated to be $1/\pi$ (see [11]).

It is known that for NLCs, the isotropic-nematic phase transition is of first order [12] and proceeds through the proccess of bubble nucleation. As a sample of NLC supercools below T_{NI} it becomes favorable for domains (bubbles) of true ground state (nematic phase) to form in the false ground state medium (isotropic phase). Very small bubbles have too much surface energy and collapse again. Bubbles beyond a certain size (the critical bubbles) gain enough bulk energy to overcome the surface energy and begin to grow (see [13] for a discussion of vortex formation by bubbles). The director field will randomly take values in $\mathcal{M}$ in different bubbles and will be roughly uniform inside a given bubble (for energetic reasons). As the bubbles collide, the actual correlation length will depend on the bubble size as well as on the rate of bubble collision and when bubble collisions are frequent, each bubble will represent an uncorrelated domain as used in the Kibble mechanism. The correlation length here is similar to the horizon in the cosmological setting in which different domains are causally disconnected and hence cannot communicate with each other. As the system cools the initially chaotic variation of the director field will smoothen as various neighboring domains coalesce and attempt to lower gradient energies by aligning their associated director fields. As described above, however, there may be topological obstructions to neighboring domains aligning completely. If there are non-trivial paths in $\mathcal{M}$ traced out by neighboring domains then π_1 (string or disclination) defects are trapped in the system somewhat like flux tubes in Type-II superconductors. This is the process we have observed in the experiment described below. One of the key quantities of interest in this mechanism of string production via bubbles is the probability of string formation per bubble. We have determined the typical domain (bubble) size and net string length produced thereby experimentally determining the value of the string formation probability.

For our observations we used the NLC $K15$ (4-cyano-4′-n-pentylbiphenyl) manufactured by BDH Chemicals, Ltd.. We employed an Olympus, Inc. model BH phase contrast microscope with a 10X objective. The microscope was equipped with a monochrome television camera and a standard videocassette recorder.

We tried two different approaches to observing the isotropic - nematic phase transition. In the first a freely suspended film of $K15$ was created by dipping a wire loop 3 mm in diameter in a sample of $K15$. This wire loop was soldered to an 18Ω resistor which acted as a heater. The loop and heater assembly were fastened to an aluminum slide with a keyhole machined to accomodate them. With this apparatus we first heated the wire loop until the film entered the isotropic phase; in this phase the microscope image was completely structureless. We then switched off the electrical heater and allowed the film to cool. A similar approach was used previously by Chuang *et al* [6] for the observation of the evolution of string defects, which formed very rapidly as the film entered its nematic phase. Our results in this regime largely reproduced this previous work.

This apparatus proved to be unsuitable for observing homogeneous bubble formation and the subsequent generation of strings. A circular front of nematic phase formed at the boundary of the film and propagated through the central region of the film before we could finish observing the evolution of the homogeneously nucleated bubbles. We tried to reduce this effect by slowing down the cooling rate by slowly reducing the current through the resistor and by shining an auxiliary illuminator during the measurements. This improved the production of bubbles but did not eliminate the problem of the circular front. We therefore tried observing the phase transitions by using a different method by using a thin drop of K15 placed on an untreated, clean microscope slide. For this arrangement heating was done using an illuminator; we were able to obtain satisfactory images of bubble formation and evolution.

One set of such images is reproduced in Fig. 1. Fig. 1(a) shows the numerous small isolated bubbles of nematic phase which form first. Fig. 1(b) and 1(c) show the images at short intervals later, as the nematic bubbles increase in size. This occurs both by natural growth and also by coalescence. In the next stage the organization of the NLC into bubbles is replaced by an image of a homogeneous medium with entangled strings (Fig. 1(d)), which further evolve by straightening, shrinking, and excising small loops of closed string (Fig. 1(e)). The string dynamics has been very well described previously [6,7]. The film becomes increasingly homogeneous in appearance on longer time scales as the string pattern coarsens.

We analyzed series of images such as Fig. 1 to estimate the string to bubble ratio. In counting the number of bubbles we have to decide which bubbles are relevant for string formation. For example between Fig. 1(b) and 1(c) small bubbles seem to just get absorbed by bigger bubbles and we do not see them contributing to any string formation. This observation is consistent with the results of the numerical simulation in [13] where the collision of two large (critical) bubbles with one small (subcritical) bubble was considered and it was found that even if a string forms in such a collision, it gets pushed out of the smaller bubble due to large momentum of the fast expanding walls of the large bubbles. Even in the collision where all the bubbles involved are very small, we have not observed any string formation. This could again be because a string will get trapped at the collision point only when the bubble walls have sufficient momentum which may not happen if bubbles are very small. Even if some strings do form in the collisions of very tiny bubbles, these strings will be very small (for isolated collisions) and will disappear quickly by shrinking. As we count the strings at a little later stage anyway (when we can clearly distinguish them), these small strings would have disappeared by that time and hence no error will be caused by the neglect of very small bubbles.

Guided by these considerations, we made a judgement of a typical size for the bubble which seems to be large enough to contribute to string formation and then consider coalescence and growth of such bubbles. Bubbles r times the average bubble size are assumed to have arisen from the coalescence of r average bubbles (as can be verified by looking at the earlier stages of the bubble formation) and are thus counted r times.

The next step is to look at a photograph corresponding to the time at which these N bubbles have expanded to fill up the entire region of the photograph (of area A). As the bubbles coalesce they do not remain spherical but actually squeeze into gaps. The final image looks roughly like a rectangular region filled by N squares (with typical side of a square d roughly equal to $\sqrt{A/N}$). We estimated a count of about 55 "effective" bubbles in Fig. 1(c); the typical length scale was about 80 μm.

Our estimate of the probability that each of these bubbles will lead to a string upon final coalescence was made as follows. Imagine that the rectangular image above is a lattice of area A with m rows and n columns such that $N = mn$. As strings form because of mismatches of the director at bubble interfaces the maximum possible string length L_{max} is given by the total sum of the lengths of all the links on the lattice which is $2(NA)^{1/2}$. For Fig. 1(c) we estimate $L_{max} = 8.7$ mm. We then examine a photograph of the earliest possible stage when one can discern strings clearly and measure the total length of string L_s; in Fig. 1(d) we obtained 2.7 mm. The probability of string formation is then taken to be $P = L_s/L_{max} = 0.31$.

We repeated this analysis on several sequences, obtaining $P = 0.33 \pm 0.01$ for the string formed per bubble. Systematic errors are not negligible; we believe that this measurement provides a lower bound on the true value of P for two reasons. First the total string length will have decreased from its initial value by the time we are able to clearly resolve and measure L_s. Secondly strings which wander significantly in the direction normal to the plane of focus (but are still formed due to top bubble layers which we observe) will have their length underestimated. The actual underestimation may not be too bad since string wiggling will be dampened in the time between their formation and our measurement of their length. Further there is a selection effect in the experiment since we are able to see strings clearly only when they do not deviate significantly from the plane of focus. Thus we are effectively counting planar strings. This will suppress the effects of the three-dimensionality of the bubble layers. In other words, the bubbles we see are only the top few layers; similarly we see the strings at the time they are roughly confined to these top layers. The numbers we obtain, therefore, correspond to an effectively two-dimensional situation.

Note that for strictly 2 dimensional situation, the strings will form perpendicular to the plane formed by the bubble layer. In the present case, however, strings will bend around in the plane as they come towards the top (where the director will eventually become normal to the surface this being the preferred boundary condition for NLC-air interface) as well as when they squeeze through the bubble layers underneath them.

We note here that the string to bubble ratio we have measured is in good agreement with the theoretical value $1/\pi \approx 0.32$ obtained by Vachaspati [11] for the ground-state manifold $S^2/\mathbf{Z}_2$ in *two* dimensions, especially considering the above arguments that our estimates for P correspond to an effectively 2 dimensional situation.

We now turn to another interesting phenomenon we are able to observe in this experiment. We first cool our sample through the nematic phase transition so that disclinations are produced and then superheat from the nematic phase to the isotropic phase. Spherical isotropic bubbles are produced in the nematic medium. These have a somewhat different more transparent appearance than the nematic bubbles in an isotropic medium because of the different optical properties of the two phases. In this way it is quite easy to distinguish the different types of bubble. We observe very clearly in this process the heterogeneous nucleation of isotropic bubbles along disclination lines like beads on a necklace (see Fig. 2). This nucleation process is of great technological and theoretical interest [14,15]. We feel that NLCs are a good testing ground to check experimentally many of the theoretical predictions concerning the mechanism of the nucleation process such as the size distribution of bubbles.

Finally we point out that the implications of the work described here extend beyond the dynamics of the formation of disclinations in nematic liquid cystals because of the close analogies with the formation of other one-dimensional topological defects such as screw dislocations in crystals and vortices in superfluid helium [5,10].

First of all I wish to thank my collaborators L. Chandar, Eric Schiff and Ajit Srivastava. Thanks are due to Bernard Yurke for his valuable advice during his visit to Syracuse University. I would also like to express very warm thanks to Edward Lipson and his biophysics group at Syracuse University for their generosity, both for sharing their microscope apparatus and for instruction in its use. Without their support the research reported here would not have been possible. This research was supported by the Outstanding Junior Investigator Grant DOE DE-FG02-85ER40231. Finally I am indebted to Norma Sánchez for her kind invitation to lecture at Erice.

References

[1] P.G. de Gennes, *The Physics of Liquid Crystals* (Clarendon Press, Oxford, 1974).

[2] N.D. Mermin, Rev. Mod. Phys. **51** (1979) 591.

[3] Y. Bouligand, in *Les Houches Session XXXV, Physics of Defects*, R. Balian, Ed. (North Holland, Dordrecht, 1981).

[4] M. Kléman, *Points, Lines and Walls in Liquid Crystals, Magnetic Systems and Various Ordered Media* (Wiley, Chichester, 1983).

[5] S. Chandrasekhar and G.S. Ranganath, *The Structure and Energetics of Defects in Liquid Crystals*, Adv. Phys. **35** (1986) 507-596.

[6] I. Chuang, R. Durrer, N. Turok and B. Yurke, Science **251** (1991) 1336; B. Yurke, A.N. Pargellis, I. Chuang and N. Turok, Princeton University preprint PUPT-91-1274 (1991).

[7] I. Chuang, N. Turok and B. Yurke, Phys. Rev. Lett. **66** (1991) 2472.

[8] M. J. Bowick, L. Chandar, E. A. Schiff and A. M. Srivastava, *The Cosmological Kibble Mechanism in the Laboratory: String Formation in Liquid Crystals*, Syracuse/Minnesota preprint SU-HEP-4241-512/TPI-MINN-92/35-T.

[9] T. Kibble, J. Phys. **A9** (1976) 1387.

[10] W.H. Zurek, Nature **317** (1985) 505.

[11] T. Vachaspati, Phys. Rev. **D44** (1991) 3723.

[12] L. Onsager, Ann. N.Y. Acad. Sci. **Vol. 51** (1949) 627.

[13] A.M. Srivastava, Phys. Rev. **D45** (1992) R3304; Phys. Rev. **D46** (1992), in press.

[14] J.D. Gunton, M.San Miguel and P.S. Sahni, in *Phase Transitions and Critical Phenomenon*, Eds. C. Domb and J.L. Lebowitz (Academic Press, New York, 1983), Vol.**8**, pp. 269-482.

[15] L. Monette and W. Klein, Phys. Rev. Lett. **68** (1992) 2336.

Figure Captions

Fig. 1. A series of five images showing the isotropic to nematic phase transition in a drop of K15 on an untreated microscope slide; the images show a width of 0.7 mm of the drop. Note the stages of bubble nucleation and growth, bubble coalescence and string formation, and string coarsening. The delay times for each image (referred to the first frame showing discernible bubbles) are (a) 2 s, (b) 3 s, (c) 5 s, (d) 11 s and (e) 23 s. The scale of the pictures is given in (f).

Fig. 2. This series of three pictures shows the heterogeneous nucleation of isotropic bubbles along disclination lines. The delay times for each image, referred to the picture in (a), are (b) 2 s and (c) 4 s. The scale is given in (d).

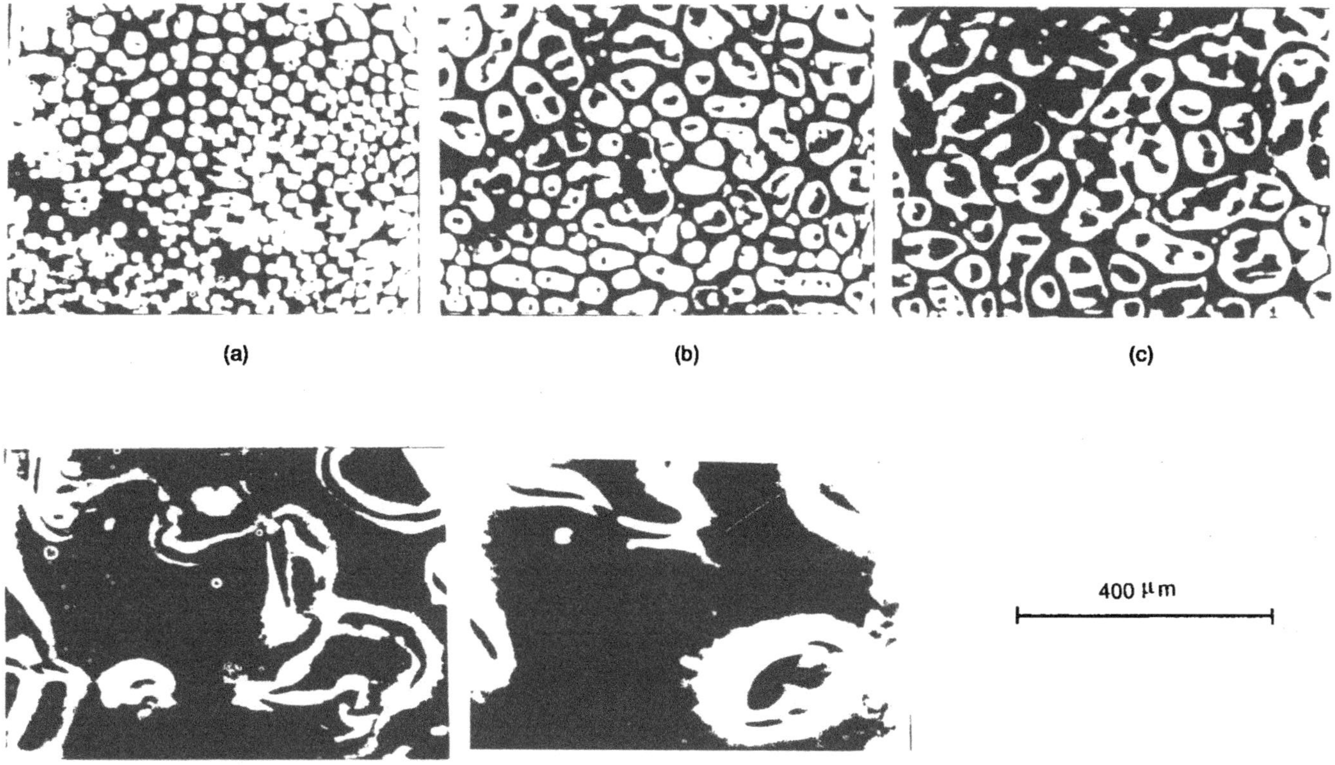

(a) (b) (c)

(d) (e) (f)

Fig. 1

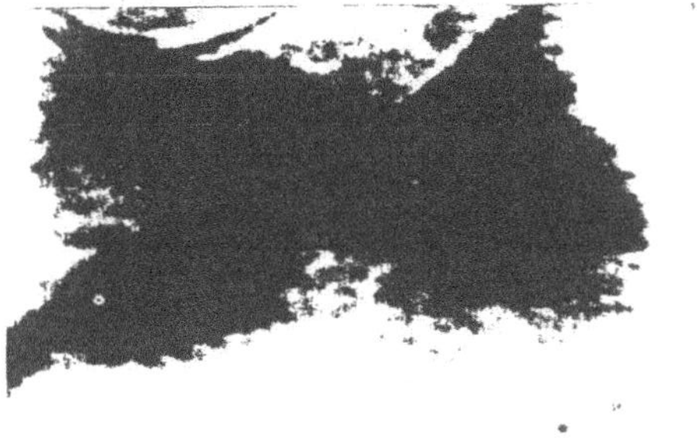

(a)

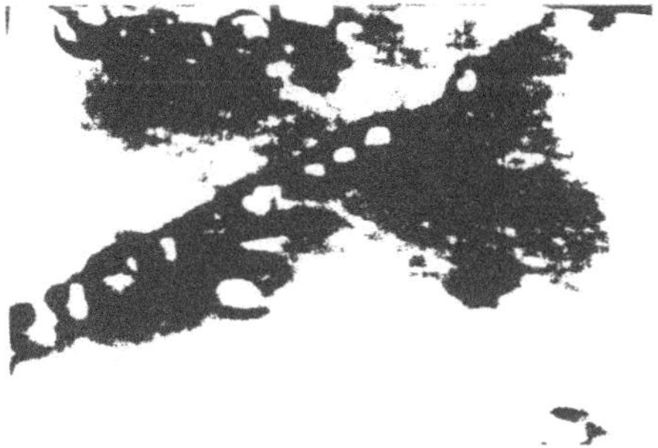

(b)

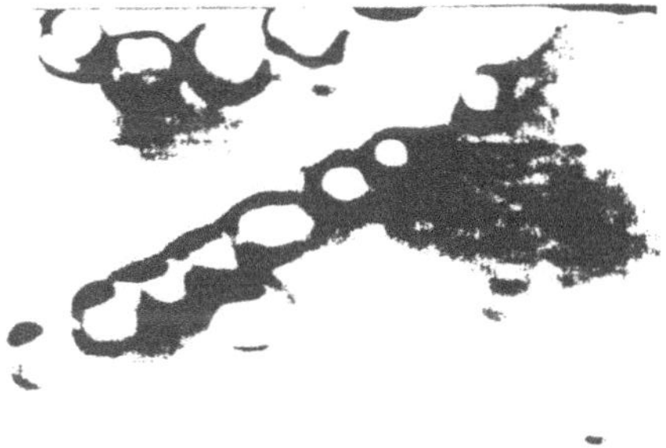

(c)

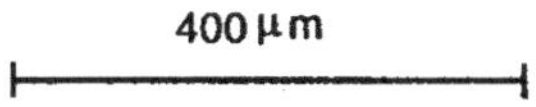

(d)

Fig. 2

The Phase Structure of Strings with Extrinsic Curvature

Mark Bowick*, Paul Coddington, Leping Han,
Geoffrey Harris and Enzo Marinari(†)

Dept. of Physics and NPAC,
Syracuse University,
Syracuse, NY 13244, USA

bowick@suhep.phy.syr.edu paulc@sccs.syr.edu han@sccs.syr.edu
gharris@sccs.syr.edu marinari@roma2.infn.it

(†): and Dipartimento di Fisica and INFN,
Università di Roma *Tor Vergata*
Viale della Ricerca Scientifica, 00173 Roma, Italy

December 21, 1992

*Invited talk delivered by Mark Bowick at the Workshop "String Quantum Gravity and Physics at the Planck Energy Scale," Erice, June 21-28, 1992 and at the 1st Iberian Meeting on Gravity, Évora, Portugal, September 21-26, 1992.

Abstract

We examine a model of non-self-avoiding, fluctuating surfaces as a candidate continuum string theory of surfaces in three dimensions. This model describes Dynamically Triangulated Random Surfaces embedded in three dimensions with an extrinsic curvature dependent action. We analyze, using Monte Carlo simulations, the dramatic crossover behaviour of several observables which characterize the geometry of these surfaces. We then critically discuss whether our observations are indicative of a phase transition.

1 Introduction

In this work, we investigate a theory of fluid, fluctuating random surfaces embedded in three dimensions. Various theories of fluctuating surfaces (string theories) arise in the description of many physical systems. Among these are the 3-dimensional Ising model, the strong interaction (QCD) in the infrared limit and fundamental unified theories of all interactions including gravity. Natural biological membranes, such as lipid bilayers, and artificial membranes, such as micelles and vesicles, also form a rich class of fluctuating surfaces together with interfaces, such as those in microemulsions, between two distinct three-dimensional bulk phases [1]. These latter systems are fluid because their component 'molecules' are loosely bound. Their constituents are arranged so that their net surface tension nearly vanishes; thus these membranes undergo large thermal fluctuations. These biological and chemical membranes exhibit self-avoidance, which we do not take into account in our simulations.

Just as field theories are described by sums over paths, string theories are formally characterized by a functional integral Z which can be written as a sum over surfaces weighted by $\exp(-S)$. Here S denotes the action associated with a particular surface. Note that we work in Euclidean space. To write down our action we introduce an explicit parametrization of a generic surface $\mathcal{M}$ in R^3 with coordinates (σ_1, σ_2) and the embedding $X^\mu(\sigma_i)$. μ runs from 1 to 3 (since we only study the case of a $3d$ embedding space). The induced metric (the pullback of the Euclidean R^3 metric via the embedding) is given by

$$h_{ij} = \partial_{\sigma_i} X^\mu \partial_{\sigma_j} X_\mu \; . \tag{1}$$

We will use Greek letters for the embedding space indices; Roman letters label the coordinate basis on each surface. A particularly natural choice for the action is the Nambu-Goto action, which is proportional to the area of the surface in the embedding space, and is given by

$$S = \mu \int \sqrt{\det(h)}. \tag{2}$$

This action is difficult to quantize, though, since it is non-polynomial. A further complication is that the measure in the path-integral constructed with this action is quite subtle, especially when the theory is discretized. By introducing additional degrees of freedom, in the form of an intrinsic metric g_{ij} on each surface, one can write down the Polyakov action [1].

$$S = \int \sqrt{|\det g|}(g^{ij}\partial_i X^\mu \partial_j X^\mu). \tag{3}$$

This action has proven to be much more tractable analytically. Note that it possesses a large degree of gauge freedom, associated with its invariance under reparametrizations of the metric and the intrinsic surface coordinates. Often, one gauge-fixes the metric to the form $g_{ij} = \exp(\rho)\delta_{ij}$; ρ is referred to as the Liouville mode. The Polyakov action is in fact independent of ρ [2], but this 'Weyl' symmetry is anomalous in general. The anomaly is expressed in terms of the central charge c. By carefully gauge-fixing the entire path-integral one can derive the anomaly condition

$$c_{\text{matter}} + c_\rho - 26 = 0. \tag{4}$$

The first term depends on the space in which the surfaces are embedded, or, in other words, on the particular field theory that lives on the two-dimensional surface. For instance, the central charge associated with D bosons is D (1 per boson); this characterizes a string embedded in R^D [3] The -26 arises from the anomaly contribution from ghosts needed for gauge-fixing reparametrization invariance. Note that when $D = 26$, c_ρ, the Liouville central charge, vanishes and the Liouville mode effectively decouples from the theory. For $D < 26$ one can integrate the matter (X^μ) fields out and obtain an effective Liouville theory. The Liouville theory is highly non-linear; furthermore, its analysis is complicated by the fact that the two-dimensional metric on the surface in which the Liouville field lives depends on the field itself! Nevertheless, with a few general assumptions about the path integral measure, the spectrum

[1] An additional term, which is a topological invariant proportional to the Euler character of the surface is added to both the Polyakov and Nambu-Goto actions. The coefficient of this term is referred to as the 'string coupling'. It is the only dimensionless parameter in the theory; therefore it serves as the perturbative expansion parameter for string theory.

[2] It follows that substituting the solution to the equations of motion back into the Polyakov action yields the Nambu-Goto action. At the quantum level, though, it has only been shown that these actions are equivalent for $D = 26$[2]. In lower dimensions this equivalence has been questioned, for example, by Distler [3]. The work of Polchinski and Strominger also suggests that there are alternate quantizations [4].

[3] A fermion has a central charge of 1/2, so that by studying an Ising model on random fluctuating surfaces, we are effectively studying strings imbedded in 'fractional' dimensions.

of the theory at each order in perturbation theory can be computed when $c_{\text{matter}} \leq 1$. Additional exponents, such as the Hausdorff dimension, which characterize both the intrinsic and extrinsic geometry of these surfaces, have also been computed from Liouville theory. These theories are often analyzed on surfaces of fixed area; in this case, the string susceptibility γ_h is determined by the dependence of the generating functional for surfaces of genus h on the area A

$$\ln(Z_h) = \exp(\mu_c A) A^{\gamma_h - 3}, \tag{5}$$

where μ_c is a non-universal constant. An analysis of the Liouville model yields

$$\gamma_h = 2 - \frac{(1-h)}{12}(25 - c_{\text{matter}} + \sqrt{(1 - c_{\text{matter}})(25 - c_{\text{matter}})}). \tag{6}$$

These computations break down in the regime $1 < c_{\text{matter}} < 25$, where exponents such as the string susceptibility given in (6) become imaginary.

Further progress in understanding string theory in a low number of dimensions has been made by mapping the sum over surfaces onto an integral over matrices. This mapping is realized by replacing continuum surfaces by their discrete cellular decompositions, such as their triangulations. In this construction, each triangle is equilateral with area 1 in the intrinsic metric; the coordination number at each vertex determines the intrinsic curvature of the surface. The coordinates i label the vertices of the triangulation. Then the discrete analogue of the intrinsic metric is the adjacency matrix C_{ij} whose elements equal 1 if i and j label neighbouring nodes of the triangulation, and vanish otherwise. Two-dimensional diffeomorphism invariance reduces, at this discrete level, to the permutation symmetry of the adjacency matrix. One of the keys, in fact, to the power of this construction is the preservation of this symmetry. The triangulation of a surface of genus h is then dual to a phi-cubed diagram of genus h. The large N ('t Hooft) expansion of an integral over $N \times N$ Hermitian matrices generates these Feynman graphs. In this case the area of the surfaces is not fixed, since all graphs of fixed genus are summed over. The action then acquires a contribution (from Legendre transformation) proportional to the product of the cosmological constant with surface area A. Orthogonal polynomial techniques can then be applied to analyze the matrix integral. Indeed the theory can be exactly solved in the 'double scaling limit' [5] in which the string coupling, cosmological constant and matter couplings (which are determined by the matrix 'potential') are tuned together. These solutions are exact, and thus include non-perturbative information about these string theories, albeit with ambiguities in certain cases. Unfortunately, the matrix models that represent theories with $c_{\text{matter}} > 1$ are too difficult to solve exactly.

It has generally been suspected that these analytic techniques fail for $c_{\text{matter}} > 1$ because the string theory becomes pathological. In this regime, a negative mass-squared particle, which is sometimes referred to as the 'tachyon', comes on shell [4]. It is thought that its presence might engender the proliferation of long tubes with thickness of the scale of the ultraviolet cutoff on surfaces that dominate the string functional integral. Cates, by analyzing the Liouville action, has put forth another perhaps related explanation of this pathology [7]. He points out that in the presence of vortex configurations of the form $\rho \sim -\ln(r)$, reducing the world-sheet cutoff to zero no longer causes the spacetime cutoff to vanish. He then shows that these vortices have negative free-energy when $c_{\text{matter}} > 1$ and proliferate to the extent that the Liouville partition function becomes ill-defined with a finite spacetime cutoff and zero cosmological constant. For positive cosmological constant, it then appears plausible that these vortex configurations would still appear, though perhaps the partition function will be finite. The predominant surfaces in the functional integral would thus be subject to large fluctuations of their internal geometry. Recent simulations of multiple Potts models with c_{matter} less than, equal to, and greater than 1 coupled to gravity have been performed [8]. These simulations, however, do not show any dramatic changes in the behaviour of the internal geometry of the dominant surfaces as c_{matter} is increased beyond 1. More analytic and numerical work clearly needs to be done to determine what exactly happens as c_{matter} becomes greater than 1.

Monte Carlo simulations of strings embedded in flat space for $D > 1$ do indicate that these theories fail to describe the fluctuations of two dimensional *smoothly embedded* surfaces in the continuum limit. The normals to the surfaces dominating the simulations are uncorrelated over the distance of a few lattice spacings. These surfaces also appear to have a large (greater than 8), or perhaps infinite, extrinsic Hausdorff dimension; they resemble

[4]More precisely, one encounters these instabilities in Liouville theory when the quantity $c - 24\Delta > 1$, where c denotes the central charge of the matter theory which describes the embedding of the surfaces and Δ is the conformal weight of the lowest weight state in this theory [6]. Since here we are considering flat space, $c = D$ and $\Delta = 0$.

branched polymers [5]. Such configurations should not describe, for instance, the domains of the 3d Ising model or QCD strings. We would thus like to find a modified string theory that is dominated by smoother surfaces.

The tachyon, and apparently these related instabilities, can be eliminated in particular cases by introducing fermionic coordinates and supersymmetry on the worldsheet, and implementing an appropriate projection of states. The fermions, presumably, effectively smooth out the surfaces. This would be consistent with what has been observed for one-dimensional geometries; the random walk of a spin one-half particle has Hausdorff dimension one and thus appears to be smooth [9]. Many authors have proposed an alternative modification of the string action [10, 11, 12, 13] via the addition of a term that directly suppresses extrinsic curvature[6]. We shall examine this class of theories in this talk.

To characterize the geometry of our surfaces further, we associate with each point in our generic surface $\mathcal{M}$ tangent vectors ($t_i^\mu \in T\mathcal{M}$) and a normal vector $n^\mu \in T\mathcal{M}^\perp$. The extrinsic curvature matrix K_{ij} (the second fundamental form) can be defined by

$$\partial_i n^\mu = -K_{ij} t^{\mu j} \ . \tag{7}$$

The eigenvalues of this matrix are the inverses of the radii of curvature of $\mathcal{M}$. One usually describes the geometry of these surfaces in terms of the mean curvature [15, 16]

$$H = \frac{1}{2} h^{ij} K_{ij} \ , \tag{8}$$

and the Gaussian curvature

$$K = \epsilon^{ik} \epsilon^{jl} K_{ij} K_{kl} \ . \tag{9}$$

One can show that the Gaussian curvature can be computed solely from the metric h_{ij}, while the mean curvature depends explicitly on the embedding X^μ.

---

[5] As above, singular configurations also dominate the Gaussian theory, which is essentially a theory of free random walks, rather than surfaces.

[6] In fact, integrating the fermions out of the Green-Schwarz superstring yields an action similar to the one we consider, but with the addition of a complex Wess-Zumino type term [14].

Our lattice model is constructed by triangulating each surface, as we discussed above in the context of matrix models. Each node of the triangulation is embedded in R^3 by the functions X_i^μ; i labels the ith node and μ runs from 1 to 3. We also associate a normal vector $(n^\mu)_{\hat{k}}$ with each triangle (indices with hats label the triangles). We shall study the theory defined by the action

$$S = S_G + \lambda S_E = \sum_{i,j,\mu} C_{ij}(X_i^\mu - X_j^\mu)^2 + \lambda \sum_{\hat{k},\hat{l},\mu} C^{\hat{k}\hat{l}}(1 - n_{\hat{k}}^\mu \cdot n_{\hat{l}}^\mu). \tag{10}$$

This model has been examined in references [17, 18, 19, 20, 21, 22, 23]. For $\lambda = 0$ this is simply a discretization of the Polyakov string action. The final term, which depends on the discretized extrinsic curvature, introduces a ferromagnetic interaction between surface normals, which one might hope would cause smoother surfaces to dominate the partition function. We would like to know if there is a smooth phase and a phase transition (at some finite λ_c) between this phase and the crumpled phase observed at $\lambda = 0$. If this were so, an interesting continuum limit of this lattice model could perhaps be constructed at this phase transition point, yielding a new continuum string theory.

The action we simulate is in fact a particularly natural discretization of

$$S = \int \sqrt{|\det g|}(g^{ij}\partial_i X^\mu \partial_j X^\mu + \frac{\lambda}{2} g^{ij} h^{kl} K_{ik} K_{jl}) \ . \tag{11}$$

Note that the second term in the action is manifestly positive, Weyl and reparametrization invariant, and that λ is a dimensionless coupling. So, it is not clear whether or not this term is relevant. Mean field (large D) and perturbative RG calculations have been performed using similar actions, such as

$$S = \int d^2\sigma \ (\mu_0 \sqrt{\det h} + \frac{1}{\alpha}\sqrt{\det h}(h^{ij} K_{ij})^2) \ , \tag{12}$$

in the regime in which the string tension μ_0 is small (unlike the usual particle physics limit of string theory, which is characterized by large μ_0). After integrating out fluctuations of the embedding X^μ between momentum scales Λ and $\tilde{\Lambda}$, it is found that the renormalization of the extrinsic curvature coupling is given to one-loop order by

$$\beta(\alpha) \equiv \Lambda \frac{d\alpha}{d\Lambda} = -\frac{3}{4\pi}\alpha^2 \ , \tag{13}$$

so that α is driven to infinity in the infra-red. This theory thus exhibits asymptotic freedom. Surfaces are smooth (the normals are correlated) below a persistence length [24]

$$\xi_p \sim \exp(\frac{4\pi}{3\alpha_{bare}}) \tag{14}$$

and are disordered above this scale. Some intuition into this result can be gained by observing that this theory is similar to the $O(3)$ sigma model, which is asymptotically free [9]. The normals to M are the analogues of $O(3)$ vectors, though in this case they are constrained to be normal to a surface governed by the action (11).

The analytic results, therefore, do not indicate that we should anticipate a finite λ phase transition. Note, though, that the RG calculations are based on a Nambu-Goto type action although we simulate an extension of the Polyakov action. Since the two actions are not clearly equivalent, *particularly* when extrinsic curvature dependent terms are added, we cannot simply assume that these analytic and our numerical results should agree.

2 The Simulation

We have considered triangulations with the topology of the torus, to minimize finite size effects. The above action was used, with the BRST invariant measure utilized also by Baillie, Johnston and Williams [18], so that

$$Z = \sum_{G \in T(1)} \int \prod_{\mu,i} dX_i^\mu \prod_i q_i^{\frac{D}{2}} \exp(-S_G - \lambda S_E) , \tag{15}$$

where $D = 3$, q_i is the connectivity of the ith vertex and $T(1)$ refers to the set of triangulations of genus 1. We used the standard Metropolis algorithm to update our configurations. To sweep through the space of triangulations we performed flips on randomly chosen links. Flips were automatically rejected if they yielded a degenerate triangulation. After a set of $3M$ flips was performed, $3M$ randomly selected embedding coordinates were updated via random shifts from a flat distribution. Most of the Monte Carlo simulations were performed on HP-9000 (720 and 750 series) workstations; we also collected some data by simulating lattices on each of the 32 nodes of a CM-5.

N	$C(\lambda)^{max}$	λ_c
36	3.484(8)	1.425(35)
72	4.571(15)	1.410(15)
144	5.37(14)	1.395(30)
288	5.55(7)	1.410(25)
576	5.81(17)	1.425(30)

Table 1: The maximum of the specific heat and its position, with errors, for different lattice sizes.

We ran on lattices ranging in size from $N = 36$ to 576 (N signifies the number of vertices) with 4 to 7 different values of λ for each N. Most of the data was this data was taken in the region $\lambda \in (1.325, 1.475)$. For small N, the runs consisted of 3×10^6 sweeps, while we performed longer runs (of up to 27×10^6 sweeps for $N = 576$) for larger lattices, because the auto-correlation times for our simulations were very large. (The correlation time for the radius of gyration was greater than 10^6 sweeps for $N = 576$!) To determine our observables as a function of λ we used a histogram reconstruction procedure. We patched different histograms by weighting them with the associated statistical indetermination (which was estimated by a jack-knife binned procedure). Various consistency checks indicate that this procedure is very reliable.

3 Observables

We measured the edge action S_E and the associated specific heat $C(\lambda) \equiv \frac{\lambda^2}{N}(< S_E^2 > - < S_E >^2)$. In Fig. 1 we plot the specific heat curve (constructed via the histogram procedure) and we tabulate its maximum and peak position for various lattice sizes in Table 1.

We see that the specific heat peak grows vigorously with N for small lattices, but that this growth quickly levels off for larger N. These observations agree fairly well with previous work [20, 21, 22]. For the larger lattices it appears that the peak position shifts very slowly towards higher values of λ, though this increase is not statistically significant. The shape of the peak does not change dramatically with N; it narrows perhaps a bit between $N = 144$ and $N = 576$.

To determine how the mean size of the dominant surfaces depend on λ, we measured the squared radius of gyration R_G;

$$R_G \equiv \frac{1}{N}\sum_{i,\mu}(X_i^\mu - X_{\text{com}}^\mu)^2 , \tag{16}$$

where the com subscript refers to the center of mass of the surface. By measuring the N dependence of the gyration radius, one can extract a value for the extrinsic Hausdorff dimension, which is given by

$$R_G \sim N^\nu \sim N^{\frac{2}{d_{extr}}} . \tag{17}$$

We plot R_G in Fig. 2(a), clearly the size of the mean surface size increases dramatically for λ near 1.4. In Fig. 2(b), we plot the effective Hausdorff dimension, given by

$$\nu(N) \equiv \frac{\log \frac{R_G(N)}{R_G(\frac{N}{2})}}{\log(2)} . \tag{18}$$

In the large λ limit $\nu \to 1$ and $d_{extr} \to 2$, as expected for flat surfaces. In the low λ limit d_{extr} becomes very large. In the pseudo-critical region ν is a linear function of λ. Curiously enough, the latter curve yields a Hausdorff dimension of 4, a value characteristic of branched polymers, near the location of the specific heat peak. This value is not particularly reliable though because of finite-size effects and also because it changes rapidly in this region. In ref. [22] a value compatible with ours ($D_H(\lambda_c) > 3.4$) is quoted for the critical theory. We stress however (and also here we are in complete agreement with [22]) that the Hausdorff dimension in the pseudo-critical region depends heavily and quite unusually on N.

In both the high and low λ regions finite size effects are quite small (compatible with zero to one standard deviation). In the pseudo-critical region, on the contrary, finite size effects are large. This effect cannot be explained by the shift in λ which one gets from the shift of the peak of the specific heat, which is far too small.

We also measured the magnitude of the extrinsic Gaussian curvature, $\int | K | \sqrt{| h |}$, given by

$$| \mathcal{K} | = \frac{1}{N}\sum_i | 2\pi - \sum_j \phi_i^j | . \tag{19}$$

Here ϕ_i^j denotes the angle subtended by the jth triangle at the ith vertex. This quantity, plotted in Fig. 3, measures the magnitude of the deficit angle in the embedding space averaged over all vertices. Note that the mean Gaussian curvature decreases rapidly in the neighborhood of $\lambda = 1.4$, indicating that a sharp crossover is occuring in this system. From this plot we can see that finite size effects increase with λ. They do not appear to peak in the region about $\lambda = 1.4$ as one might expect for a typical phase transition.

The magnitude of the intrinsic Gaussian curvature, $|\mathcal{R}|$, given by

$$|\mathcal{R}| = \frac{\pi}{3N}\sum_i |6 - q_i| \,, \tag{20}$$

is shown in Fig. 4. When the intrinsic and extrinsic metrics are equal, the intrinsic and extrinsic deficit angles are identical and $K = R/2$. Both extrinsic and extrinsic curvatures behave in a qualitatively similar manner; $|\mathcal{R}|$ drops off rapidly, just as $|\mathcal{K}|$ does. Through the peak region, though, $|\mathcal{K}|$ decreases by about a factor of 5 while $|\mathcal{R}|$ diminishes to only about .6 of its value on the left-hand side of the peak. Since the action explicitly suppresses mean curvature, and the mean and extrinsic Gaussian curvature are closely related (for instance, $H^2 > \frac{K}{2}$), we would expect that for large λ extrinsic fluctuations would be suppressed much more than fluctuations of intrinsic geometry.

The question of the equivalence of the Nambu-Goto and Polyakov actions motivated us to study the correlations between intrinsic geometry (which is not introduced independently in the Nambu-Goto formulation) and extrinsic geometry. We measured the quantity which we refer to as $\mathcal{K} * \mathcal{R}$

$$\mathcal{K} * \mathcal{R} \equiv \frac{\int KR}{\sqrt{\int K^2 \int R^2}} = \frac{\sum_i (2\pi - \sum_j \phi_i^j)(6 - q_i)}{\sqrt{\sum_i (2\pi - \sum_j \phi_i^j)^2 \sum_i (6 - q_i)^2}} \,. \tag{21}$$

This quantity is 1 when the metrics are equal, 0 if they are un- correlated, and negative when these curvatures are anti-correlated. The plot of $\mathcal{K} * \mathcal{R}$ in Fig. 5 indicates that intrinsic and extrinsic geometry are strongly correlated for small λ, but as one passes through the peak region they become decorrelated. This is not particularly surprising, given that the action directly suppresses only extrinsic fluctuations. Note that RG calculations based on the Nambu-Goto action plus an extrinsic curvature term perturb about a background that is both intrinsically and extrinsically flat. Given the observed decorrelation between intrinsic and extrinsic geometry, we would not anticipate that this background appears in the low- temperature limit of the model which we simulate.

We also measured various other observables which characterize both the intrinsic and extrinsic geometry of these surfaces. These measurements are discussed in another write-up of this work [23]. They all exhibit sharp crossover behaviour in the region near $\lambda = 1.4$. We found that the autocorrelation times of these observables grew rapidly as λ increased, but we did not note any maximum in these times in the region about $\lambda = 1.4$.

The crossover behaviour became also quite apparent when we examined typical snapshots of our simulated surfaces for various values of λ. In Figs. 6 (a)-(d) we present pictures of typical surfaces for $\lambda = .8, 1.3, 1.5$, and 2.0. Note that the surfaces rapidly become smoother and the normal-normal correlation length increases significantly as one passes from the second to the third of these pictures, which correspond to only slightly different values of λ.

4 Interpretation

This model of crumpled surfaces appears to exhibit sharp crossover behaviour in the region around $\lambda = 1.4$. The sharp change in the magnitude of the Gaussian curvature, the radius of gyration and other observables indicates that the normals acquire long-range correlations, up to the size of the systems we examine. The zero string tension measurement of [22] also shows that the disordered regime differs from the regime in which the surfaces are ordered (up to scale of the lattices that are simulated) by only a small shift in λ. This evidence might indicate the presence of a phase transition at this point. Since the peak growth rapidly diminishes for large N, such a phase transition would likely be higher than second order. Still, the apparent absence of diverging correlation times and, in some cases, increasing finite size effects in the peak region leads us to question whether we are actually observing a typical phase transition.

There are indeed other possible interpretations of our data. Note that the surfaces which we simulate are quite small. For instance, if the surfaces in our simulations had an intrinsic dimension of 2.87 (characteristic of $D =$

0 gravity), they would have roughly a linear size of fewer than 10 lattice spacings [7].

Perhaps the simplest alternative explanation for the presence of this peak is suggested by the arguments of Kroll and Gompper [25]. They argue that the peak occurs when the persistence length of the system approaches the size of the lattice ($\xi_p \sim N^{\frac{1}{d}}$); d denotes the intrinsic Hausdorff dimension. Fluctuations on a larger scale become more important. When this scale is greater than the lattice size these fluctuations are suppressed. Thus one might surmise that the specific heat will drop for large λ. Typically, the persistence length grows as $\xi_p \sim \exp(C\lambda)$; C is inversely proportional to the leading coefficient of the beta function. We would expect that the peak position should shift to the right with increasing N in this scenario as

$$\lambda_{peak}(N') - \lambda_{peak}(N) = \frac{\ln(\frac{N'}{N})}{dC}. \tag{22}$$

Quite a large value of C is needed to explain the rapid crossover; roughly values of $C \sim 10, d_{intr} \sim 3$ are more or less consistent with the magnitude of the peak shift and crossover width. The RG calculations using different forms of the action yield $C = \frac{4\pi}{3}$ (see equation 14), but this may not apply to the action we simulate. This reasoning also indicates that the peak should widen as the lattice size increases; we do not observe this at all. It seems plausible though that these arguments, based only on the leading term of the high λ expansion, are too naive.

An alternative scenario, which builds on the ideas in the above paragraph, is suggested by the tantalizing similarities between the results of our fluid surface simulations and what has been observed for the $d = 4$ $SU(2)$ Lattice Gauge Theory [26] and for the $d = 2$ $O(3)$ model.

The $O(3)$ model, which is thought to be asymptotically free, exhibits a specific heat peak near $\beta = 1.4$ (first measured via Monte Carlo simulations by Colot [27]). The origin of this peak is understood [28]; it is due to the fluctuations of the sigma particle, a low-mass bound state of the massless $O(3)$ pions. The sigma induces short-range order and contributes to the specific heat as a degree of freedom only at high temperatures (when the correlation length in the system becomes smaller than its inverse mass). The peak thus

[7] Of course, our lattices are too small, by one or two orders of magnitude, to really exhibit a convincing fractal structure.

occurs at the beginning of the crossover regime, when the correlation length is several lattice spacings.

According to the low temperature expansion, the correlation length grows as $\xi \sim \exp(2\pi\beta)/\beta$. Thus one would expect a fairly rapid crossover in the $O(3)$ model; the correlation length should increase by roughly a factor of 9 when β is shifted by about 0.35. Such a crossover is indeed observed, though it is not so apparent that it is as dramatic as the crossover behaviour observed for fluid surfaces. [8]

Recent simulations of the $O(3)$ model [29] indicate that the specific heat peak grows significantly when the lattice size L is increased from 5 to 15, and that virtually no growth in peak height is evident as L is increased further up to 100. Furthermore, the peak position shifts to the right as L grows and then appears to stabilize for large L. This is more or less what we observe in our simulations of fluid surfaces, on lattices of small size. We point out these similarities largely to emphasize that there does exist an asymptotically free theory (with low mass excitations) which exhibits crossover behaviour qualitatively similar to that observed in our simulations. The analogy is perhaps deeper, though, since the fluid surface action (with extrinsic curvature) in certain guises looks like a sigma model action. It would not therefore be so surprising from this point of view to find a sigma particle in these theories, perhaps associated with $(\hat{n}^2 - 1)$, in which $\hat{n}$ denotes the unit normal to our surfaces.

Another additional possibility is that fluctuations of the intrinsic geometry (the Liouville mode) are responsible for short-range order and contribute to the specific heat peak.

5 Conclusion

We have introduced a model of fluid random surfaces with an extrinsic curvature dependent action and explored its phase diagram. Unfortunately, we have been unable to determine if our model undergoes a phase (crumpling) transition at finite coupling. We have observed dramatic crossover behavior for particular observables in our Monte Carlo simulations, but on the other

[8] To compare quantitatively the width of the crossover regimes for these two models it would be necessary to measure a correlation length (perhaps extracted from the normal-normal correlation function) in these random surface simulations.

hand, the correlation times and certain finite-size effects do not behave as one would expect in the presence of a phase transition. The behavior of other lattice models also indicates that it is possible that we are observing the effects of finite-mass excitations on small lattices, rather than a phase transition. We hope that future work will clarify this murky state of affairs, to determine if there indeed exists a crumpling transition for fluid surfaces.

6 Acknowledgments

This work has been done with NPAC (Northeast Parallel Architectures Center) and CRPC (Center for Research in Parallel Computing) computing facilities. The research of MB was supported by the Department of Energy Outstanding Junior Investigator Grant DOE DE-FG02-85ER40231 and that of GH by research funds from Syracuse University. We gratefully acknowledge discussions, help and sympathy from Jan Ambjørn, Kostas Anagnostopoulos, John Apostolakis, Clive Baillie, Mike Douglas, David Edelsohn, Geoffrey Fox, Volyosha Kazakov, Emil Martinec, Alexander Migdal, David Nelson, Giorgio Parisi, Bengt Petersson, Steve Shenker, and Roy Williams. Deborah Jones, Peter Crockett, Mark Levinson and Nancy McCracken provided invaluable computational support.

References

[1] See F. David, in *Two Dimensional Quantum Gravity and Random Surfaces*, eds. D.J. Gross, T. Piran, and S. Weinberg, World Scientific (1992) pp. 80-124.

[2] T. Morris, Nucl. Phys. **B341** (1990) 443; J. Govaerts, Int. J. Mod. Phys. **A4** (1989) 173.

[3] J. Distler, *A Note on the 3D Ising Model as a String Theory*, Princeton University preprint PUPT-1324 (hep-th/9205100).

[4] J. Polchinski and A. Strominger, Phys. Rev. Lett. **67** (1991) 1681.

[5] E. Brézin and V.A. Kazakov, Phys. Lett. **236B** (1990) 144; M.R. Douglas and S.H. Shenker, Nucl. Phys. **B335** (1990) 635; D. J. Gross and A. A. Migdal, Phys. Rev. Lett. **64** (1990) 127.

[6] D. Kutasov and N. Seiberg, Nucl. Phys. **B358** (1991) 600.

[7] M.E. Cates, Europhys. Lett. **7** (1988) 719.

[8] C. Baillie and D. Johnston, U. of Colorado preprint COLO-HEP-276, 1992; S. Catterall, J. Kogut, and R. Renken, Univ. of Illinois preprint ILL-TH-92-7; S. Catterall, presented at Int. Symposium on Lattice Field Theory, Amsterdam 1992; G. Thorleiffson, presented at Int. Symposium on Lattice Field Theory, Amsterdam 1992 (to appear in Nucl. Phys. Proc. Suppl.).

[9] A. M. Polyakov, *Gauge Fields and Strings* (Harwood Academic Publishers, Chur, Switzerland 1987).

[10] P. B. Canham, J. Theor. Biol. **26** (1970) 61.

[11] W. Helfrich, J. Naturforsch. **28C** (1973) 693.

[12] A. M. Polyakov, Nucl. Phys. **B268** (1986) 406.

[13] H. Kleinert, Phys. Lett. **174B** (1986) 335.

[14] P. Wiegmann, Nucl. Phys. **B323** (1989) 330.

[15] Y. Choquet-Bruhat, C. DeWitt-Morette and M. Dillard-Bleick, *Analysis, Manifolds, and Physics* (Elsevier, Amsterdam 1987).

[16] M. P. Do Carmo, *Differential Geometry of Curves and Surfaces* (Prentice-Hall, Englewood, N.J., USA 1976).

[17] S. Catterall, Phys. Lett. **220B** (1989) 207.

[18] C. Baillie, D. Johnston and R. Williams, Nucl. Phys. **B335** (1990) 469.

[19] C. Baillie, S. Catterall, D. Johnston and R. Williams, Nucl. Phys. **B348** (1991) 543.

[20] S. Catterall, J. Kogut and R. Renken, Nucl. Phys. **B** (Proc. Suppl.) **99A** (1991) 1.

[21] J. Ambjørn, J. Jurkiewicz, S. Varsted, A. Irbäck and B. Petersson, Phys. Lett. **275B** (1992) 295.

[22] J. Ambjørn, A. Irbäck, J. Jurkiewicz and B. Petersson, Niels Bohr Institute Preprint NBI-HE-92-40 (hep-lat/9207008).

[23] M. Bowick, P. Coddington, L. Han, G. Harris and E. Marinari, Syracuse and Rome preprint SU-HEP-4241-517, SCCS-3557, ROM2F-92-48 (hep-lat/9209020) submitted to Nucl. Phys. **B**.

[24] L. Peliti and S. Leibler, Phys. Rev. Lett. **54** (1985) 1690.

[25] D. Kroll and G. Gompper, Europhys. Lett. **19** (1992) 581.

[26] M. Falcioni, E. Marinari, M. L. Paciello, G. Parisi and B. Taglienti, Phys. Lett. **102B** (1981) 270.

[27] J. L. Colot, J. Phys. **A16** (1983) 4423.

[28] J. Orloff and R. Brout, Nucl. Phys. **B270** (1986) 273.

[29] K. Anagnostopoulos, private communication.

7 Figure Captions

Fig. 1 The specific heat $C(\lambda)$ as a function of λ. As in all other pictures, fille circles and a dotted line correspond to $N = 144$, crosses and a dashe line indicate $N = 288$, and empty squares and a solid line represen $N = 576$.

Fig. 2a The gyration radius R_G defined in (16), plotted as in Fig. 1.

Fig. 2b The effective inverse Hausdorff dimension ν as a function of λ, as de fined in (17). The filled dots and the dashed curve are from a fit t the $N = 288$ and $N = 144$ data, while the empty dots and solid curv represent the fit to $N = 576$ and $N = 288$.

Fig. 3 The extrinsic Gaussian curvature $\mid \mathcal{K} \mid$ defined in (19), plotted as in Fig. 1.

Fig. 4 The intrinsic curvature $\mid \mathcal{R} \mid$ defined in (20), plotted as in Fig.1.

Fig. 5 The intrinsic extrinsic curvature correlation, as defined in (21), plotte as in Fig.1.

Fig. 6a A snapshot of a 576 node surface of toroidal topology for $\lambda = 0.8$.

Fig. 6b As in Fig. 6a for $\lambda = 1.3$.

Fig. 6c As in Fig. 6a for $\lambda = 1.5$.

Fig. 6d As in Fig. 6a for $\lambda = 2.0$.

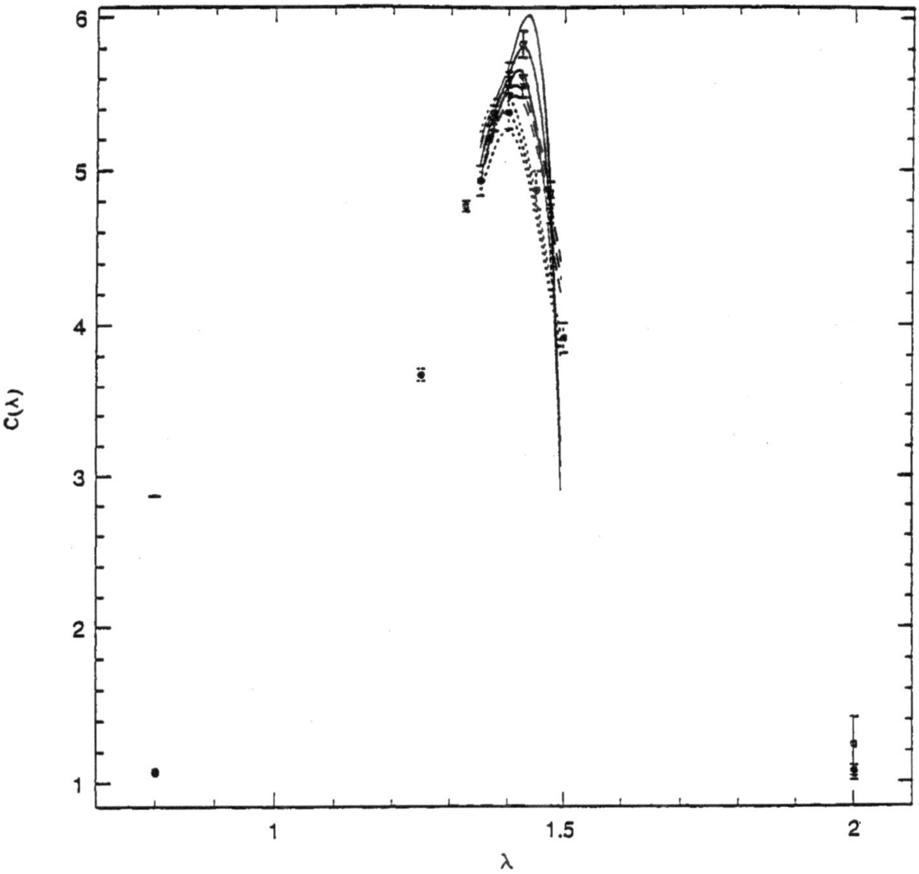

Fig. 1

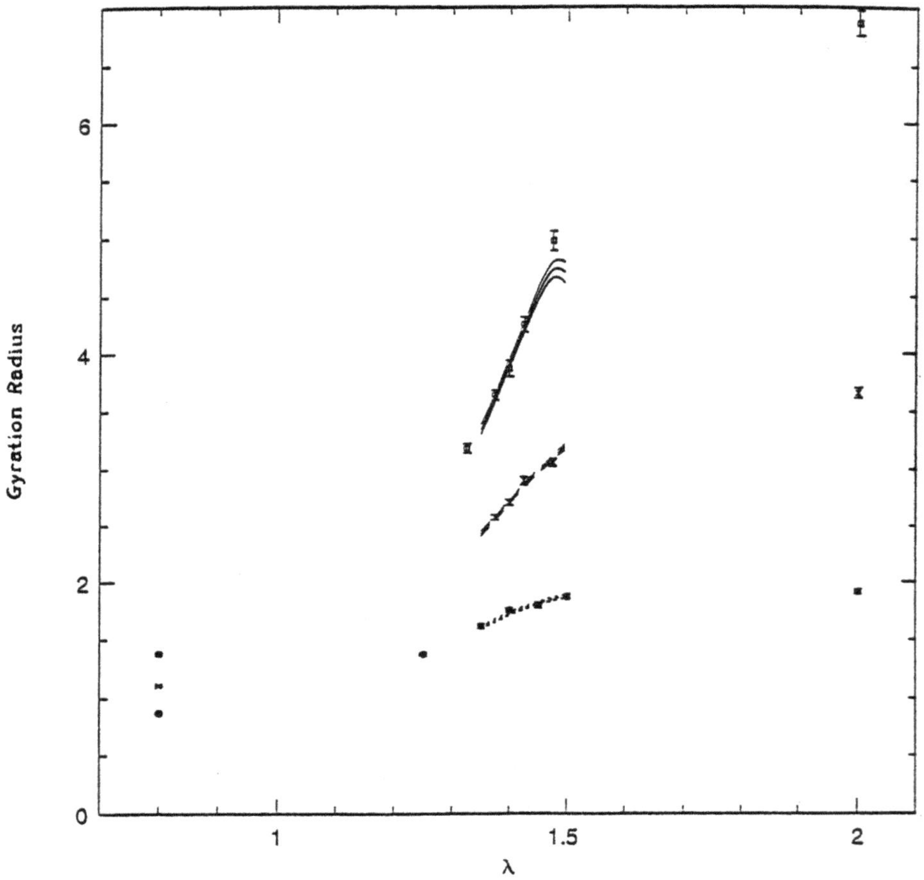

Fig. 2a

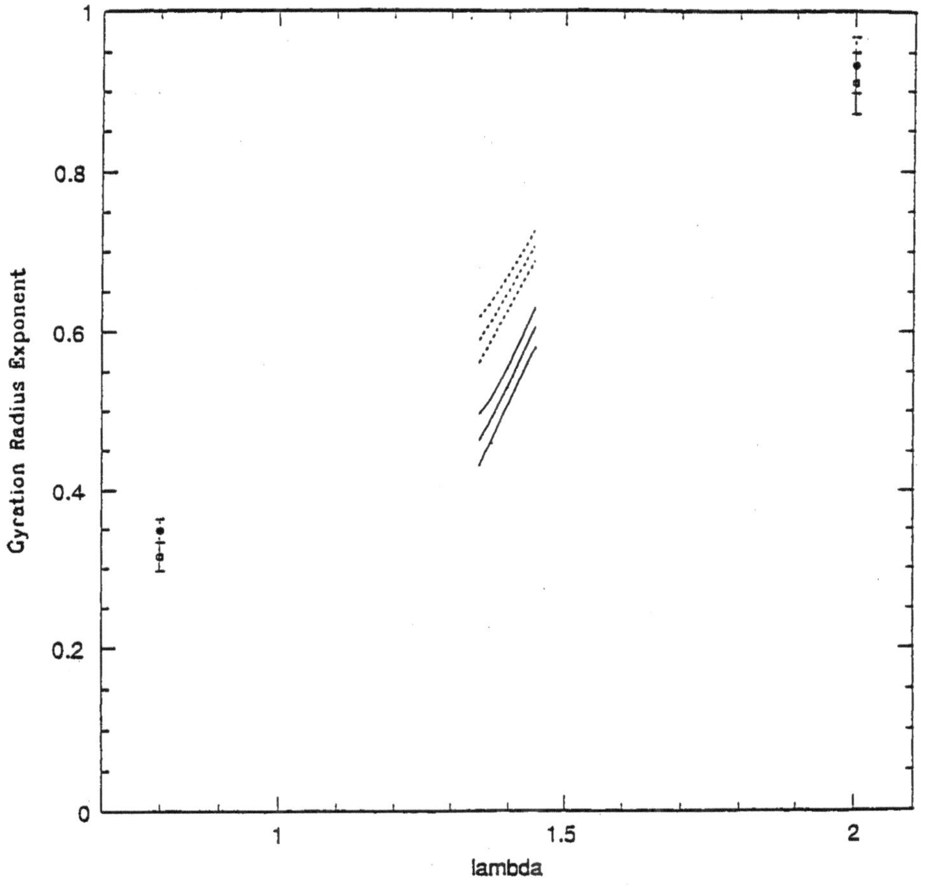

Fig. 2b

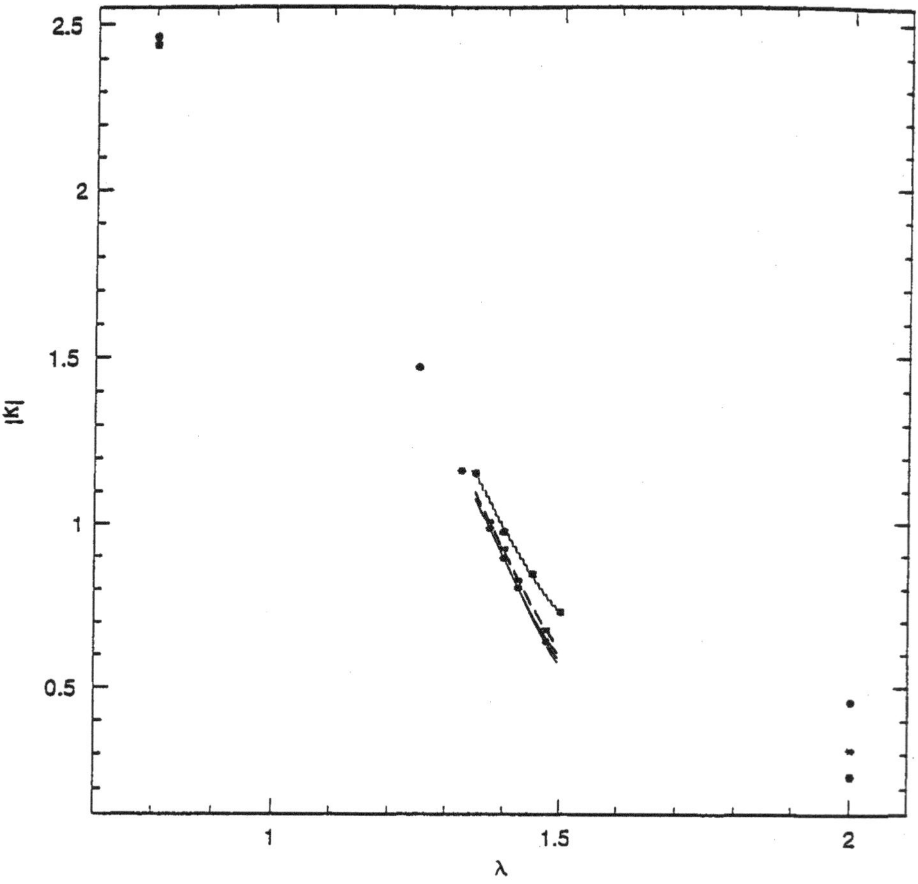

Fig. 3

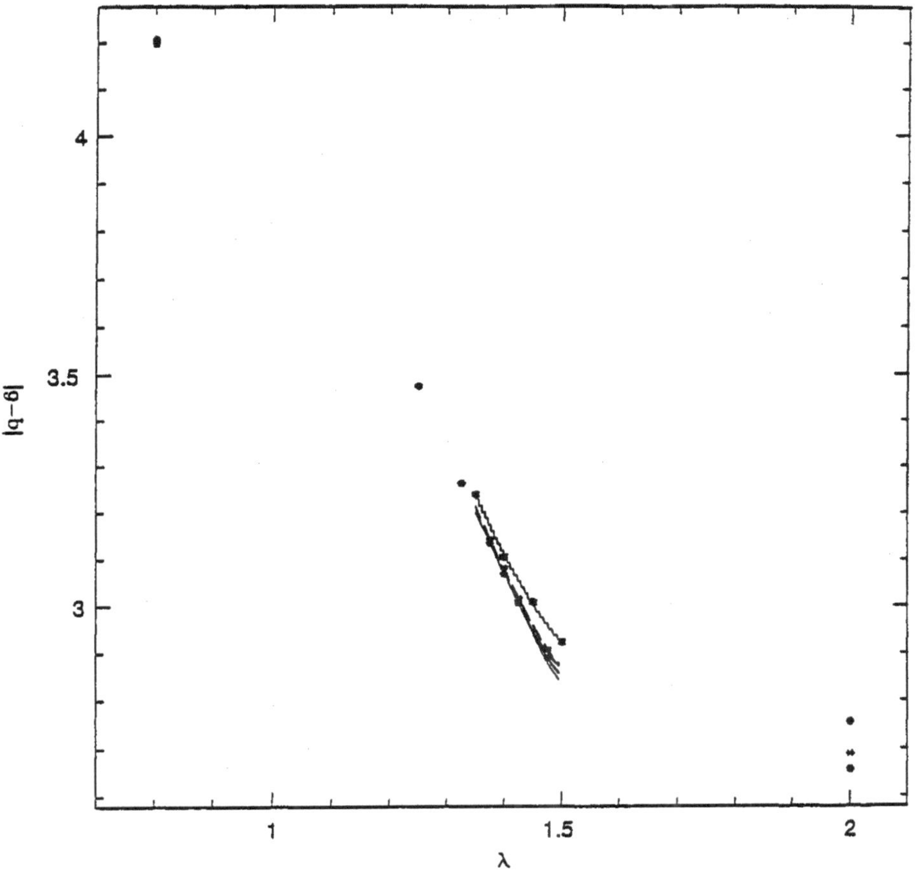

Fig. 4

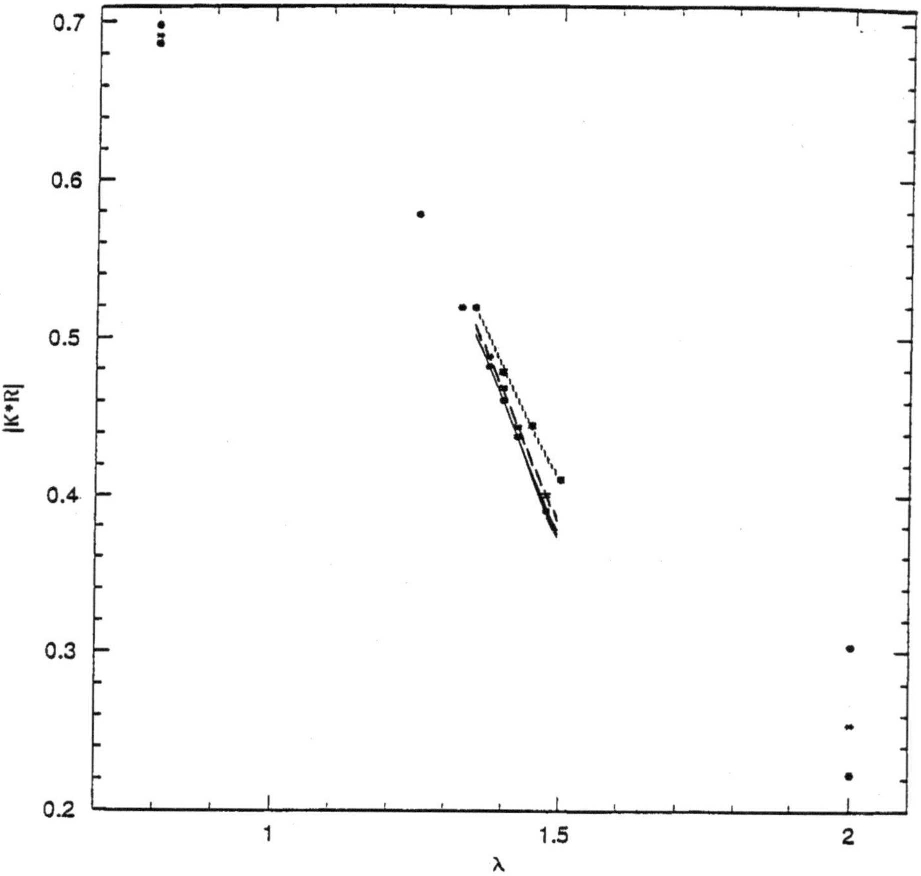

Fig. 5

Fig. 6a

Fig. 6b

Fig. 6c

Fig. 6d

CLASSICAL THEORY OF TWO-DIMENSIONAL BLACK HOLES

VALERY P. FROLOV*

The Copenhagen University Observatory,
Øster Voldgade 3, DK-1350 Copenhagen K, Denmark†

Abstract

The classical theory of two-dimensional black-hole is discussed. It is shown that the main results of the four-dimensional black holes physics (such as the uniqueness theorems, mass formulas and thermodynamical analogy) have their counterparts in the two-dimensional case. It is proved the existence of static solutions describing black holes with tachyon hairs. The entropy of a charged two-dimensional black hole is shown to be $2\pi \exp(\Phi_H)$, where Φ_H is the value of the dilaton field at the horizon. Four laws of two-dimensional black holes physics are formulated.

1 Introduction

There are several reasons why the theory of two-dimensional black holes attracted recently such a great attention. Quite long time ago it was understood that quantum corrections must necessarily modify Einstein equations when the spacetime curvature becomes comparable with the Planckian curvature l_{Pl}^{-2} ($l_{Pl} = \sqrt{(\hbar G/c^3)}$). The fundamental string theory allows one in principle to obtain the effective equations which include the necessary modifications. The case when the target space for the string theory has only two dimensions is the simplest one. The string motivated effective two-dimensional gravitational theory is in many aspects simpler then the full four-dimensional theory. It is natural to use two-dimensional gravity as a toy model for the latter. The existence of a black-hole like solutions in a string motivated two-dimensional gravity made it possible to analyze many of unsolved problems of

*On leave of absence from P.N.Lebedev Physical Institute, Leninsky prospect 53,Moscow 117924 Russia
†E-mail address: frolov@astro.ku.dk

four-dimensional black hole physics. Among these problems we mention the following: i. internal structure of black holes and the singularities inside a black hole; ii. final state of an evaporating black hole; iii. loss of quantum coherence in the proccess of quantum evaporation of a black hole.

The most consistent way to deal with these problems is to try to solve them in the framework of the complete string theory. It has not been done. Instead of this much simpler semiclassical approximation was developed in which the gravitational field is descibed as classical and quantum corrections are taken into account. The natural first step in this direction is to develop the general theory of classical two-dimensional black holes and to understand up to what extend it is possible to reproduce in it the standart basic results of four-dimensional black-hole's physics. The main purpose of this paper is to discuss the classical theory of two-dimensional black holes.

We begin with quite general remarks. The black holes in general case are defined as regions in asymptotically-flat spacetime where the gravitational field is so strong that no information-carrying signals can escape from these regions to infinity. The idea of black holes has proved to be highly fruitful. Originally black holes arose as special solutions of the Einstein equations. Later it was understood that the black holes may play an important role in the Universe. The theoretical discovery by Hawking of quantum radiation from black holes provided us with understanding that black holes may also play the role of a 'Rosetta stone' to relate gravity, quanta and thermodynamics.

Many properties of black holes are directly connected with the non-trivial causal structure of the spacetime in the presence of a black hole. There are also a number of fundamental results (such as the increasing of the black holes surface [1], uniqueness theorems [2]-[6], mass formulas and thermodynamic analogy [7]) which for their proof require the Einstein equations[1]. An interesting general problem is how far these results can be extended to other versions of the theory of gravity. The theory of black holes in the string-motivated two-dimensional gravity may be considered as an intersting and important example.

The evolution of a classical string is described by a 2-dimensional sheet in a d-dimensional target space. If g is an external metric in the target space then the conditions of self-consistency of quantum strings equations require that the metric g must obey special relations. These relations determined by setting the β-function to zero can be also obtained by varying some effective action [9]. This effective action contains the curvature. It describes the theory of gravity which is a generalization of the Einstein theory and which (for bosonic strings) besides the gravitational field g contains also the dilaton field Φ and the tachyon field T (as well as some other fields).

In this construction one can choose the number d of target space dimensions arbitrarily. The case when $d = 2$ plays an important role. It is singled out by the property that due to the absence of the transversial directions a string can not vibrate. As a consequence the string-motivated two-dimensional gravity contains only

[1]For the general discussion see [8] and references therein.

one dynamical degree of freedom which is connected with the tachyon field. It was shown by Witten [10] and by Mandel, Sengupta, and Wadia [11] that in the absence of the tachyon field the string-motivated two-dimensional gravity allows black-hole like solutions. These solutions, their properties and quantum effects in their presence appeared to be important for the string theory [10]-[14]. On the other hand in the framework of these two-dimensional toy models the most intriguing problems of black holes physics (singularities, loss of coherence and final state of evaporating black holes) can be analysed. Many papers on this subject appeared recently. Without pretending to be complete, we refer only to some of them [15]-[32].

The main aim of the paper is to develop the general theory of two-dimensional black holes and to compare its results with the results of the 'standard' theory of four-dimensional black holes.

The sign conventions of MTW [33] and the units in which $G = \hbar = c = 1$ are used.

2 Equations of String Two-Dimensional Gravity

We consider two-dimensional gravitational theory described by the effective action

$$S = S[g, \Phi, T, A] \equiv \frac{1}{2}\int_V d^2x\sqrt{-g}e^{\Phi}(R+L) - \int_{\partial V} dx\sqrt{|h|}e^{\Phi}K, \tag{2.1}$$

where

$$L = (\nabla\Phi)^2 - (\nabla T)^2 + \mu^2T^2 + \lambda - \frac{1}{2}F_{\mu\nu}F^{\mu\nu}, \tag{2.2}$$

$$F_{\mu\nu} = \nabla_\mu A_\nu - \nabla_\nu A_\mu. \tag{2.3}$$

The fields Φ, T and A_μ which enter the action are the dilaton, the tachyon and the electromagnetic field, respectively. The surface term is added in Eq.(2.1) in order to make the variation procedure self-consistent. The quantity h is the induced metric on ∂V and K is its extrinsic cuvature.

In the absence of the electromagnetic field the action Eq.(2.1) coincides with the effective action for closed bosonic strings [9, 10, 11, 28, 29]. We include the electromagnetic field into the action (2.1) because many results of two-dimensional black-holes physics can be easily obtained in the general case of charged black holes. On the other hand the consideration of charged black holes allow us to trace more deeply the analogy between two-dimensional black hole solutions and 4-d ones. This action naturally arises in the two-dimensional heterotic string theory [27].

The following system of equations is obtained by varying the action (2.1) :

$$T_{\mu\nu} \equiv e^{\Phi}[\Phi_{;\mu\nu} + T_{,\mu}T_{,\nu} - \frac{1}{2}g_{\mu\nu}(R + 2F^2)] = 0, \tag{2.4}$$

$$\Box\Phi + (\nabla\Phi)^2 - \frac{1}{2}(R+L) = 0, \tag{2.5}$$

$$e^{-\Phi}\nabla^{\mu}(e^{\Phi}\nabla_{\mu}T)+\mu^{2}T=0, \tag{2.6}$$

$$(e^{\Phi}F)_{,\alpha}=0, \tag{2.7}$$

where $F_{\mu\nu}=Fe_{\mu\nu},\quad e_{\mu\nu}=e_{[\mu\nu]},\quad e_{01}=(-g)^{1/2}$.

Instead of Eq.(2.5) one may use the relation which follows from it and from the trace equation $T^{\mu}_{\mu}=0$ and which is of the form

$$(\nabla\Phi)^{2}-(\nabla T)^{2}+R-\mu^{2}T^{2}-\lambda+3F^{2}=0. \tag{2.8}$$

It should be emphasized that it is the presence of the dilaton field which makes the considered two-dimensional theory of gravity nontrivial. The dilaton field which enters all the gravitationalequations and which violates the equivalence principle is to be considered as fundamental as the gravitational field itself.

3 Static Solutions

In the absence of the tachyon field the solutions for the action (2.1) allow a Killing vector. In order to prove this we note that for $T=0$ the two-dimensional gravity field equations are of the form

$$\Phi_{;\mu\nu}=\frac{1}{2}g_{\mu\nu}(R+2F^{2}). \tag{3.1}$$

If $\Phi_{;\alpha}\neq 0$[2] then the equation (3.1) implies that the vector field

$$\xi^{\mu}\equiv -e^{\mu\alpha}\Phi_{;\alpha} \tag{3.2}$$

obeys the Killing equation $\xi_{(\mu;\nu)}=0$. Moreover Eq.(2.7) and Eq.(3.2) show that

$$\xi^{\mu}\Phi_{;\mu}=0,\qquad \xi^{\mu}F_{;\mu}=0, \tag{3.3}$$

and hence the dilaton and electric fields are constant along the Killing trajectories. This result may be considered as a two-dimensional version of the Birkhoff's theorem.

In a two-dimensional space the tensor $\xi_{[\mu}\xi_{\nu;\lambda]}$ vanishes identically and according to the Frobenius' theorem one has

$$\xi_{\mu}=\beta\eta_{,\mu}. \tag{3.4}$$

The scalar functions $t=\alpha\eta$ and $r=\alpha^{-1}\Phi$ where $\alpha=\lambda^{1/2}$ can be chosen as coordinates. In these coordinates [27, 34, 35]

$$ds^{2}=-Adt^{2}+\frac{dr^{2}}{A}, \tag{3.5}$$

[2] In the degenerate case when $\Phi_{;\alpha}=0$ the solution does not contain arbitrary parameters, the spacetime metric coinsides with anti-de Sitter metric, and $F=\alpha$ [35].

$$A = 1 - \frac{2M}{\alpha} e^{-\alpha r} + \frac{Q^2}{\alpha^2} e^{-2\alpha r}, \tag{3.6}$$

$$\Phi = \alpha r, \qquad F = Qe^{-\alpha r}. \tag{3.7}$$

For $M \geq Q$ this solution describes a charged (with charge Q) two-dimensional black hole, the coordinates t and r being valid out of the horizon defined by the relation

$$r = r_{\pm} \equiv \alpha^{-1} \ln[\alpha^{-1}(M \pm \sqrt{M^2 - Q^2})]. \tag{3.8}$$

The global structure of this solution is the same as for the 4-dimensional charged black hole. For $Q > M$ the horizons are absent and instead of a black hole one has a naked singularity. For $Q = 0$ the solution (3.5)–(3.7) coinsides with the black-hole solution in the string theory [10, 11].

Two-dimensional black holes allow tachyon hairs. The solution describing an uncharged black hole with tachyon hairs is of the form [35]

$$ds^2 = -e^{2U} dt^2 + \alpha^{-2} e^{2x} U'^2 dx^2, \tag{3.9}$$

$$\Phi = x - U + \ln \frac{M}{\alpha}, \tag{3.10}$$

$$U(x) = -\int_x^{\infty} \left[\frac{1 - T'^2}{1 + e^{2x}(1 + \beta^2 T^2)} \right]^{1/2} dx. \tag{3.11}$$

Here the coordinate x is connected with the values of the acceleration w of the Killing observor ($x = -\ln(w/\alpha)$), and M is the integration constant (mass). The static tachyon field which enter these relations can be found as the solution of the following *master equation* [35]

$$(e^{-2x} + 1 + \beta^2 T^2) T'' + (1 - T'^2)(1 + \beta^2 T^2) T' + \beta^2 (1 - T'^2) T = 0. \tag{3.12}$$

The boundary condition of regularity of the tachyon field at the black hole horizon ($x = -\infty$)

$$\lim_{x \to -\infty} T' = 0. \tag{3.13}$$

singles out a solution of the master equation which is regular at the horizon and for $\beta < 1$ is decreasing at space infinity. This solution is unambiguously specified by its value at the horizon $T_H = T(-\infty)$. The metric and the dilaton field defined by this solution are regular everywhere and they describe a static black hole with the tachyon hairs. The general static uncharged black-hole solution contains two parameters: the value of the tachyon field at the horizon T_H and the mass of a black hole M.

4 Surface Gravity, Mass Formula

One of the important characteristics of a static black hole is its surface gravity κ which can be defined as the limiting value at the horizon of red-shifted acceleration of the Killing observer $\kappa = w\xi$. The helpful representation for the surface gravity can be obtained by using the following relation

$$\kappa = \frac{1}{2}\int_{\Sigma} R\xi^{\alpha} d\Sigma_{\alpha}. \qquad (4.1)$$

The integration in the right-hand side is taken over the exterior of the black hole.

By using the field equations and the Stocks' theorem one can transform Eq.(4.1) into the form

$$\kappa = \frac{\alpha}{2} + \frac{1}{2}\int_{\Sigma}[(\nabla T)^2 - F^2]\xi^{\alpha} d\Sigma_{\alpha}. \qquad (4.2)$$

This relation shows that in the absence of the electromagnetic field the surface gravity of the black hole with tachyon hairs is always higher then $\alpha/2$, the value of the surface gravity for tachyon free solutions. The presence of the electromagnetic field decreases the surface gravity.

An important role in the standard black hole physics is played by the so called *mass formula* which relates the asymptotic properties of the black-hole metric with the parameters of the event horizon. To obtain the mass formula for two-dimensional black holes we ussume that all the field equations except the gravitational one (Eq.(2.4)) are satisfied. The coordinate invariance of the action (2.1) implies that

$$T^{\mu\nu}{}_{;\nu} = 0, \qquad (4.3)$$

where $T^{\mu\nu}$ is defined by Eq.(2.4). We assume that the gravitational field is static and allows a timelike Killing vector ξ^{μ}. Then one has

$$(T^{\mu\nu}\xi_{\mu})_{;\nu} = 0. \qquad (4.4)$$

In two dimensions Eq.(4.4) implies that there exist such a scalar function m that

$$m_{,\alpha} = -e_{\alpha\mu}T^{\mu\nu}\xi_{\nu}. \qquad (4.5)$$

(Eq.(4.4) plays the role of the integrability condition for Eq.(4.5).) Straightforward calculations show [34] that the "mass function" m at a point x can be written in the following explicit form

$$m(x) = e^{\Phi(x)}w(x)\xi(x) + \varphi(x)Q(x), \qquad (4.6)$$

where

$$\xi = |\, \xi_{\alpha}\xi^{\alpha}\, |^{1/2}, \quad w = |\, w_{\alpha}w^{\alpha}\, |^{1/2}, \quad w_{\alpha} = \nabla_{\alpha}\ln\xi, \qquad (4.7)$$

$\varphi(x) \equiv A_{\alpha}(x)\xi^{\alpha}(x)$ is the value of the electric potential at a given point and $Q(x) \equiv e^{\Phi(x)}F(x)$ is the electric charge 'inside' this point. In what follows we assume that the Killing vector is normalized by the condition $\xi(x=\infty)=1$ and we choose electric potential to vanish at infinity.

Eq.(4.6) is valid for any static gravitational field. In the case when the field equations $T_{\mu\nu}=0$ are satisfied Eq.(4.5) implies that $m=const$. In other words Eq.(4.6) gives the first integral of the gravitational field equations. This first integral can be used to relate the mass of a black hole seen by a distant observer with the parameters of the black-hole's horizon. The asymptotic value of $e^{\Phi}w\xi$ at infinity coincides with the ADM mass M of the black hole and hence $m=M$. On the other hand the limiting value of $w\xi$ at the horizon gives the surface gravity of the black hole

$$\kappa=[w\xi]_H. \tag{4.8}$$

Hence we have

$$M=e^{\Phi_H}\kappa+\varphi_H Q, \tag{4.9}$$

where φ_H is the electric potential and $Q=[e^{\Phi}F]_H$ is the charge of the black hole. The obtained relation Eq.(4.9) is the desired *mass formula* for two-dimensional black holes.

For tachyon free solutions (3.5)–(3.7) the values of Φ_H, κ, and φ_H are

$$\Phi_H=\ln\left[\frac{M+\sqrt{M^2-Q^2}}{\alpha}\right]; \tag{4.10}$$

$$\kappa=\frac{\alpha\sqrt{M^2-Q^2}}{M+\sqrt{M^2-Q^2}}; \tag{4.11}$$

$$\varphi_H=\frac{Q}{M+\sqrt{M^2-Q^2}}. \tag{4.12}$$

These formulas resemble the analogous formulas for a 4-*d* Reissner-Nordström black hole.

5 Euclidean Two-dimensional Black Hole, Entropy

The study of the analytic continuation of a two-dimensional black-hole spacetime and its Euclidean section is important for understanding quantum aspects of the black-hole physics and for development of the thermodymanical analogy. The metric of the Euclidean black hole can be by making the Wick's rotation $\tau=it$ and it can be written in the form:

$$ds_E^2\equiv g_{\mu\nu}^E dx^\mu dx^\nu=\xi_E^2 d\tau^2+\frac{dx^2}{P}. \tag{5.1}$$

Denote by l a proper distance from the Euclidean horizon $\xi_E^2=0$ ($dl=dx/P^{1/2}$) then the metric Eq.(5.1) near the horizon has the following form

$$ds_E^2\approx\kappa^2 l^2 d\tau^2+dl^2, \tag{5.2}$$

where

$$\kappa = \lim_{l \to 0}(w\xi) = \left[P^{\frac{1}{2}}\frac{d\xi}{dx}\right]_{l=0} \tag{5.3}$$

is a surface gravity of a black hole. The Euclidean black hole is defined as a regular manifold with the metric (5.1). The regularity of the metric at $l = 0$ implies that τ is periodic with the period $2\pi/\kappa \quad (\tau \in (0, 2\pi/\kappa))$.

Various thermodynamical quantities for a system containing a black hole can be obtained by differentiating of the Massieu function

$$\Omega(\beta, \Phi_B, \varphi_B) = -\frac{1}{\beta}\ln Z_{GC}, \tag{5.4}$$

where Z_{GC} is the grand canonical partition function. It is assumed that a black hole is inside a cavity with the boundary B and β, Φ_B and φ_B are the inverse temperature, the value of the dilaton field, and the chemical potential at the boundary. The chemical potential for a charged black hole is conjugate to the electric charge and hence it is an electrostatic potential energy per unit charge. The electric charge Q and the value of the dilaton field at the horizon Φ_H are not fixed and enter the Massieu function as parameters. In Refs.[34-37] there was proposed a method for obtaining an approximation to $\ln Z_{GC}$ based upon finding the stationary points of the Euclidean action [44] from which the constraints have been eliminated. We use this approach in order to show that the entropy of a charged two-dimensional black hole is[3]

$$S_H = 2\pi e^{\Phi_H}. \tag{5.5}$$

Following [39] we rewrite the metric in the form

$$ds^2 = Bd\eta^2 + \frac{dx^2}{P}, \tag{5.6}$$

where η is periodic with the period 2π and $x \in [0, 1]$. The boundary surrounding a black hole is located at $x = 1$. Denote

$$B(1) = B_B, \qquad \Phi(1) = \Phi_B, \qquad \Phi(0) = \Phi_H, \tag{5.7}$$

then the inverse temperature β measured at the boundary is

$$\beta = \int_0^{2\pi} B_B^{1/2} d\eta = 2\pi B_B^{1/2}. \tag{5.8}$$

The regularity condition of the metric and the dilaton field at the origin give

$$\left[B^{1/2}(P^{1/2})'\right]_{x=0} = 1, \qquad \left[(\nabla\Phi)^2\right]_{x=0} \equiv \left[P\Phi'^2\right]_{x=0} = 0. \tag{5.9}$$

[3]It can be shown that for an uncharged black hole the entropy coincides with the value of the action evaluated for the Euclidean black hole solution after subtruction of the vacuum contribution from it.

Here $(\ )' = \partial/\partial x$.

The vector-potential for the electromagnetic field can be chosen in the form

$$A_\mu dx^\mu = -i\varphi(x)d\eta, \tag{5.10}$$

where $\varphi(x)$ is a real function obeying the boundary conditions

$$\varphi(0) = 0, \qquad \varphi(1) = \frac{\beta\varphi_B}{2\pi}. \tag{5.11}$$

The electromagnetic field constraint $(e^\Phi F^{\eta\mu})_{;\mu} = 0$ and the gravitational constraint $T_{\eta\eta} = 0$ at $\eta = const$ give

$$\varphi' = Q\left(\frac{B}{P}\right)^{1/2} e^{-\Phi}, \tag{5.12}$$

$$W \equiv P\Phi'^2 = \alpha^2 - De^{-\Phi} + Q^2 e^{-2\Phi}. \tag{5.13}$$

where Q is the charge of the black hole and the integration constant D connected with the mass of a black hole M

$$D \equiv 2\alpha M = \alpha^2 e^{\Phi_H} + Q^2 e^{-\Phi_H}. \tag{5.14}$$

Denote by I the reduced action which is obtained from the Euclidean action by substituting the constraints solutions and using the boundary conditions (5.7-5.9), (5.11). The calculations give

$$I \equiv I(\beta, \Phi_B, \varphi_B; \Phi_H, Q) = -\beta e^{\Phi_B}\sqrt{W_B} - 2\pi e^{\Phi_H} - \beta Q\varphi_B, \tag{5.15}$$

where

$$W_B = \left(1 - e^{\Phi_H - \Phi_B}\right)\left(\alpha^2 - Q^2 e^{-\Phi_B - \Phi_H}\right). \tag{5.16}$$

The reduced action I diverges at large values of Φ_B. In order to make it finite the vacuum contribution $I_{vac} = -\alpha\beta\exp(\Phi_B)$ must be subtracted

$$I_* \equiv I - I_{vac} = \beta e^{\Phi_B}(\alpha - \sqrt{W_B}) - 2\pi e^{\Phi_H} - \beta Q\varphi_B. \tag{5.17}$$

The stationary points of the reduced action I_* with respect to the parameters Φ_H and Q are defined by a differentiation with β, Φ_H and φ_H fixed. By using these relations we get

$$\varphi_B = Q\frac{e^{-\Phi_H} - e^{-\Phi_B}}{\sqrt{W_B}}, \tag{5.18}$$

$$\beta = \frac{4\pi\sqrt{W_B}e^{\Phi_H}}{\alpha^2 e^{\Phi_H} - Q^2 e^{-\Phi_H}} = \frac{\sqrt{W_B}}{\alpha}\beta_H, \tag{5.19}$$

where

$$\beta_H = \frac{1}{T_H} = \frac{2\pi}{\kappa}, \tag{5.20}$$

with κ given by Eq.(4.11).

We denote by $\tilde{I} = \tilde{I}(\beta, \varphi_B, \Phi_B)$ the value of I_* calcilated for the stationary points. The entropy of the two-dimensional black hole[4] defined by the relation

$$S_H \doteq \left[\beta \frac{\partial \tilde{I}}{\partial \beta}\right]_{\Phi_B, \varphi_B} - \tilde{I}. \tag{5.21}$$

is

$$S_H = 2\pi e^{\Phi_H}. \tag{5.22}$$

By using the reduced action $\tilde{I}$ one can also find the mean values of the charge $\langle Q \rangle$ and the thermal energy $\langle E \rangle$ of a charged black hole

$$\langle Q \rangle = -\beta^{-1} \left[\frac{\partial \tilde{I}}{\partial \varphi_B}\right]_{\beta, \Phi_B} = Q, \tag{5.23}$$

$$\langle E \rangle = \left[\frac{\partial \tilde{I}}{\partial \beta}\right]_{\beta \varphi_B, \Phi_B} = e^{\Phi_B}(\alpha - \sqrt{W_B}). \tag{5.24}$$

In the limit $\Phi_B \to \infty$ one has $\langle E \rangle = M$.

The above approach based on the reduced action allows one to get the complete thermodynamical description of a system containing a black hole. If one is interested only in the calculation of the black hole's entropy the following approach proposed recently in Ref.[40] can also be used.

The grand canonical partition function Z_{GC} in the presence of concerved charges C_i and their related chemical potentials is

$$Z_{GC} \equiv \exp(-\beta\Omega) = Tr \exp[-\beta(H - \mu_i C_i)]. \tag{5.25}$$

As we mentioned earlier the grand canonical partition function Z_{GC} (for an appropriate choice of boundary conditions at the cavity) can be written in the form

$$Z_{GC} = \exp(-I^B), \tag{5.26}$$

where I is the value of the Euclidean action. We use here the superscript B to emphasize that the surface term in the action integral containing the extrinsic curvature is to be calculated at the external boundary (cavity). This surface term is not needed at the Euclidean horizon because the manifold is regular there.

On the other hand the mean value of the energy E is directly related with the amplitude of the imaginary time evolution between two surfaces τ_1 and τ_2 with given boundary conditions [41, 40]

$$\langle \tau_1 | \tau_2 \rangle = \exp[-(\tau_2 - \tau_1)(E - \mu_i C_i)]. \tag{5.27}$$

---

[4]Another approach to the calculation of the two-dimensional black hole's entropy was proposed in [43].

In the same as earlier quasi-classical approximation one has

$$\beta(E - \mu_i C_i) = I_H^B, \tag{5.28}$$

where I is again the Euclidean action, and the indices B and H indicate that now besides the external boundary B the contribution of the surface in the action at the Euclidean horizon has to be taken into account.

By using the relations (5.26) and (5.28) one can get for the entropy of the two-dimensional black hole the following expression

$$S_H = \beta(E - \mu_i C_i) - \beta\Omega = I_H^B - I^B. \tag{5.29}$$

The difference of the Euclidean actions standing in the right-hand side of Eq.(5.29) contains only the surface term at the Euclidean horizon and is of the form

$$S_H = -\int_H dx\sqrt{|\, h\,|}e^{\Phi}K. \tag{5.30}$$

Simple calculations give the expression (5.22) for the entropy of a charged two-dimensional black hole.

The entropy of a 4-d black hole is proportional to its surface and in this sence it is of pure geometrical origin. One may expect that in a general non-Einsteinian theory of gravity which allows black holes the entropy of a black hole besides the pure geometrical part (surface of the black-hole's area) contains also a part connected with other (non-metric) physical fields. In the two-dimensional gravity the 'surface' of a black hole is null-dimensional (a point) and there is no sence to speak about its area. We have seen that the entropy of a 2-d black hole in the absence of a pure geometrical part is determined by the value of a dilaton field which is present in the theory in addition to the metric. The general connection between the entropy and the surface terms in action [40] indicates that only those fields which accompany the highest (second) derivatives of metric in the action will contribute to the entropy of a black hole. The analoguos situation takes place in the theory of gravity with higher in curvature corrections which requires additional (with respect to the Einstein theory) surface terms.

6 Four Laws of Two-Dimensional Black-Holes Physics

In this Section we make some remarks concerning the thermodynamical analogy in two-dimensional black-holes physics. We begin by considering general (not necessary static) solutions describing black holes and prove some results on the behavior of the event horizon.

We assume that a spacetime described by a solution of the Eqs.(2.4) -(2.7) is asymptotically flat. As usual one may define the event horizon H as a line $\dot{J}^-(\mathcal{J}^+)$.

The standard arguments used for the proof of the Penrose theorem show that this line at its regular points is to be null. Any null line in two dimensions is geodesic. Denote by V an affine parameter along the horizon and let $l^\mu \partial_\mu \equiv \partial_V$ be its tangent vector.

The results of the previous section allow us to conclude that the entropy of a black hole is to be defined as follows

$$S_H = 2\pi e^{\Phi_H}. \tag{6.1}$$

The general equation which describes the evolution of Φ_H can be obtained by using the gravitational field equation (2.4). These equations imply the following relation which must be fulfilled at the horizon

$$\frac{d^2\Phi_H}{dV^2} \equiv [l^\mu(l^\nu \Phi_{;\nu})_{;\mu}]_H = - \mid l^\mu D_\mu T \mid^2_H \equiv -\Pi(V). \tag{6.2}$$

The subscribe H in these relations indicates that the corresponding quantities are calculated at the horizon. The solution of the Eq.(6.2) under the assumption that the black hole becomes static at the late time can be written as follows

$$\Phi_H(V) = \Phi_H(\infty) - \int_V^\infty dV' \int_{V'}^\infty dV'' \Pi(V''), \tag{6.3}$$

where $\Phi_H(\infty)$ is the future asymptotic value of the dilaton field at the horizon. For our choice of the action the local 'energy flux' of the tachyon field through the horizon is non-negative ($\Pi(V) \geq 0$) and Eq.(6.3) implies that $\Phi_H(V)$ (and hence the entropy of the black hole S_H) is a monotonically non-decreasing function of time. This result resembles the Hawking theorem for 4-d black holes.

Eq.(6.3) show us also that for any moment of time before the moment when the black hole's parameters reach their asymptotic values the value of $d\Phi_H/dV$ is strictly negative provided there was any flux of the tachyon field through the horizon. This and another properties of time-dependent solutions can be illustrated by using the two-dimensional analogue of Vaidya solutions described in the Section 3.

After the discussion of the general behavior of the event horizons we are able to formulate the four laws of two-dimensional black hole physics. The analogue of the zeroth law of thermodynamics is quite evident. For a static equilibrium configuration the black hole's temperature $\theta = \kappa/2\pi$ is constant at the horizon.

For formulation of the first law we assume that the system incorporating a black hole switches from one static state to another. The mass formula (4.9) shows that its mass changes by

$$\delta M = \delta(e^{\Phi_H})\kappa + e^{\Phi_H}\delta\kappa. \tag{6.4}$$

(For simplicity we consider the uncharged black hole.) The variation of κ can be found by using Eq.(4.2)

$$\delta\kappa = \frac{1}{2}\delta \int_\Sigma (\nabla T)^2 \xi^\alpha d\Sigma_\alpha. \tag{6.5}$$

Using the definitions of the entropy $S_H = 2\pi e^{\Phi_H}$ and the temperature $\theta = \kappa/2\pi$ of a black hole we get

$$\delta M = \theta \delta S_H + \frac{1}{2} e^{\Phi_H} \delta \int_\Sigma (\nabla T)^2 \xi^\alpha d\Sigma_\alpha. \tag{6.6}$$

The last equation ('differential mass formula') relating the change of the energy of a black hole to the change of its entropy plays the role of the first law in the thermodynamical analogy. The second term in the right-hand side of Eq.(6.6) describes the change of the contribution of the matter (the tachyon field) outside the black hole to the total energy.

In the beginning of this Section it was shown that S_H is a non-decreasing function of time

$$\delta S_H \geq 0 \tag{6.7}$$

provided $| l^\mu D_\mu T |_H^2 \geq 0$. The latter inequality ('weak energy condition') is valid for classical tachyon field but it can be violated due to the quantum effects (e.g. during the process of quantum evaporation of a black hole). Nevertheless one may expect that the generalized entropy (the entropy S_H of a black hole plus the entropy S_r of the quantum radiation) is to be a non-decreasing function of time

$$\delta S_H + \delta S_r \geq 0. \tag{6.8}$$

This inequality is the second law.

The third law, i.e. the statement that the temperature of a black hole cannot be reduced to zero by a finite number of operations, is evidently valid for uncharged black holes because for them $\theta \geq \alpha/4\pi$. For charged tachyon free black holes Eq.(4.11) shows that any attempt to reach $\theta = \kappa/2\pi = 0$ must destroy the black hole in the same manner as it happens in the 4-d case. It means that the same arguments as in the 4-d case [45] can be applied to the two-dimensional black holes. It would be interesting to obtain the rigorios general proof of the third law for two-dimensional black holes.

In conclusion we make a remark concerning the tachyon hairs. The positive integral $I[T] = \int_\Sigma (\nabla T)^2 \xi^\alpha d\Sigma_\alpha$ standing in the right-hand side of Eq.(6.6) vanishes only for the tachyon free solutions. It may be used to charaterize the deflection of a solution descibing a black hole with tachyon hairs from a tachyon free one. The classical radiation of the tachyon field from a black hole carries away energy and hence it decreases the mass of a black hole. On the other hand the second law (6.7) implies that the entropy of the black hole cannot decrease and hence for this process $\delta M - \theta \delta S_H \leq 0$. The first law allows one to conclude that $\delta I[T] < 0$. In other words by radiating the tachyon field the black hole becomes closer to its 'no-hair state'. It means that one may expect classical instability of tachyon hairs unless there exist some conservation law preventing this process.

Acknowledgements: The work was supported by the Danish Natural Science Research Council, grant 11-9524-1SE.

References

[1] S.W.Hawking. *Commun.Math.Phys.* , **25** (1972) 152.

[2] W.Israel.*Phys.Rev.*, **164** (1967) 1776.

[3] W.Israel. *Commun.Math.Phys.* , **8** (1968) 245.

[4] B.Carter.*Phys.Rev.Lett.*, **26** (1971) 331.

[5] D.C.Robinson.*Phys.Rev.Lett.*, **34** (1975) 905.

[6] P.O.Mazur. *J.Phys.*, **A15** (1982) 3173.

[7] J.M.Bardeen, B.Carter, and S.W.Hawking. *Commun.Math.Phys.* , **31** (1973) 161.

[8] I.D.Novikov and V.P.Frolov. *Physics of Black Holes.* (Kluwer Acad. Publ., Dordrecht, Boston, London, 1989).

[9] C.G.Callan, D.Friedan, E.J.Martinec, and M.J.Perry. Nucl.Phys. **B262** (1985) 593.

[10] E.Witten,*Phys.Rev.*, **D44** (1991) 314.

[11] G.Mandal, A.M.Sengupta, and S.R.Wadia, *Mod.Phys.Lett.*,**6** (1991) 1685.

[12] S.de Alwis and J.Lykken. *Phys.Lett.* **B269** (1991) 264.

[13] J.Ellis, N.E.Mavromatos, and D.V.Nanopoulos. *Phys.Lett.* **B267** (1991) 465.

[14] R.Dijkgraaf, H.Verlinde, and E.Verlinde. *Nucl.Phys.* **B371** (1992) 269.

[15] C.G.Callan, Jr., S.B.Giddings, J.A.Harvey, and A.Strominger.*Phys.Rev.* **D45** (1992) R1005.

[16] R.B.Mann. *GRG* **24** (1992) 433.

[17] R.B.Mann and T.G.Steele. *Class.Quantum Grav.* **9** (1992) 475.

[18] J.G.Russo, L.Susskind, and L.Thorlacius. *Phys.Lett.* **B292** (1992) 13.

[19] S.Hawking. *Phys.Rev.Lett*, **69** (1992) 406.

[20] E.Witten. *Two-dimensional string theory and black holes*, Preprint IASSNS-HEP-92/24 (1992).

[21] L.Susskind and L.Thorlacius. *Nucl.Phys.* **B382** (1992) 123.

[22] B.Birnir, S.B.Giddings, J.A.Harvey, and A.Strominger. *Phys.Rev.* **D46** (1992) 638.

[23] S.B.Giddings and W.M.Nelson. *Phys.Rev.* **D46** (1992) 2486.

[24] A.Bilal and C.Callen. *Liouville models of 1+1 black hole evaporation*, Princeton preprint PUPT-1320 (1992).

[25] A.Strominger. *Phys.Rev.* **D46** (1992) 4396.

[26] J.G.Russo, L.Susskind, and L.Thorlacius. *Phys.Rev.* **D46** (1992) 3444.

[27] M.D.McGuigan, C.R.Nappi and S.A.Yost. *Nucl.Phys.* **B375** (1992) 421.

[28] A.A.Tseytlin. *On the thachyon terms in the string effective action*, Preprint JHU-TIPAC-91004 (1991).

[29] S.Elitzur, A.Forge, and E.Rabinovici. *Nucl.Phys.* **B359** (1991) 581.

[30] J.G.Russo and A.A.Tseytlin. *Scalar-tensor quantum gravity in two dimensions*,Preprint SU-ITP-92-2, DAMTP-1-1992 (1992).

[31] A.A.Tseytlin. *String cosmology and dilatonInt.J.Modern.Phys.* (1992) (to appear).

[32] A.A.Tseytlin and C.Vafa.*Nucl.Phys.* **B372** (1992) 443.

[33] C.W.Misner, K.S.Thorne, and J.Wheeler. *Gravitation*, Freeman, San Francisco, 1973.

[34] V.P.Frolov and N.Sánchez (unpublished).

[35] V.P.Frolov. *Phys.Rev.* **D46** (1992) 5383.

[36] J.W.York, Jr.*Phys.Rev.* **D33** (1986) 2092.

[37] J.M.York, Jr. *Physica* **A158** (1989) 425.

[38] J.M.York, Jr. In *Conceptual Problems of Quantum Gravity* (Eds. A.Ashtekar and J.Stachel), Birhäuser (1991).

[39] H.W.Braden, J.D.Brown, B.F.Whiting, and J.W.York, Jr. *Pfys.Rev.* **D42** (1990) 3376.

[40] R.Kallosh, T.Ortin, and A.Peet. *Entropy and Action of Dilation Black Holes* , Stanford preprint SU-ITP-92-29 (1992)

[41] S.W.Hawking, in *General Relativity, An Einstein Centenary Survey,* eds. S.W.Hawking and W.Israel Cambridge, 1979), Chapter 15.

[42] W.Israel.*Phys.Rev.*Lett. **57** (1986) 397.

[43] G.W.Gibbons and M.J.Perry. *The physics of 2-d Stringy Spacetimes* Preprint HEP-TH/9204090 (1992).

[44] G.W.Gibbons and S.W.Hawking.*Phys.Rev.* **D15** (1977) 2751.

[45] W.Israel.*Phys.Rev.*Lett. **57** (1986) 397.

TOPOLOGY, CAUSALITY, AND CHRONOLOGY PROTECTION

VALERY P. FROLOV*

The Copenhagen University Observatory,
Øster Voldgade 3, DK-1350 Copenhagen K, Denmark†

Abstract

The paper contains discussion of causality properties of locally static multiply-connected spacetimes. It is shown that the graviational field in such spaces is generically non-potential. As the result of non-potentiality of the gravitational field closed timelike curves are created. The renormalized stress-energy tensor of a quantum field diverges near a Cauchy horizon. The general problems connected with chronology protection are briefly discussed.

1 Introduction

One of the most intriguing features of the General Relativity is possible existence of spacetimes with non-trivial topological and causal structure. Wormhole solutions described by Wheeler [1, 2] are well known examples of spaces with non-trivial topology. Black hole solutions give example of situations when the causal structure of spacetime is non-trivial because of the existence of regions (lying inside a black hole) which cannot be seen by a distant observer. The loss of the possibility of free information exchange in a spacetime with a black hole results in a number of important physical consequences. Thermal properties of the black holes are among them. There are also known solutions of Einstein equations (the Gödel solution [3] describing a stationary rotating universe is one of them) which allow closed time-like curves. According to the standard (naive) interpretation the existence of closed timelike curves allows one to return into one's own past. This creates highly non-trivial problems connected with causality.

*On leave of absence from P.N.Lebedev Physical Institute, Leninsky prospect 53,Moscow 117924 Russia

†E-mail address: frolov@astro.ku.dk

It was shown [4, 5] that a spacetime with a traversible wormhole (if it only exists) can be transformed into a spacetime with closed timelike curves. Namely it was shown that closed timelike curves may arise as a result of the relative motion of the wormhole's mouths. This principal possibility to create closed timelike curves in a spacetime where they originally were absent attracted new interest to the wormhole-like solutions. From more general point of view the described model of a 'time machine' might be considered as an indication that there exist non-trivial relations between topology and causality.

There are at least four quite general physical problems connected with these results:

1. Is it possible to create and to preserve a traversible wormhole without violating known physical laws?
2. Is it possible to have a non-contradictory physics in a spacetime with closed timelike curves?
3. Do there exist any physical laws which may forbid closed timelike curves formation?
4. Are there other cases when the regions with closed timelike curves can be generated in an initially regular spacetime?

We are able to make here only brief comments on the results connected with these problems (mainly by giving the corresponding references).

In the framework of classical General Relativity it is impossible to change the space topology without violating weak energy condition (WEC) and creation of closed timelike curves [7]. The wormholes (if only they exist) may be long living (traversible) only if the averaged weak energy condition is violated [6, 4]. The averaged weak energy condition was shown to be valid for scalar and electromagnetic field in the Minkowski spacetime [8, 9, 10, 11]. On the other hand the general prove of the validity of the averaged weak energy condition in a curved four-dimensional spacetime is not known.

The scattering theory for a non-interacting (free) classical and quantum field can be constructed in a spacetime with closed timelike curves [12, 13]. An interaction resulting in non-linearity of physical equations changes the situation drastically. Simple examples (e.g. so called 'billiard ball problem' [14, 18]) show that because of self-interaction trivial extension of a local solution to the global one is as the rule inconsistent in a spacetime with closed timelike curves. It looks like that non-contradictory physics in a spacetime with a closed timelike curves is possible only if one imposes self-consistency as additional principle [15, 16, 17]. This principle plays the role of global boundary conditions. It allows to single out such local solutions to the equations of physics which can occur in the Universe, which can be extended to be part of a global solution, which is well defined everywhere in the spacetime [14].

In Ref.[18] it was demonstrated that there are infinite numbers of self-consistent solutions in the billard-ball problem. But in general case it is not clear whether the measure (defined in any reasonable way) of solutions of non-linear equations obeying the self-consistency principle in a spacetime with closed timelike curves is large enough to allow any interesting non-trivial physics. It should be also stressed that the self-consistency principle means in particular that it is impossible to change one's own past by free will. In this sence the existence of closed timelike curves does not allow to construct what usually (e.g. in science fiction) is understood under the time machine. In other words the action of such a 'time machine' will be essentially restricted by the requirement of self-consistency (for more detailed dicussion, see [14, 19]).

It is also unclear whether the self-consistency principle is effectively working in quantum domain. Hartle [20] has shown that in a spacetime with closed timelike curves there are problems with unitarity already in the quantum mechanics. He used the decoherence approach developed in Ref.[21] to study this problem. The lack of the unitarity for interacting quantum fields in a spacetime with closed timelike curves was demonstrated in Ref.[13].

In the case when a 'time machine' is not eternal and closed timelike curves are created in a spacetime, where they originally were not present, the region with closed timelike curves is restricted in the past by the Cauchy horizon. The existence of closed null geodesics lying at the Cauchy horizon or in its vicinity immideately results in the divergency of quantum fluctuations and the renormalized stress-energy tensor near the Cauchy horizon[1]. The structure of the renormalized stress-energy tensor divergency in the one-loop approximation was analysed for a model of a wormhole with moving mouths by Kim and Thorne [22] and for a wormhole with a locally-static gravitational field by Frolov [23]. Later Hawking [24] proved the existence of a closed null geodesic lying on the compactly generated horizon and the existence of the divergency of renormalized stress-energy tensor in the one-loop approximation near this geodesic. These results indicate that quantum gravity effects are necessarily to be taken seriously when discussing the 'time machine' problem. One may expect the cut-off of one-loop divergencies near the Cauchy horizon due to these effects but without developted theory of quantum gravity the corresponding calculations were not done. Kim and Thorne [22] and Hawking [24] gave different estimations for the possible value of the cut-off. For more invariant cut-off proposed by Hawking the resulting energy density due to the vacuum polarization reaches the Plankian values and as the result the Cauchy horizon must be destroyed. The assumption that it really takes place every time when according to the classical description the closed timelike curves are to be formed was called by Hawking the 'chronology protection conjucture'.

As for the other possibilities to create closed timelike curves the proposed by Gott [25] example recently attracted a lot of interest. Gott has shown that spacetimes with two relatively moving infinitely long straight strings can possess closed timelike curves. The study of solutions proposed by him show that such spacetimes cannot

[1]The arguments in favor of existence of these divergencies were given by a number of people (independent unpublished talks by Gnedin, Kim and Thorne, Kompaneets, and Frolov, presented at the NORDITA Workshop, 1990.)

arise from a regular at infinity asymptotically flat spacetime which initially contains no closed timelike curves and has a positive-definite total energy [26, 27]. On the other hand the renormalized stress-energy tensor is also divergent near the Cauchy horizon [28] as in the wormhole's models.

In this paper we consider physical properties of spacetimes with wormholes. The aim of the paper is to present some ideas concerning deep relations between the topology and causality and to discuss the possible role of quantum effects for the chronology protection. To be more concrete we consider so called locally static multiply-connected spacetimes and show that:

1. the gravitational field in a locally static multiply-connected spacetime is generically non-potential;
2. the existence of closed timelike curves is a generic property of a spacetime with non-potential gravitational field;
3. quantum effects connected with the vacuum polarization are highly important near the Cauchy horizon so that they may prevent closed timelike curves formation.

In the paper the units in which $G = c = \hbar = 1$ and sign conventions of MTW [29] are used.

2 Locally Static Multiply Connected Spacetimes

We begin by explaining why the gravitational field in a locally static multiply-connected spacetime is generically non-potential. A spacetime is called locally static if in any its simply connected region there exist a uniquely defined (up to normalization) timelike Killing vector ξ, obeying the relations

$$\xi_{(\mu;\nu)} = 0, \qquad \xi_{[\mu;\nu}\xi_{\lambda]} = 0, \tag{2.1}$$

The first equation is the definition of a Killing vector, while the second equation means that ξ is locally 'surface orthogonal'. In the general case in a multiply-connected spacetime (e.g. when a wormhole is present) it is impossible to define such a Killing vector globally. The existence of non-contractible closed paths (i.e. the paths which passes through a wormhole and cannot be contracted to a point by continious transformations) may create an obstacle for the existance of a global Killing vector field. In general case after continious propagation of a Killing vector field along a closed non-contractible path it returns back to the initial point having the same direction as the initial vector but with a different norm. In order to be able to deal with these kind of situations it is convenient to introduce a unit timelike vector $u^\mu \equiv \xi^\mu/|\xi^2|^{1/2}$ instead of ξ. This vector of four-velocity of a Killing observer as well as its four-acceleration $w^\mu \equiv u^\nu u^\mu{}_{;\nu}$ do not depend on the norm of ξ and hence they are well defined globally. It is easy to verify that the following set of equations

$$u_\mu u^\mu = -1, \qquad u_{\mu;\nu} = -w_\mu u_\nu, \qquad w_{[\mu;\nu]} = 0 \tag{2.2}$$

is locally equivalent to Eq.(2.1). A locally static spacetime is a spacetime which admits two globally defined vector fields u^μ and w^μ obeying the relations (2.2).

It is convenient to consider a locally static spacetime M as the collection W of Kiling trajectories [30], i.e. the set of integral lines $X^\mu(\tau)$ of u^μ $(dX^\mu/d\tau = u^\mu)$. We assume that a unique Killing trajectory passes through each point of the spacetime M and that $M = T \times W$, where T topologically is R^1. One may refer to W as to the three-dimensional space under consideration. In the presence of a wormhole this space has non-trivial topology. A non-contractible loop passing through a wormhole n-times in the chosen direction can be characterized by its winding number n. Two loops with the same winding number n are connected by a continiuos transformation (are homotopic). These loops are also homological, i.e. they are boundaries of some two-dimensional surface. In other words the first fundamental group π_1 and the first homology group H_1 calculated for W (as well as for the spacetime M itself) coincide with the group of integer numbers **Z**.

The last equation in (2.2) shows that the acceleration form $w \equiv w_\mu dX^\mu$ is closed $(dw = 0)$. According to the Stokes' theorem the integral of this form over any closed path C_n

$$I_n[w] = \oint_{C_n} w \tag{2.3}$$

depends only on the winding number n of the path. One has $I_n = nI$. We chose a direction of the path with $n = 1$ in such a way that $I[w] \equiv I_1[w] \leq 0$.

If the period $I[w]$ vanishes then the form w is exact, i.e. there exist a globally defined scalar function φ (gravitational potential) such that $w = d\varphi$. For such potential gravitational field the vector field $\xi^\mu = e^\varphi u^\mu$ is a global Killing vector field.

In the general case (when $I[w] \neq 0$) the gravitational potential φ cannot be defined as a single valued function. Such a gravitational field is called non-potential. The period $I[w]$ may be considered as the measure of non-potentiality of the field. The important physical property of a non-potential gravitational field is that the work done by such a field over a particle moving along a closed countour C_n passing throuth a wormhole does not vanish. This work is proportional to $e^{-I_n[w]}$.

3 Clocks Synchronization

Before considering properties of spacetimes with a non-potential gravitational field we discuss a quite general problem of clocks synchronization in a multiply-connected spacetime. The presence of non-contractible paths makes this problem non-trivial already in a potential gravitational field. The reason is very simple. In a static gravitational field it is possible to synchronize the clocks along any path lying in a simply connected region. But for the process of clocks' synchronization along the closed path passing through a wormhole one cannot guarantee that the 'synchronized clocks' shows the same time as the initial ones. In general case there will be a non-vanishing time gap for the clocks' synchronization along a non-contractible path.

For potential gravitational field the value of time gap for the clocks' synchronization in a multiply-connected spacetime is a topological invariant. In order to show this we introduce the 1-form

$$\eta \equiv e^{-\varphi} u_\mu dX^\mu. \tag{3.1}$$

One can verify that this globally defined 1-form is closed ($d\eta = 0$). That is why its integral over a closed path C_n

$$J_n[\eta] = \oint_{C_n} \eta \tag{3.2}$$

is a topological invariant depending only on the winding number. We show now that the value of $J_n[\eta]$ coincides with the time gap for the clocks' synchronization along a closed path with a chosen winding number n.

For this purpose we chose a point X_0^μ and denote by $X_0^\mu(\tau)$ a Killing trajectory passing through this point ($X_0^\mu = X_0^\mu(\tau_0)$). Consider now a curve γ determined by the equation $X^\mu = Y^\mu(\lambda)$ ($\lambda \in (0,1)$), which begins at X_0^μ, passes once through the wormhole, and ends at the point X_1^μ lying at the same Killing trajectory ($X_1^\mu = X_0^\mu(\tau_1)$). We assume that γ is everywhere orthogonal to the four-velocity of a Killing observer: $u_\mu dY^\mu/d\lambda = 0$. The latter condition means that the curve γ is formed by the events which are simultaneous (in the reference frame of Killing observers) with the initial event X_0^μ. We add the part of a Killing trajectory lying between τ_0 and τ_1 to this curve to make it closed. The evaluation of the integral (2.3) for this closed path with a winding number $n = 1$ is quite simple. Only part of the path lying along the Killing trajectory gives non-vanishing contribution. As the result the value of the integral $J_1[\eta]$ calculated for the chosen path coincides with the time gap for the clocks' synchronization along this path. It is evident now that $J_1[\eta]$ does not depend on the particular choice of a path with the given winding number $n = 1$ and that $J_n[\eta] = nJ_1[\eta]$.

In the non-potentialgravitational field the time gap $\Delta_1(\tau_0) \equiv \tau_0 - \tau_1$ for the clocks' synchronization is not any more time independent. It can be shown [31] that it obeys the equation

$$\frac{d\Delta_1(\tau_0)}{d\tau_0} = 1 - e^{I[w]}. \tag{3.3}$$

This relation shows that the time gap for the clocks' synchronization along any closed path with a winding number $n = 1$ is growing with time and the value $I[w]$ may be considered as the measure of this growing. As soon as the gap becomes greater than the minimal time needed for a light signal to propagate along a closed path passing through the wormhole and to return back, the closed timelike curve can arise.

In [31] it was argued that the interaction of a wormhole with classical matter generically generates a non-potential component of the gravitational field. That is why a locally static wormhole is generically unstable with respect to the processes which transform it into a 'time machine'. The relative motion of a wormholes mouths also generates a time gap for the clocks' synchronization [4]. One can interprete the above results in the following way. There exist internal relation between topological and causal properties of a spacetime. The existence of closed timelike curve is a generic property of multiply connected locally static spacetimes.

In any asymptotically flat spacetime a Cauchy horizon separates a past region without closed timelike curves from a future region which possesses them. In general case the Cauchy horizon is a null surface generated by null geodesics with no end points [32]. In the case under consideration the Cauchy horizon is compactly generated, i.e. the past-directed generators of the Cauchy horizon will enter and remain within a compact set [24]. In principle there may be no closed null geodesics lying on the Cauchy horizon. But this property is unstable and any small perturbations of metric results in the arizing closed null geodesics on the Cauchy horizon [24].

In a locally static multiply-connected spacetime one can prove quite general result concerning the behavior of null geodesics in the vicinity of the Cauchy horizon. Namely for a chosen number n and for any chosen Killing trajectory there exist such a point p_n lying on it and such a null geodesic which begins at p_n, passes through the wormhole exactly n times in a chosen direction and returns back to the initial point p_n [23]. It should be emphasized that the initial and final tangent vectors for these null geodesics non necessarily are the same. The points p_n for a chosen value n form a so called n'th polarized hypersurface H_n. The hypersufaces H_n lie above the Cauchy horizon and for $n \to \infty$ the set of H_n is concentrated near it [22].

4 Quantum Effects

The Cauchy horizon seems to be stable against classical perturbations [4, 14]. For quantum fields due to the existence of zero-point fluctuations, which cannot be isolated or suppressed, the situation is quite different [22, 23, 24]. We show that the renormalized stress-energy tensor in a locally static spacetime with a wormhole is divergent near the Cauchy horizon and describe the structure of the divergency.

Consider an amplitude $G(X, X')$ of a quantum particle propagation from a point X to a point X'. This amplitude $G(X, X')$ can be presented as a Feynman integral over all trajectories connecting these points. For close points in a multiply-connected spacetime besides trivial paths there exist homotopically non-equivalent classes of paths numerated by an integer winding number n. In the geometric optics approximation the main contribution to the Feynman propagator is given by narrow tubes of paths located near geodesics connecting X and X'. In the limit $X' \to X$ the trivial direct path contribution diverges. This results in the divergecy of the vacuum expectation value of a stress-energy tensor $\langle T_{\mu\nu}\rangle$. This divergency exists already in a topologically trivial flat spacetime. According to the standard renormalization scheme this divergency it is to be subtracted. Besides the regular part remaining after subtractions made from $n = 0$ contribution, the renormalised stress-energy tensor $\langle T_{\mu\nu}\rangle^{ren}$ in a multiply-connected spacetime in the geometric optics approximation containes also the contributions of the closed geodesics with non-vanishing winding numbers. In the region lying to the past from the Cauchy horizon these additional geodesics (with $n \neq 0$) are spacelike and there contribution generally remains finite. The existence of closed null geodesics with $n \neq 0$ near the Cauchy horizon results in the divergency of $\langle T_{\mu\nu}\rangle^{ren}$ near the Cauchy horizon [22, 23].

For simplicity we restrict ourselves by considering the quantum theory of a scalar massless field Φ obeying the equation

$$\Box\Phi - \xi R\Phi = 0, \tag{4.1}$$

where R is a scalar curvature and a parameter ξ for a conformal invariant field takes the value 1/6. There are two equivalent ways to quantize the field Φ in a multiply-connected spacetime. The first is to work directly with solutions of Eq.(4.1) in the spacetime M. The second approach uses the universal covering space $\tilde{M}$ [33, 34, 35, 36]. The discrete group of isometries Γ which is isometric to the fundamental group $\pi_1(M)$ acts on $\tilde{M}$ from the left and the quotient $\Gamma\backslash\tilde{M}$ coincides with the original spacetime M. The universal covering space $\tilde{M}$ by definition is simply connected. That is why $\tilde{M}$ allows a global Killing vector field which did not exist in the multiply-connected locally static spacetime M. The existence of a global Killing vector field in $\tilde{M}$ makes it possible to separate naturally positive and negative frequencies and to develope the standard scheme of quantization. The physically interesting results in the spacetime M can be obtained by projecting the field theory on $\tilde{M}$ onto the field theory on M.

The basic idea of this projection is to identify M with a fundamental domain $\Gamma\backslash\tilde{M}$ and regard a field theory on M as a field theory on $\tilde{M}$ obeying certain conditions. Because the changing of the fundamental domain is the symmetry transformation and the metric $d\tilde{s}^2$ on $\tilde{M}$ is invariant under these transformations $\gamma \in \Gamma$, the invariance of the field action implies

$$\Phi(\gamma X) = a(\gamma)\Phi(X). \tag{4.2}$$

The group property of Γ gives $a(\gamma_1\gamma_2) = a(\gamma_1)a(\gamma_2)$. The fields Φ on $\tilde{M}$ obeying the Eq.(4.2) are known as automorphic fields.

The Hadamard function $G^1(X, X')$ in the physical spacetime M is connected with the Hadamard function $\tilde{G}^1(X, X')$ in the universal covering space $\tilde{M}$ by the relation [36]

$$G^1(X, X') = \sum_{\gamma\in\Gamma} a(\gamma^{-1})\tilde{G}^1(X, \gamma X'). \tag{4.3}$$

The renormalized stress-energy tensor $\langle T_{\mu\nu}\rangle^{ren}$ in a physical spacetime can be obtained by applying the operator

$$D_{\mu\nu} = \frac{1}{2}(\frac{1}{2} - \xi)(\nabla_{\mu'}\nabla_{\nu} + \nabla_{\mu}\nabla_{\nu'}) + (\xi - \frac{1}{4})g_{\mu\nu}\nabla_{\rho}\nabla^{\rho'} - \frac{1}{2}\xi(\nabla_{\mu}\nabla_{\nu} + \nabla_{\mu'}\nabla_{\nu'})$$
$$+ \frac{1}{8}\xi g_{\mu\nu}(\nabla_{\rho}\nabla^{\rho} + \nabla_{\rho'}\nabla^{\rho'}) + \frac{1}{2}\xi[R_{\mu\nu} - \frac{1}{2}(1 - 3\xi)g_{\mu\nu}R] \tag{4.4}$$

to $G^1(X, X')$ and after subtracting the necessary divergences taking the coincidence limit $X' = X$ [37, 38]. For simplicity we restrict ourselves by considering a non-twisting conformall massless field and put $\xi = 1/6$ and $a(\gamma) = 1$.

The renormalized stress-energy tensor in the physical spacetime M can be written as

$$\langle T_{\mu\nu}\rangle_M^{ren} = \langle T_{\mu\nu}\rangle_{\tilde{M}}^{ren} + T_{\mu\nu}, \tag{4.5}$$

where

$$T_{\mu\nu}(X) = \sum_n{}' T^n_{\mu\nu}(X), \qquad T^n_{\mu\nu}(X) = \lim_{X'\to X} D_{\mu\nu}\tilde{G}^1(X, \gamma^n X'). \tag{4.6}$$

The prime in the sum indicates that the term with $n = 0$ is omitted from the summation. The universal covering space $\tilde{M}$ is a regular static spacetime. That is why generally $\langle T_{\mu\nu}\rangle_{\tilde{M}}^{ren}$ is regular and finite. On the other hand $T^n_{\mu\nu}(X)$ is divergent near the n'th polarized hypersurface. To show this we choose a point X_0 on H_n and denote by C_n a null geodesic in $\tilde{M}$ connecting X_0 and $X_{0,n} \equiv \gamma^n X_0$. We assume that C_n lies in some causal domain $\Omega[C_n]$, so that the Hadamard function $\tilde{G}^1$ allows the following expansion [39, 40, 41]

$$\tilde{G}^1(X, X') = \frac{\Delta^{1/2}}{4\pi^2}\left[\frac{1}{\sigma(X, X')} + v(X, X')\ln|\sigma(X, X')| + w(X, X')\right], \tag{4.7}$$

where $\sigma(x, X') = \frac{1}{2}s^2(X, X')$, $s(X, X')$ is a geodesic interval, and $\Delta = \Delta(X, X')$ is a Van Vleck-Morette determinant. The functions v and w are finite in the coincidence limit.

Consider now points $X^\mu = X_0^\mu - \delta t\xi^\mu(X_0)$ and $X_n^\mu \equiv \gamma^n X^\mu = X^\mu_{0.n} - \delta t A^n \xi^\mu(X_{0,n})$ located on the Killing trajectories passing through X_0 and $X_{0,n}$. For these points $\sigma(x, X_n) \simeq \delta t$. The leading (divergent at $\delta t \to 0$) part of $T^n_{\mu\nu}$ is of the form

$$T^n_{\mu\nu} \simeq \frac{\Delta_n^{1/2}}{12\pi^2\sigma^3(X, X_n)}[\sigma_{\mu'}\sigma_\nu + \sigma_\mu\sigma_{\nu'} - \frac{1}{2}g_{\mu\nu}\sigma_\rho\sigma^{\rho'} - \frac{1}{2}(\sigma_\mu\sigma_\nu + \sigma_{\mu'}\sigma_{\nu'})]. \tag{4.8}$$

This relation shows that $T^n_{\mu\nu}$ is divergent near n'th polarized surface and this divergency is of the form $\sim (\delta t)^3$. Near the Cauchy horizon (for large n) the renormalized stress-energy tensor desribes null fluid propagation with the negative infinitely growing at the horizon energy density. (For more detailed discussion of the divergency structure, see [23].)

To summarize, we show that in a non-potential gravitational field of a locally static spacetime with a wormhole the renormalized stress-energy tensor $\langle T_{\mu\nu}\rangle^{ren}$ is divergent near each of the polarized hypersurfaces H_n. This result is in a complete agreement with the result of calculations of Kim and Thorne [22] who obtained the renormalized stress-energy tensor near a Cauchy horizon for the model of moving wormhole's mouths in a flat spacetime.

5 Summary and Concluding Remarks

The divergence of the renormalized stress-energy tensor near a Cauchy horizon means that the backreaction of the quantum effects on the background geometry is to be taken into account and one needs to consider a self-consistent problem which includes quantum effects. One cannot expect to solve this problem. Nevertheless it is possible to make quite general remarks concerning the possible properties of such solutions. First of all if we really believe in one-loop result and assume that there exist only one massless scalar field then we may conclude that the 'time machine' formation is forbidden by quantum effects. In our concrete model according to classical theory a 'time machine' is formed due to the non-potentiality of the gravitational field of a wormhole. The negative energy flux due to the vacuum polarization is directed in such a way that it automotically reduces the non-potentiality of the field. If only no singularity arises then this effect is acting until the gravitational field becomes potential. In other words the back reaction either prevents the formation of the Cauchy horizon or destroys it by creating a singularity.

In reality there exist many physical fields of different spins. The general conclusion about the divergence of the renormalized stress-energy tensor near the Cauchy horizon evidently remains valid also for massive fields because the leading (divergent at small geodesic distances) part of the massive propagator has the same singularity as for massless case. The sign of the vacuum energy density depends on the spin of field and has opposite sign for fermions. Nevertheless the mutual cancellation of divergences for bosons and fermions is impossible [42]. If the contribution of fermions is dominated the resulting flux of positive mass increases the non-potentiality of the gravitational field until a singularity is created.

One may expect a cut-off of the divergency due to quantum gravity effects but no calculations were done to prove it. Different estimations for this cut-off were proposed by Kim and Thorne [22] and by Hawking [24]. For more invariant cut-off proposed by Hawking the curvature of the spacetime due to quantum effects must reach Plankian value which effectively in the framework of the classical theory is to be considered as singularity.

All the above discussion of quantum effects was inexpicitely based on the assumption that before the Cauchy horizon formation one can use the standard local causal quantum field theory. In principle one cannot exclude the possibility that the non-contradictory quantum mechanics in a spacetime with closed timelike curves is possible only if there exist some global acausal restrictions on the choice of states and their evolution which modify the standard equations non only in the regions with closed timelike curves, but also in the regions lying to the past of the Cauchy horizon. In this case the above conclusions concerning the behavior of the renormalized stress-energy tensor near the Cauchy horizon may fail. In any case the proof of the 'chronology protection conjucture' might require further development of the theory.

Acknowledgements: The work was supported by the Danish Natural Science Research Council, grant 11-9524-1SE.

References

[1] J.A.Wheeler, *Geometrodynamics* (Academic, New York, 1962).

[2] C.W.Misner and J.A.Wheeler, *Ann.Phys.* **2** (1957) 525.

[3] K.Gödel, *Rev.Mod.Phys.* **21** (1949) 447.

[4] M.S.Morris, K.S.Thorne, and U.Yurtsever, *Phys.Rev.Lett.***61** (1988) 1446.

[5] I.D.Novikov, *JETP* **95** (1989) 769.

[6] M.S.Morris and K.S.Thorne, *Am.J.Phys.* **56** (1988) 395.

[7] R.Geroch, *J.Math. Phys.* **8** (1967) 782.

[8] U.Yurtsever, *Class.Quant.Grav.* **7** (1990) L251.

[9] R.M.Wald and U.Yurtsever, *Phys.Rev.* **D44** (1991) 403.

[10] G.Klinkhammer, *Phys.Rev.* **D43** (1991) 2542.

[11] A.Folacci, *Phys.Rev.* **D46** (1992) 2726.

[12] J.Friedman and M.S.Morris, *Phys.Rev.Lett.* **66** (1991) 401.

[13] J.L.Friedman, N.J.Papastamatiou, and J.Z.Simon, *Phys.Rev.* **D46** (1992) 4456.

[14] J.Friedman, M.S.Morris, I.D.Novikov, F.Echeverria, G.Klinkhammer, K.S.Thorne, and U.Yurtsever, *Phys.Rev.* **D42** (1990) 1915.

[15] I.D.Novikov *Evolution of the Universe*, (Cambridge University Press, p.169) (1983).

[16] Ya.B.Zel'dovich and I.D.Novikov, *Structure and Evolution of the Universe*, (Nauka, Moscow) (1975) (in Russian).

[17] I.D.Novikov and V.P.Frolov, *Physics of Black Holes*, (Kluwer Acad.Publ.) p.109, (1989)

[18] F.Echeverria, G.Klinkhammer, and K.S.Thorne, *Phys.Rev.* **D44** (1991) 1077.

[19] I.D.Novikov, *Phys.Rev.* **D45** (1992) 1989.

[20] J.B.Hartle, Univ.of California at Santa Barbara Report No. UCSBTH-92-04.

[21] M.Gell-Mann and J.B.Hartle, em J.Phys. **A24** (1991) L159.

[22] S.W.Kim and K.S.Thorne, *Phys.Rev.* **D43** (1991) 3929.

[23] V.P.Frolov, *Phys.Rev.* **D43** (1991) 3878.

[24] S.W.Hawking,*Phys.Rev.* **D46** (1992) 603.

[25] J.R.Gott III, *Phys.Rev.Lett.* **66** (1991) 1126.

[26] S.M.Carroll, E.Farhi, and A.H.Guth, *Phys.Rev.Lett.* **68** (1992) 263.

[27] S.Deser, R.Jackiw, and G. 't Hooft, *Phys.Rev.Lett.* **68** (1992) 267.

[28] D.G.Boulware,*Phys.Rev.* **D46** (1992) 4421.

[29] C.W.Misner, K.S.Thorne, and J.Wheeler. *Gravitation*, Freeman, San Francisco, 1973.

[30] R.Geroch, *J.Math. Phys.* **12** (1971) 918.

[31] V.P.Frolov and I.D.Novikov, *Phys.Rev.* **D42** (1990) 1057.

[32] S.W.Hawking and G.F.R.Ellis, *The Large Scale Structure of Space Time* (Cambridge University Press, Cambridge, England, 1973).

[33] L.S.Schulman, *Phys.Rev.* **176** (1968) 1558.

[34] J.S.Dowker, *J.Phys.* **A5** (1972) 936.

[35] J.S.Dowker and R.Banach, *J.Phys.* **A11** (1978) 2255.

[36] R.Banach and J.S.Dowker and, *J.Phys.* **A12** (1979) 2255, 2545.

[37] S.M.Christensen, *Phys.Rev.* **D14** (1976) 2490; **D17** (1978) 946.

[38] N.D.Birrell and P.C.W.Davies, *Quantum Fields in Curved Space.* (Cambridge University Press, Cambridge, England. 1982).

[39] J.Hadamard, *Lectures on Cauchy's Problem in Linear Partial Differential Equations* (Yale University Press, New Haven, 1923).

[40] B.S.DeWitt and R.W.Brehme, Ann. Phys.(N.Y.) **9** (1960) 220.

[41] F.G.Friedlander, The Wave Equation in Curved Space-Time (Cambridge University Press, Cambridge, England, 1975).

[42] G.Klinkhammer, *Phys.Rev.* **D46** (1992) 3388.

FOUR DIMENSIONAL QUANTUM GRAVITY: THE COVARIANT PATH INTEGRAL AND QUANTIZATION OF THE CONFORMAL FACTOR

Emil Mottola
Theoretical Division, Los Alamos National Laboratory
T-8, M. S. B285, Los Alamos, NM 87545, USA

ABSTRACT

After first reviewing the issue of vacuum energy (the cosmological constant problem) in the Einstein theory, the covariant path integral for gravity in four dimensions is constructed. The problem of vacuum energy requires determining the correct ground state of the quantum theory of gravity, and as such is an *infrared* problem, arising prior to and independently of the physics of the Planck scale. It is addressed in these lectures by studying the infrared fixed point of the low energy effective action of the conformal factor generated by the quantum trace anomaly in four dimensions. The infrared fixed point of this effective theory describes a conformally invariant phase of gravity with a vanishing effective cosmological term.

1. Introduction: The Problem of Vacuum Energy

The cosmological constant Λ was introduced into Einstein's equations,

$$R^a{}_b - \tfrac{1}{2}R\delta^a{}_b + \Lambda\delta^a{}_b = \kappa T^a{}_b, \tag{1.1}$$

by Einstein himself in 1917, when he realized that the original form of the equations did not admit a static cosmological solution in the presence of matter and/or radiation. In this prejudice that the universe should be static, quiescent and eternal, Einstein was following a long tradition at least as old as Newton. Just a few years later, in the early twenties, Slipher and Hubble revolutionized cosmology by showing that the systematic red-shifting of the spectral lines of distant nebulae could only be explained if the entire universe is expanding, like the fragments of a colossal primordial explosion or "big bang." The "nebulae," Hubble realized, must be star systems as vast and as complex as our own Milky Way, but very much farther away and receeding from earth at speeds approaching the speed of light. The observational basis for Einstein's prejudice was removed, and he is quoted as reacting: "If there is no quasi-static world, then away with the cosmological term!"

With only a few exceptions, "away with the cosmological term" describes pretty well the practice of cosmologists for the better part of the next half century. There is a very good observational reason for setting $\Lambda \approx 0$. If transposed to the right side of the Einstein equations the Λ term corresponds to a constant energy density $\rho_v = \frac{\Lambda}{\kappa}$, and isotropic pressure $p_v = -\frac{\Lambda}{\kappa}$ permeating all of space *uniformly*, and *independently* of any localized matter sources. Hence, even if $T^a_b = 0$, such a cosmological term results in spacetime becoming curved with a radius of curvature of order $|\Lambda|^{-\frac{1}{2}}$. From the mere fact that the universe is at least 10^{27} cm. large (and perhaps a great deal larger), we immediately infer an extremely stringent bound on Λ from observations,[1] *viz.*

$$|\Lambda_{obs}| < 10^{-54}\ \text{cm}^{-2}. \tag{1.2}$$

In purely classical physics there is no natural scale for Λ, it is perfectly consistent to set it equal to zero, and this is precisely what is done in the standard Friedmann-Robertson-Walker cosmological model.

That might be the end of the story if it were not for the other revolution in theoretical physics that took place in the twenties, *i.e.* quantum mechanics. When special relativity is combined with quantum mechanics, the result is quantum field theory, and an intrinsic and necessary feature of quantum field theory is quantum fluctuations. Despite leading to the difficulty of infinities in some quantities, the physical reality of quantum fluctuations is beautifully demonstrated by the Lamb shift and the Casimir force between two conducting plates.[2] In 1957 this force was measured by Sparnaay and found to agree with Casimir's formula for the force per unit area,

$$\frac{F}{A} = \frac{\hbar c \pi^2}{240\ d^4} = \frac{1.3 \times 10^{-18}}{d^4}\ \text{dyne/cm}^2, \tag{1.3}$$

if d is measured in cm. Spaarnay found a value between 1 and 4×10^{-18} dyne/cm^2. Since that time the experiments have been refined and repeated, and yet better agreement has been found. In addition to these direct checks every success of quantum theory serves as indirect verification for its underpinnings, and there can be little doubt that quantum fluctuations must be taken seriously.

As soon as the observational limit on the cosmological constant is reconsidered from the point of view of quantum theory, we are confronted with a startling paradox. The point about quantum fluctuations is that they carry energy, and that this fluctuation energy is not associated with localized quasi-classical matter sources, but with the *vacuum* itself. Since gravity weighs all forms of energy, quantum fluctuations should contribute to the cosmological term in Einstein's equations. Moreover, with the introduction of Planck's constant into physics there is now a "natural" scale for the cosmological term. It is the Planck scale, $(\hbar\kappa)^{-1}$. Converting the observational bound of (1.2) above to these units, we find that this limit is some 118 orders of magnitude smaller than unity in Planck units! If Λ is not *identically* zero for some hitherto undiscovered reason, then its finite value would represent the tiniest known constant of nature, which we would seem to have no hope of computing with our present understanding of physics.

The issue is complicated by the divergences of quantum field theory: each mode of a field theory contributes $\frac{1}{2}\hbar\omega$ to the vacuum energy, and the sum over modes is quartically divergent. Not only is there no apparent reason for the effective value of Λ to be as tiny in Planck units as the observational bound (1.2) requires, but a naive calculation of Λ produces an *infinite* answer!

Infinities are a reflection of our ignorance of physics at ultra-short distance scales. In quantum electrodynamics, we learn how to absorb all of our ignorance of ultraviolet physics into a small number of physical parameters, chiefly the electric charge and mass of the electron. When this renormalization procedure is carried out, we obtain an *effective* theory with a great deal of predictive power in terms of the renormalized parameters. The price is that we forego any hope of calculating those parameters. The infinite contribution to Λ is similar: it may be removed formally by renormalization, at the price of introducing a free parameter of the quantum theory which cannot be set equal to zero without fine tuning. Fine tuning is a technically precise term, in the sense that there is no apparent symmetry forbidding the generation of a Λ term of order unity in Planck units. As Steven Weinberg has remarked: "Just because something is infinite, does not necessarily mean that it's zero." If there is no reason for it to be zero, then why should it be so tiny in Planck units? This conundrum is often called the "cosmological constant problem," although as I shall argue that this name is somewhat misleading.

The point is brought home most forcefully if one considers a quantum theory with symmetry breaking, such as the Weinberg-Salam theory of electroweak interactions. The energy density difference between the symmetric vacuum of the theory, in which it finds itself in the very early universe at temperatures above 1TeV, and the symmetry broken vacuum of the present epoch is of order $1\text{TeV}^4 \approx 10\text{cm}^{-2}$, or some 55 orders of magnitude above the observational bound (1.2). In other words, if the standard model is correct, the universe passed through a phase transition in which a latent heat 55 orders of magnitude larger than the present vacuum energy density was dumped into the system, so efficiently that there is no presently observable effect on the curvature of spacetime! If only a few parts in 10^{55} of this latent heat did not convert to ordinary matter and radiation but remained as vacuum energy density today, then it would exceed the observational bound. If the universe passed through several such phase transitions at still higher temperatures, the fine tuning becomes even more incredible.

With both quantum fluctuation vacuum energy and symmetry breaking seemingly indisputable features of microphysics, this fine tuning problem is a serious problem for theoretical cosmology. The embarassment is particularly acute for the inflationary universe scenario which explicitly relies upon vacuum energy to drive expansion of the universe in the inflationary era, but then offers no explanation as to why there isn't even a trace of such vacuum energy in the present epoch. Simply assuming that Λ is zero (or nearly zero) in the present universe would seem to distinguish the present state of the universe in a special way, and violates the whole spirit of the inflationary scenario.

Since the vacuum energy problem (as I would prefer to call it) does not arise in purely classical general relativity, one might be tempted simply to lay the problem at the door of quantum gravity, by which is usually meant the physics of the Planck scale. However, the problem with energy densities enormously larger than the observational bound (1.2) appears also at distance scales very much larger than the Planck scale, as the exercise with the Weinberg-Salam phase transition above demonstrates. There is a problem at distance scales where quantum gravitational effects (in the sense of graviton loops, for example) ought to be negligible, and where it should be perfectly permissable to treat spacetime as a classical continuous manifold, described by a low-energy effective field theory in terms of a classical metric. Moreover, since vacuum energy is shifted by spontaneous symmetry breaking, it is clear even in semi-classical gravity that the spacetime curvature in Einstein's equations depends on the vacuum state of all quantum fields. The problem of vacuum energy is indeed a problem of *quantum* gravity, but *not* one that should depend on the physics of the Planck scale. Rather, it is the physics of the correct vacuum state of quantum gravity that is really the issue, and (if strong interactions, electroweak symmetry breaking, and a host of other physical systems are any guide at all) this is controlled by the long-distance or *infrared* properties of an effective field theory.

My goal in these pedagogical lectures is twofold: first I would like to define and develop the techniques necessary for the covariant path integral quantization of gravity, transcribing much of what has been learned in two-dimensional quantum gravity (otherwise known as non-critical string theory) directly to four dimensions. Secondly, I would like to apply these techniques to the quantization of the conformal part of the spacetime metric, the low-energy effective action for which is determined by the form of the conformal trace anomaly. It is this effective action and, in particular, its infrared fixed point that I would like to argue is relevant to the question of the ground state of quantum gravity in four dimensions. A good deal of the second and third lectures (appearing here as sections 4 to 6 and 7 to 10 respectively) is technical, so it important to keep in mind throughout the motivation for this program: the problem of vacuum energy.

The lectures are organized as follows. After surveying the various approaches to the vacuum energy problem, and sketching the proposal for its solution suggested by an analogy with electromagnetism in this first lecture, I turn to the formal aspects of path integral quantization in the second lecture. The path integral for gravity will be constructed in a way that preserves manifest coordinate invariance at the quantum level and applied to the stability of flat Minkowski spacetime. Finally in the third lecture I describe the effective theory of the conformal factor generated by the quantum trace anomaly, and quantize this theory by the path integral method. I discuss there the infrared conformal fixed point of this modified theory of gravity and the behavior of correlation functions in de Sitter spacetime, which strongly suggest that the *effective* cosmological constant in the true vacuum state of quantum gravity in four dimensions is identically zero.

2. A Brief Survey of Approaches

The introduction was meant to convince you that the problem of vacuum energy is real and, moreover that it should have a solution within the confines of a low-energy effective theory of gravity, without direct reliance on Planck scale phenomena. In order to reinforce this point and to see how the approach I shall follow fits into the historical background, I would like to review briefy other approaches that have been suggested up until now. The general outline is borrowed from S. Weinberg's very clear recent review article,[3] to which you are referred for an extensive bibliography and further reading.

2.1. Supersymmetry, Supergravity, Superstrings

One way to set a constant of nature to zero that appears to have no reason to be small is to postulate a new symmetry. Supersymmetry would seem to be a very attractive idea in this regard, because fermionic degrees of freedom are related to spacetime transformations by the fundamental algebra of supersymmetry, *viz.*

$$\{Q_\alpha, Q_\beta\}_+ = (\sigma_\mu)_{\alpha\beta}P^\mu \ . \tag{2.1}$$

Hence if the vacuum has zero fermionic charge, $Q_\alpha|\ \rangle = 0$, it automatically has vanishing energy-momentum: $P^\mu|\ \rangle = 0$. This is simply a reflection of the cancellation between positive vacuum energy of bosons with the negative vacuum energy of their supersymmetric partners. This cancellation is exact and supersymmetry would forbid a vacuum energy term only if the supersymmetry is unbroken. The problem is that we are sure that supersymmetry *is* broken (if it exists at all in nature). Otherwise, there would be degenerate mass supermultiplets which are not observed. Since it must be broken to agree with particle physics as we know it, the problem of a non-zero and generally large vacuum energy reappears.

When supergravity theories are considered the problem becomes more subtle. One can construct models with zero vacuum energy, even with supersymmetry broken. However, all known examples of this kind are based on specific choices of superpotentials, or tied in some way to a tree-level or low loop order perturbative cancellation. The broken supersymmetry does not forbid higher loop violations of the cancellation, and a fine tuning problem remains. Superstring theories seem to have exactly the same problems as supergravity models in this respect. Even if a way around large contributions to Λ in a fundamental supersymmetric or superstring theory could be found, the mechanism would have to survive all the way down to electroweak and QCD phase transition scales. Such an extreme sensitivity of low energy phenomenology to an ultra-short distance theory runs counter to all of our experience in physics.

2.2. Wormholes

Some years ago Hawking proposed a Euclidean formulation of the path integral for quantum gravity.[4] In analogy with the partition function of other systems, he suggested that one consider the Euclidean path integral,

$$Z \sim \int [\mathcal{D}g_{ab}] \ \exp(-S_E[g_{ab}]) \ , \tag{2.2}$$

where S_E is the Euclidean signature version of the Einstein-Hilbert action functional. If one interprets this expression as a probablity functional on the space of Euclidean metrics, then it is sharply peaked around $\Lambda = 0$. Hence if Λ becomes dynamical in some way, it may seem plausible to argue that the preferred value of Λ_{eff}

is zero. This is the original Baum-Hawking proposal.[5] Later Banks and Coleman suggested,[6] in effect that the way Λ could become dynamical is if the path integration over metrics includes summation over non-trivial topologies, *i.e.* wormholes or "baby universes." Then the effect of this sum over wormholes is to yield an effective wormhole-free theory of the same form as (2.2) but with an effective shifted value of the cosmological term, $\Lambda_{eff} = \Lambda(\alpha)$ that depends on wormhole coherent state parameters α with some probability measure. If this measure is reasonable smooth, then integrating over α implies that the total wave functional is sharply peaked at the value of Λ_{eff} where (2.2) peaks, namely zero. In an extremely simplified form this is Coleman's argument for $\Lambda_{eff} = 0$.

The problems with this argument are many, and it has been criticized from several different points of view in the literature.[7] The most glaring difficulty with the entire Baum-Hawking-Coleman approach is that the expression (2.2) above is completely ill-defined. The Euclidean Einstein-Hilbert action is not a bounded functional. Therefore $\exp(-S_E[g_{ab}])$ cannot be interpreted as a probability measure of anything in its present form. This problem is called the conformal factor problem, because the unboundedness arises in the integration over the conformal part of the metric σ, when g_{ab} is expressed in the form $e^{2\sigma}\bar{g}_{ab}$. Gibbons, Hawking and Perry subsequently proposed to "fix" the unboundedness of the Euclidean action by "rotating the contour" of the σ integration to lie along the imaginary axis.[8] In general this makes the metric complex and/or changes its signature, and does not help to elucidate the physical meaning of (2.2) as a probability density.

The difficulty with the unboundedness of the Euclidean action appears even when expanding about *flat* Minkowski spacetime with no cosmological term. But in this case, the positive energy theorem for asymptotically flat spacetimes applies and the physical spin-2 excitations can only raise the energy, not lower it. Hence flat space is definitely *not* unstable to conformal deformations of the metric which vanish at infinity, in pure Einstein gravity, notwithstanding the unboundedness of the Euclidean action. This is a very confusing state of affairs and one that casts doubt on the whole procedure of Euclidean continuation in quantum gravity. The Coleman argument simply ignores the unboundedness difficulty and *assumes* that (2.2) is sensible. Even if we swallow hard and accept this point of view, there are more problems. Coleman's analysis borrows from instanton physics in the sense that classical solutions to Euclidean field equations are posited to dominate the path integral. In field theory one can calculate certain exponentially suppressed tunnelling amplitudes by a dilute gas of such configurations. This makes sense if their contributions are exponentially small in coupling, and there are suitable cut-offs in integrations over scale size that keep the contributions exponentially small. In the wormhole case, neither of these conditions is satisfied. There is no "dilute gas" of wormholes because the contribution of one wormhole is exponentially *large*, not small. This is related to the unboundedness of the Euclidean action, since the value of the action is *negative* on the wormhole solution. This is no mere technical difficulty that one may imagine removing in a more refined calculation, since the sign of the action is crucial to the entire argument about peaking of the wavefunction at

$\Lambda_{eff} = 0$, so the problem is intrinsic to the whole approach. If this is not enough, there is no obvious infrared cut-off in wormhole size, and we are in danger of being confronted by closely packed wormholes of all sizes, up to macroscopic distance scales.[7] Needless to say this is in conflict with observation.

In all, the wormhole proposal raises many more questions than it answers. No one has ever shown convincingly that summing over arbitrary topologies is even necessary, let alone how to evaluate the sum reliably. Yet, the wormhole suggestion does illustrate an important point that is independent of summing over topologies. That is, if by *whatever* mechanism the effective cosmological term becomes dynamical, then it may be that zero is its preferred value in the ground state, and this conclusion depends only on low-energy considerations in an effective field theory (large wormholes in Coleman's proposal). Wormholes simply provide a definite way for this to happen and a fairly specific suggestion for what the ground state wave functional might be, but any low-energy effective field theory in which the effective value of Λ is not constrained but dynamical should do just as well. We shall see that fluctuations in the conformal factor of the metric provide just such an alternative.

2.3. Dynamic Adjustment Mechanisms

The basic idea of this variety of approaches to the vacuum energy problem is to introduce some new degree(s) of freedom (usually a scalar field) into the gravitational Lagrangian. If this scalar field ϕ couples to the trace of the energy-momentum tensor T_{ab}, in such a way that T^a_a vanishes at $\phi = \phi_0$ where $\Lambda_{eff}(\phi_0) = 0$, then the effective cosmological term vanishes in the vacuum state. This approach tries to build upon the right physical intuition (in my opinion) of finding the right vacuum state through a dynamical principle, rather than relying on an exact symmetry of the fundamental theory of gravity at the Planck scale, whatever it may be.

The trouble is that there is generally no way to arrange $\Lambda_{eff}(\phi_0) = 0$ except through fine tuning some parameter(s) of the scalar Lagrangian. In his discussion of this class of models, Weinberg points out that fine tuning can be avoided only if the Lagrangian depends on ϕ and the metric g_{ab} solely through the combination, $\hat{g}_{ab} = e^{2\phi} g_{ab}$. In that case the vanishing of the trace of T_{ab} and therefore Λ_{eff} follows immediately from the ϕ eqs. of motion. With standard kinetic terms for the scalar field ϕ it is impossible to arrange this at constant ϕ_0 without fine tuning.[3]

From the point of view we shall adopt later in these lectures, we shall see that the covariant quantization of gravity *requires* a scalar field, the conformal part of the metric itself, which *automatically* appears in the right combination with g_{ab}, and moreover that the conformal anomaly determines the form of the effective action for this field which is *not* that of a conventional scalar field. Also, the ground state of the resulting effective field theory is not necessarily one with constant ϕ_0. The route to a solution of the problem we'll be considering in the last lecture may be regarded as a variation of a dynamic adjustment mechanism which slips through the loopholes in the no-go argument above.

2.4. Anthropic Principle

This suggestion comes down to the statement that the universe could not be other that as it is, because if it were there would be no one here to observe it. So perhaps it can be argued that Candide was right after all that "all is for the best in this best of all possible worlds," and Λ must be small enough to allow Candide[9] and the rest of us to exist and behold its beauty.

One can attempt to make this slippery statement more quantitative by asking what are the features of our universe that are absolutely necessary to intelligent life, and how small must Λ be in order for these features (like stars and planets) to have formed.[11] This sort of argument is difficult to dispute on purely scientific grounds. The sceptic can only ask politely how one defines intelligent life, or how one knows *a priori* what constants of nature and/or laws of physics are necessary for its emergence. We not only lack any sound information on these questions, but even (it seems to me) any possibility of ever obtaining such information. We can no more answer such questions as we can answer the trick question on the poet's physics final exam that asks: "Define the Universe. Give three examples."

So my objection to an anthropic argument is not scientific but philosophical. It seems to me more like a statement of ignorance rather than a scientific principle. Medieval scolastics, if pressed could always argue that what is is the way it is by the will of God, but this explains nothing, and (if taken seriously) has the effect of cutting off further discussion. Would it not be more intellectually honest to leave the door open for further inquiry by saying simply, "I don't know?"

2.5. Modifying Gravity

Once one discards general relativity there are very few guidelines of how to replace it. An interesting suggestion involving minimal alteration of Einstein's theory has been discussed by several authors.[10] The proposal is that only the spin-two modes of the metric field should really be treated as physical degrees of freedom which must be varied dynamically. This is equivalent to the ordinary unconstrained variation of the metric in the Einstein-Hilbert action together with a Lagrange multiplier to enfore the constraint on the determinant of g_{ab}. In this way one finds only the ***traceless*** part of Einstein's field equations, in which the cosmological term drops out,

$$R^a{}_b - \tfrac{1}{4}R\delta^a{}_b = \kappa(T^a{}_b - \tfrac{1}{4}T\delta^a{}_b)\ . \tag{2.3}$$

Now if one takes the covariant derivative of this equation and uses the Bianchi identity and covariant conservation of T_{ab}, one finds that $\kappa T + R$ is a constant, which we may call 4Λ if we like. Then Einstein's equations are recovered in their original form, except that Λ has appeared as a constant of integration of the classical eqs. of motion, not as a fundamental parameter of the Lagrangian. Hence, the field equations (2.3) admit a flat space solution when T_{ab} vanishes, by suitable choice of the integration constant. Of course, this doesn't solve our problem, since now the question becomes: why should one prefer the flat space solution? It does recast the issue in an interesting way, however, and again serves to emphasize that Λ depends on the state of the system (the quantum analog of an integration constant), rather than being determined by a fundamental Planck scale formulation of quantum gravity. Next I would like to point out in some detail an analogy between this formulation of cosmological constant problem and a similar situation in electromagnetism.

3. An Electromagnetic Analogy

An analogy to the vacuum energy problem exists also in electromagnetism. I will call it the "cosmological electric field problem." It consists of the elementary observation that Maxwell's equations *in vacuo* admit a solution with constant, uniform electric field $\mathbf{E}_{cosm}$ of arbitrary direction and magnitude E_{cosm}. Why then are all electric fields found in nature always associated with localized electrically charged sources? Why do we not observe some huge uniform electric field pointing in the direction of the magellanic clouds, for example? Classically, we just set $E_{cosm} = 0$ by a boundary condition, but this explains nothing. Is there a deeper reason in the quantum theory?

The answer is almost certainly yes, because of vacuum polarization effects. Suppose that E_{cosm} were not zero. Then the quantum vacuum of any charged matter fields interacting with it is polarized. This is described by a polarization tensor,

$$\Pi^{ab}(x, x'; E) = i\langle \mathcal{T} j^a(x) j^b(x') \rangle_E \; , \tag{3.1}$$

where j^a is the charge current operator and the expectation value is evaluated in a vacuum state with classical background field E. The time ordering symbol $\mathcal{T}$ enforces the Feynman boundary conditions on the polarization operator, *i. e.* positive frequency modes propagate forward in time, while negative frequency modes propagate backward in time. Notice that these boundary conditions are *not* time symmetric. Time-ordering (with an $i\epsilon$ prescription in the propagator) defines a different polarization tensor or Feynman Green's function from anti-time-ordering (with a $-i\epsilon$ prescription). This time-asymmetry is built into quantum theory by the demands of causality and the existence of a lowest energy ground state, or vacuum.

The polarization operator Π^{ab} contains two pieces, an even and an odd piece under time reversal. The even piece describes the polarizability of the vacuum, since the vacuum fluctuations of virtual charge pairs may be thought of as giving rise to an effective polarizability of the vacuum (dielectric constant). The time reversal odd piece describes the creation of *real* particle anti-particle pairs from the vacuum by the Schwinger mechanism.[12] The creation of real charge pairs means that a real current flows $j \geq 0$ in the direction of the electric field at later times, even if none existed initially. This implies that the electric field does work at a rate, $j \cdot E$. To the extent that this power cannot be recovered because the created particles interact and lose the coherence they may have had in their initial state, this is the rate of energy *dissipation*. The vacuum behaves like a normal conductor with a finite conductivity and resistivity, due to the random, uncorrelated motions of its fluctuating charge carriers. At the same time, the electric field is diminished by the Maxwell equation,

$$\frac{\partial E}{\partial t} = -\langle j \rangle, \tag{3.2}$$

in the case of exact spatial homogeneity of the average current. If the expectation value of j points in the direction of E, then we should expect the magnitude of E to decrease with time, and finally approach zero. This is certainly what occurs if j contains an Ohmic dissipation term σE. In other words, there is a dynamical reason why zero cosmological electric field is preferred: any non-zero field eventually decays to zero when interactions with quantized charged matter fields is allowed.

The Einstein gravitational field equations may be cast into a form precisely analogous to the spatially homogeneous form of Maxwell's eqs. as follows. Consider the Einstein equation for a spatially homogeneous and isotropic spacetime with Robertson-Walker scale factor, $a(t)$:

$$H^2 \equiv \left(\frac{\dot{a}}{a}\right)^2 = \frac{\kappa}{3}\rho, \tag{3.3}$$

where any cosmological term is included in the right hand side as a vacuum energy contribution with equation of state,

$$\rho_v = -p_v. \tag{3.4}$$

Eq. (3.3), together with the equation of covariant energy conservation,

$$\dot{\rho} + 3\ H\ (\rho + p) = 0, \tag{3.5}$$

imply that:

$$\dot{H} = -\frac{\kappa}{2}\ (\rho + p). \tag{3.6}$$

Compare this with Eq. (3.2) above. If $\rho + p$ is strictly non-negative, and contains a dissipative term proportional to H in a curved background, then H will decrease monotonically, *independently* of the value of Λ in the Einstein Lagrangian. Whether or not such a dissipative term exists is a *dynamical* question, depending on the behavior of the non-local polarization operator,

$$\Pi^{abcd}(x, x'; \bar{g}) = i\langle T T^{ab}(x) T^{cd}(x')\rangle_{\bar{g}}, \tag{3.7}$$

at *large* separations of the points x and x' in a curved background $\bar{g}_{ab}$. The purely local counterterms of the effective action needed to absorb the ultraviolet infinities in Π as $x \to x'$ have nothing to do with Ohmic dissipation in QED, and do not generate any dissipative term proportional to H in (3.6). Such terms can only be generated at long distances in the far infrared.

This idea of vacuum fluctuations giving rise to the mechanism of dissipating vacuum energy into other modes, thereby effectively screening any fundamental cosmological term at large distance or time scales, requires a formalism to treat the effect of particle creation back on the metric of spacetime (dynamical back reaction).

Actually this is an attractive line of thought which I have pursued for several years now. However, to describe this work in detail would take me far afield of the main focus of these lectures, so I simply refer you to the literature,[13–16] and turn to a different methodology in the next lecture.

4. Basics of Path Integral Quantization

I wish to turn now from general physical considerations to the formal aspects of path integrals, because we'll need some basic mathematical tools in order to analyze quantum gravity from a covariant point of view. The path integral is the best route because (unlike canonical methods, for example) symmetries can be preserved in a natural geometric way, and general covariance will play a crucial role in our subsequent analysis.

Let us begin with a non-relativistic quantum mechanical Hamiltonian,

$$H = \frac{p^2}{2m} + V(x) \,, \tag{4.1}$$

where $p = \frac{\hbar}{i}\frac{\partial}{\partial x}$ in the position representation. The evolution of the wavefunction with arbitrary initial condition at time t_0 is described succintly by the propagator or kernel K,

$$\psi(x;t) = \int_{-\infty}^{\infty} dx_0 \; K(x, x_0; t - t_0) \; \psi(x_0; t_0) \,, \tag{4.2}$$

where K satisfies the Schrödinger Eq.,

$$HK(x, x_0;\, t - t_0) = i\frac{\partial}{\partial t}K \,, \tag{4.3}$$

which has the solution,

$$K(x, x_0; t) = \langle x | \exp(-iHt) | x_0 \rangle \,. \tag{4.4}$$

The path integral representation for this quantity is easily derived by first dividing the total time interval t into N equal partitions of length ϵ (with $t_n = n\epsilon$ and $n = 0, 1, \ldots, N$), and then inserting the identity operator in the form of an integration over complete sets of position eigenstates of the particle at each of the $N - 1$ intermediate times. In this way we find that:

$$K(x, x_0; t) = \lim_{N\to\infty} \prod_{n=1}^{N} \left\{ \int dx_n \langle x_n | (1 - i\epsilon H(t_n)) | x_{n-1} \rangle \right\} \delta(x - x_N) \,. \tag{4.5}$$

Evaluating each of the N matrix elements by again inserting the identity in the form of N integrations over conjugate momentum eigenstates $|p_n\rangle$, and using $\langle p | x \rangle = \exp(-ipx)$ yields the desired phase space path integral representation, *viz.*

$$\begin{aligned} K(x, x_0; t) = & \lim_{N\to\infty} \delta(x - x_N) \times \\ & \prod_{n=1}^{N} \left\{ \int \frac{dx_n dp_n}{2\pi} \exp\left\{ -i\epsilon \left[\frac{p_n^2}{2m} + V(x_n) \right] + ip_n(x_n - x_{n-1}) \right\} \right\} \,. \end{aligned} \tag{4.6}$$

The important point that I wish to stress about this representation is that the canonical phase space measure appears with *no* arbitrariness whatsover, and the limit $N \to \infty$ exists, so Eq. (4.6) is completely well-defined mathematically.

Further, since all of the p_n integrations are Gaussian, they may be performed explicitly to arrive at the position space representation of the Feynman path integral,

$$K(x, x_0; t) = \lim_{N\to\infty} \prod_{n=1}^{N} \left\{ \int \frac{m\, dx_n}{2\pi i\epsilon} \exp\left\{ \frac{im(x_n - x_{n-1})^2}{2\epsilon} - i\epsilon V(x_n) \right\} \right\} \delta(x - x_N). \tag{4.7}$$

Since the quantity in the exponent is precisely the time-discretized form of the classical action functional evaluated on all possible paths between the initial position x_0 and the final position x in time t, Eq. (4.7) gives a completely well-defined, unambiguous meaning to the shorthand notation,

$$K(x, x_0; t) = \int_{x(0)=x_0}^{x(t)=x} [\mathcal{D}x] \exp\{iS[x]\}, \qquad S[x] = \int_0^t du \left[\frac{m\dot{x}^2}{2} - V(x) \right]. \tag{4.8}$$

If t were replaced by $-i\tau$ this is exactly the Wiener path integral which is well studied in the mathematical and statistical mechanics literature.[17]

Although the configuration space path integral is well-defined by discretization of time, such a procedure is awkward to implement, and is not well-suited to covariant theories where time should play no preferred role. Here the situation is reminiscient of the Riemann definition of ordinary integration: it's important that such a definition exists but that's not how we actually perform integrals in practice. Unfortunately, the only path integrals we know how to do analytically are Gaussian, but for these it is possible to give an alternate definition which generalizes nicely to covariant field theories.

For a harmonic oscillator action functional we may expand about the classical solution to quadratic order,

$$S_{quad} = \frac{m}{2} \int_0^t du \left[\left(\frac{dx}{du}\right)^2 - \omega^2 x^2 \right] = S_{cl}(x, x_0; t) + \frac{m}{2} \int_0^t du\, \xi \Delta_\omega \xi\,, \tag{4.9}$$

where $\xi(u) = x(u) - x_{cl}(u)$ and $x_{cl}(u)$ satisfies the boundary conditions, $x_{cl}(0) = x_0$ and $x_{cl}(t) = x$. The differential operator Δ_ω is given by

$$\Delta_\omega = -\frac{d^2}{du^2} - \omega^2\,. \tag{4.10}$$

Since this operator may be diagonalized by finding a complete set of eigenfunctions $\xi_n(u)$ on the interval $[0, t]$ satisfying vanishing boundary conditions at the endpoints, we would like to be able to define the path integral over ξ in terms of the determinant of Δ_ω which is just the product of its eigenvalues, λ_n. The problem is that this product is infinite, and requires a regulator. Just such a regulator is provided by the ζ-function technique. Define

$$\zeta(s|\Delta_\omega) \equiv \sum_n \left(\frac{\mu^2}{\lambda_n} \right)^s. \tag{4.11}$$

where μ is an arbitrary scale parameter with dimensions of ω. Eq. (4.11) defines a meromorphic function of s with simple poles on the negative integers. In particular, the sum converges for $Re(s) > 1$, and this defines a function which may analytically continued to $s = 0$. Since for a finite dimensional matrix operator,

$$-\frac{d\zeta}{ds}(s=0|\Delta_\omega) = \sum_n \ln\left(\frac{\lambda_n}{\mu^2}\right) = \ln\det\left(\frac{\Delta_\omega}{\mu^2}\right), \tag{4.12}$$

we may use Eq. (4.12) to *define* the determinant of a differential operator with an infinite spectrum. Then the Gaussian path integral (4.8) with action (4.9) is defined (up to the arbitrary parameter μ) by

$$\int_{x(0)=x_0}^{x(t)=x} [\mathcal{D}x] \exp\{iS_{quad}[x]\} = \det{}^{-\frac{1}{2}}\left(\frac{\Delta_\omega}{\mu^2}\right) \exp\{iS_{cl}(x, x_0; t)\}. \tag{4.13}$$

Since these lectures are meant to be pedagogical, let us perform the exercise of explicitly evaluating this expression by the ζ-function method.

First, the complete set of eigenfunctions of Δ_ω satisfying the vanishing boundary conditions at the endpoints of the interval $[0, t]$ are just sine functions,

$$\xi_n(u) = \sin\left(\frac{\pi n u}{t}\right) \qquad n = 1, 2, \ldots \tag{4.14}$$

with the corresponding eigenvalues of Δ_ω,

$$\lambda_n = \left(\frac{\pi n u}{t}\right)^2 - \omega^2. \tag{4.15}$$

Hence the ζ-function (4.11) may be written as

$$\begin{aligned} \zeta(s|\Delta_\omega) &= \sum_{n=1}^{\infty}\left[\frac{\mu^2}{\left(\frac{\pi n}{t}\right)^2 - \omega^2}\right]^s \\ &= \left(\frac{\mu t}{\pi}\right)^{2s} \sum_{n=1}^{\infty} n^{-2s}\left[1 - \left(\frac{\omega t}{\pi n}\right)^2\right]^{-s}. \end{aligned} \tag{4.16}$$

Using the binomial expansion for the last term in this expression, *viz.*,

$$(1 - x^2)^{-s} = \sum_{k=0}^{\infty} \frac{\Gamma(k+s)}{\Gamma(k+1)\Gamma(s)} x^{2k}, \tag{4.17}$$

and interchanging the order of the k and n sums, we find that the sum over n may be performed explicitly in terms of the Riemann zeta function ζ_R:

$$\zeta(s|\Delta_\omega) = \left(\frac{\mu t}{\pi}\right)^{2s} \sum_{k=0}^{\infty} \frac{\Gamma(k+s)}{\Gamma(k+1)\Gamma(s)} \left(\frac{\omega t}{\pi n}\right)^{2k} \zeta_R(2s + 2k). \tag{4.18}$$

This form is now suitable for analytic continuation towards $s = 0$. Evaluating the first derivative there gives

$$\left.\frac{d\zeta(s|\Delta_\omega)}{ds}\right|_{s=0} = 2\ln\left(\frac{\mu t}{\pi}\right)\zeta_R(0) + 2\zeta_R'(0) + \sum_{k=1}^{\infty}\frac{1}{k}\left(\frac{\omega t}{\pi n}\right)^{2k}\zeta_R(2k)\,. \tag{4.19}$$

The properties of ζ_R are well known and may be found in standard mathematical references. In addition to the values of the function and its derivative at $s = 0$, the relationship of $\zeta_R(2k)$ to the Bernoulli numbers B_k is needed. The remaining sum is easily evaluated using

$$\sum_{k=1}^{\infty}\frac{1}{2k}\frac{x^{2k}}{(2k)!}B_k = \ln\left(\frac{x}{2\sin(\frac{x}{2})}\right)\,. \tag{4.20}$$

We find finally the simple result

$$\left.\frac{d\zeta(s|\Delta_\omega)}{ds}\right|_{s=0} = \ln\left(\frac{\omega}{2\mu\sin\omega t}\right)\,. \tag{4.21}$$

Hence from the definition (4.12)

$$\det{}^{-\frac{1}{2}}\left(\frac{\Delta}{\mu^2}\right) = \exp\left(\frac{1}{2}\left.\frac{d\zeta(s|\Delta)}{ds}\right|_{s=0}\right) = \left(\frac{\omega}{2\mu\sin\omega t}\right)^{\frac{1}{2}}\,. \tag{4.22}$$

The unknown constant μ is somewhat analogous to a constant of integration. It can be determined from the normalization condition on K which follows from Eq. (4.4), namely

$$K(x, x_0; t = 0) = \delta(x - x_0)\,. \tag{4.23}$$

Equivalently one can compare the form of the kernal obtained from Eqs. (4.8), (4.13) and (4.22) for the harmonic oscillator with the known result for the free particle with $\omega = 0$. By either method one finds that μ must be replaced by $\frac{i\pi\hbar}{m}$ in order to give the correct answer. This is the same result for K that is obtained by the more laborious method of evaluating the discretized path integral expression Eq. (4.7) directly.

As a further exercise with the harmonic oscillator path integral, one can verify that substituting $t = -i\beta$ and $x_0 = x$, and performing the Gaussian integration over x gives the finite temperature partition function of the oscillator,

$$Z(\beta) = \int_{-\infty}^{\infty} dx\, K(x, x; -i\beta) = Tr\, e^{-\beta H} = \frac{1}{2\sinh(\frac{\omega\beta}{2})}\,. \tag{4.24}$$

This can be evaluated also by the ζ-function method directly if one computes the determinant of Δ_ω in Eq. (4.10) with periodic boundary conditions in imaginary time, $t = -i\beta$, otherwise following the same steps as Eqs. (4.14)-(4.22) above. In this case the parameter μ drops out of the final answer and the familiar result (4.24) is obtained directly.

5. Covariant Path Integral Quantization

We turn now to the path integral quantization of systems with a reparameterization or gauge invariance. In order to build confidence with path integral techniques, we shall procedure in a methodical way from simpler systems to gravity.

5.1. Relativistic Point Particle

There are several forms of the action for a relativistic point particle, all of which lead to the same answer. We shall use the form in flat Minkowski space,

$$S[x;N] = \frac{1}{2}\int_0^1 du\left(\frac{\dot{x}^a \eta_{ab}\dot{x}^b}{N} - m^2 N\right). \tag{5.1}$$

This action possesses the reparameterization gauge invariance, $\mathcal{G}$

$$\begin{aligned} x^a(u) &\to x^a(f(u)), \\ N(u) &\to \frac{df}{du}N(f(u)), \end{aligned} \tag{5.2}$$

for any differentiable function $f(u)$ that leaves the endpoints of the interval $[0,1]$ fixed, *i.e.* provided

$$f(0) = 0\,, \qquad f(1) = 1\,. \tag{5.3}$$

The appearance of one factor of the determinant of the reparameterization, $u \to f(u)$ in the transformation of N in Eq. (5.2) characterizes it as a density of weight one. The coordinate x^a is a density of weight zero, *i.e.* a scalar under coordinate reparameterizations.

The momenta conjugate to x^a and N are:

$$\begin{aligned} p_a &\equiv \frac{\delta S}{\delta \dot{x}^a} = \frac{1}{N}\eta_{ab}\dot{x}^b, \qquad \text{and} \\ p_N &\equiv \frac{\delta S}{\delta \dot{N}} = 0\,, \end{aligned} \tag{5.4}$$

respectively. Since the momentum conjugate to N vanishes identically, the Hamiltonian system is a degenerate one with a primary first class constraint, in Dirac's nomenclature. The secondary first class constraint following from the primary one $p_N = 0$ is

$$\dot{p}_N = -\frac{\delta S}{\delta N} = p_a \eta^{ab} p_b + m^2 \equiv \mathcal{H} = 0\,. \tag{5.5}$$

Because of this degeneracy, the phase space Hamiltonian path integral quantization cannot be applied without some modification. The correct modification is that of

Batalin, Fradkin and Vilkovisky.[18] Since the Hamiltonian $\mathcal{H}$ is quadratic in momenta, there is a simple coordinate space path integral obtained by integrating out the momenta in the BFV formalism. Like the non-relativistic example considered previously the canonical path integral has a completely unambiguous well-defined measure in phase space, which defines a configuration space measure after the Gaussian integration over momenta. Rather than carrying this out explicitly I'll follow a different route and construct the coordinate space path integral by geometric reparameterization invariance considerations alone.[19–22]

The space of all configurations, $\mathcal{M}$ labelled by the functions $(x^a(u), N(u))$ may be treated by methods borrowed from Riemannian geometry. At every "point" in the function space, X^i we introduce the cotangent space, $\delta\mathcal{M}$ labelled by the basis vectors $\delta X^i = (\delta x^a(u), \delta N(u))$. In Riemannian geometry one can introduce a metric on a space by defining a quadratic form, the line element, that maps $\delta\mathcal{M} \times \delta\mathcal{M}$ to the real numbers,

$$(\delta s)^2 = G_{ij}(X)\delta X^i \delta X^j \,. \tag{5.6}$$

The (infinitesimal) invariant volume measure of integration on the cotangent space, $\delta\mathcal{M}$

$$d(\delta\mathcal{V}) = \sqrt{\det G_{ij}(X)}\; d(\delta X^1) \times d(\delta X^2) \times \ldots \tag{5.7}$$

may be chosen to satisfy the Gaussian normalization condition,

$$\int d(\delta\mathcal{V}) \exp\left\{-\tfrac{i}{2}\delta X^i G_{ij}(X)\delta X^j\right\} = 1\,, \tag{5.8}$$

and immediately induces an invariant volume measure on the full space $\mathcal{M}$,

$$d\mathcal{V} = \sqrt{\det G_{ij}(X)}\; dX^1 \times dX^2 \times \ldots \tag{5.9}$$

To define the path integral for a system possessing a reparameterization gauge invariance, we have only to define a quadratic inner product on the cotangent space $\delta\mathcal{M}$. Then the construction of the invariant measure on the function space of all configurations proceeds exactly along the lines of (5.6)-(5.9). Since the path integral is specified by an invariant action functional and an invariant integration measure, this procedure preserves the classical reparameterization invariance under quantization. Then by identifying configurations which differ only by a reparameterization of coordinates, we may integrate only over *equivalence classes* of coordinates, $\mathcal{M}/\mathcal{G}$ in a manifestly gauge invariant way.

Let us illustrate this procedure for the relativistic point particle. Since both N and δN have weight one under (5.2), and

$$\int_0^1 N(u)\, du \equiv \tau \tag{5.10}$$

is invariant under reparameterizations, we can define an invariant inner product on $\mathcal{M}$ by

$$\begin{aligned} \langle \delta N, \delta N\rangle_N &\equiv \int_0^1 \left(\frac{\delta N}{N}\right)^2 N\, du = \int_0^1 \frac{(\delta N)^2}{N}\, du \qquad \text{and} \\ \langle \delta x, \delta x\rangle_N &\equiv \int_0^1 \delta x^a G_{ab} \delta x^b\, du = \int_0^1 \delta x^a \eta_{ab} \delta x^b N\, du\,, \end{aligned} \tag{5.11}$$

since η_{ab} is the only Lorentz invariant tensor possible in flat Minkowski space, up to a multiplicative constant, which may be absorbed into normalization of the measure by

$$\begin{aligned}\int[\mathcal{D}(\delta N)]\exp\left\{-\tfrac{i}{2}\langle\delta N,\delta N\rangle_N\right\} &= 1,\\ \int[\mathcal{D}(\delta x)]\exp\left\{-\tfrac{i}{2}\langle\delta x,\delta x\rangle_N\right\} &= 1.\end{aligned} \tag{5.12}$$

This defines an invariant functional measure on the full space $\mathcal{M}$ by eqs (5.6)-(5.9). In order to pull this measure back to an invariant measure on the quotient space of equivalence classes $\mathcal{M}/\mathcal{G}$, let us parameterize the gauge orbits $\mathcal{G}$ by the set of diffentiable functions $f(u)$ satisfying the endpoint conditions, (5.3). Then an arbitrary lapse function $N(u)$ may be written in the form,

$$N(u) = \tau\frac{df}{du}\,; \qquad f(u) = \frac{1}{\tau}\int_0^u N(u)\,du\,. \tag{5.13}$$

In other words, the gauge equivalence class $\mathcal{M}/\mathcal{G}$ of all the functions $N(u)$ is characterized completely by the single parameter τ defined in Eq. (5.10), and the gauge fiber $\mathcal{G}$ is coördinatized by $f(u)$. Hence the integration measure on the quotient space is given by

$$\frac{[\mathcal{D}N]}{[\mathcal{D}f]} = J(\tau)d\tau\,, \tag{5.14}$$

where J is the Jacobian of change of variable from N to (τ, f). The invariant path integral is then

$$K(x,x_0) = \int J(\tau)d\tau\int_{x(0)=x_0}^{x(1)=x}[\mathcal{D}x]e^{iS[x;\tau]}\,. \tag{5.15}$$

Our task now is to determine the Jacobian J. To do so write the cotangent space form of Eq. (5.13),

$$\delta N = \delta\tau\frac{df}{du} + \tau\frac{d\delta f}{du}\,, \tag{5.16}$$

and substitute this into the inner product definition (5.11). The cross term between $\delta\tau$ and δf vanishes by using the endpoint conditions (5.3). Changing variables in the last term from u to $v \equiv f(u)$ and defining

$$\xi(v) \equiv (\delta f)(u)\Big|_{u=f^{-1}(v)}\,, \tag{5.17}$$

we obtain

$$\langle\delta N,\delta N\rangle_N = \frac{(\delta\tau)^2}{\tau} + \tau\int_0^1 dv\,\xi(v)\left[-\frac{d^2}{dv^2}\right]\xi(v), \tag{5.18}$$

after an integration by parts. Now, it is straightforward to verify that the quantity $\xi(v)$ is invariant under diffeomorphism group transformations operating from the right, *i.e.*,

$$\xi(v) \to \xi(v)\,, \qquad f(u) \to f(\alpha(u))\,, \tag{5.19}$$

but that it transforms as a density of weight -1 under the *inverse* diffeomorphism group transformations operating from the left, *i.e.*,

$$\xi(w) \to \frac{1}{\left(\frac{d\beta^{-1}(w)}{dw}\right)}\xi(\beta^{-1}(w)) , \qquad f(u) \to \beta(f(u)) \equiv w . \tag{5.20}$$

The only quadratic form in ξ that is invariant under both of these two transformations is[20]

$$\langle \xi, \xi \rangle \equiv \int_0^1 dv\, \tau^3\, \xi^2(v) , \tag{5.21}$$

provided that τ remains invariant under the first transformation, but

$$\tau \to \tau \left(\frac{d\beta^{-1}(w)}{dw}\right) \tag{5.22}$$

under the second. The geometric meaning of these transformations is not difficult to visualize. In the case of (5.19), $N(u)$ as parameterized by (5.13) transforms according to Eq. (5.2) under right multiplication of the fiber coordinate $f \to f \circ \alpha$, while the coordinate in the quotient space $\mathcal{M}/\mathcal{G}$, τ remains invariant. In the second transformation (5.20), the base space coordinate τ transforms according to (5.22), in order to cancel the change under left multiplication, $f \to \beta \circ f$ and leave the total $N(u)$ invariant.

With the fully invariant inner product on the cotangent space to $\mathcal{G}$ defined by Eq. (5.21), we may define the integration measure on $\mathcal{G}$ by the Gaussian normalization condition,

$$\int [\mathcal{D}\xi] \exp\left\{-\tfrac{i}{2}\langle \xi, \xi \rangle\right\} = 1 . \tag{5.23}$$

Returning then to our problem of evaluating the Jacobian J in Eq. (5.14), we find that

$$\langle \delta N, \delta N \rangle_N = \frac{(\delta\tau)^2}{\tau} + \frac{1}{\tau^2}\langle \xi, \Delta_0 \xi \rangle , \tag{5.24}$$

with Δ_0 given by the same second order operator (with $\omega = 0$) whose determinant we evaluated in the previous section. Therefore we have immediately

$$\begin{aligned} 1 &= \int [\mathcal{D}(\delta N)] \exp\left\{-\tfrac{i}{2}\langle \delta N, \delta N \rangle_N\right\} \\ &= \int d(\delta\tau)\, J(\tau) \int [\mathcal{D}\xi]\, \exp\left\{-\frac{i(\delta\tau)^2}{2\tau} - \frac{i\langle \xi, \Delta_0 \xi \rangle}{2\tau^2}\right\} \\ &= cJ(\tau)\, \tau^{\frac{1}{2}}\, \det{}^{-\frac{1}{2}}\left(\frac{\Delta}{\tau^2}\right) \\ &= c' J(\tau) , \end{aligned} \tag{5.25}$$

where c and c' are constants, by Eq. (4.22). Hence the Jacobian is independent of τ and we may take it to be one.

The evaluation of the propagator in Eq. (5.15) is now straightforward. The functional integration over $x^a(u)$ with fixed enpoint conditions is the same as that for the free non-relativistic particle, and we are left with only a simple integral over the proper time τ to perform:

$$\begin{aligned} K(x,x_0) &= \int_0^\infty d\tau \, \exp\Big\{i\frac{(x-x_0)^a\eta_{ab}(x-x_0)^b}{2\tau} - i\tau m^2\Big\} \\ &= \int_0^\infty d\tau \int \frac{d^4p}{(2\pi)^4} \exp\Big\{ip_a(x-x_0)^a - i\tau\,(p_a\eta^{ab}p_b + m^2)\Big\} \\ &= \int \frac{d^4p}{(2\pi)^4}\frac{e^{ip\cdot(x-x_0)}}{(p^2+m^2)}\,, \end{aligned} \tag{5.26}$$

provided that a small negative imaginary part is added to m^2 to define the τ integral. Eq. (5.26) we recognize as just the Feynman propagator for the free relativistic scalar field, which we have obtained here by the path integral treatment of the reparameterization invariant first quantized particle action (5.1). As in the non-relativistic case, the arbitrary multiplicative constant is fixed by an appropriate normalization condition.

5.2. Yang-Mills Theory

The Yang-Mills action,

$$S_{inv}[A] = -\frac{1}{4g^2}\int d^4x F^i_{\mu\nu}F^{i\mu\nu} \tag{5.27}$$

is invariant under non-abelian gauge transformations, $A^i_\mu \rightarrow B^i_\mu$ with A and B related by

$$A^i_\mu = \{\mathcal{U}^{-1}(\theta)B^i_\mu\mathcal{U}(\theta)\}^i - i\{\mathcal{U}^{-1}(\theta)\partial_\mu\mathcal{U}(\theta)\}^i\,, \tag{5.28}$$

where $\mathcal{U}$ is an arbitrary element of the gauge group at each spacetime point. For $\mathcal{U}$ close to the identity the infinitesimal form of Eq. (5.28) is:

$$A^i_\mu \rightarrow A^{i(\theta)}_\mu \equiv A^i_\mu + (\nabla_\mu\theta)^i \equiv A^i_\mu + \partial_\mu\theta^i + f^{ijk}A^j_\mu\theta^k\,, \tag{5.29}$$

where f^{ijk} are the structure constants of some non-abelian gauge group.

We may regard Eq. (5.28) as setting up a coordinate system in the space of all gauge fields $\mathcal{M}$ with gauge fiber coordinates on $\mathcal{G}$ specified by $\mathcal{U}(\theta)$ and base space coordinates in the quotient space $\mathcal{M}/\mathcal{G}$ specified by B^i_μ, analogous to (5.13) of the relativistic point particle example. Then we consider the only Poincare invariant quadratic inner product on the cotangent space, namely

$$\langle\delta A, \delta A\rangle = \int d^4x\ \delta A^i_\mu(x)\ \eta^{\mu\nu}\ \delta A^i_\nu(x). \tag{5.30}$$

This natural, invariant quadratic form is completely analogous to the invariant line element ds^2 of finite dimensional Riemmanian manifolds, and therefore induces a natural, invariant volume form on the space of gauge fields. In the present case this metric is flat, being independent of $A^i_\mu(x)$. Hence the invariant volume form is just the product of coordinate differentials. We may fix the irrelevant normalization of this functional measure by a Gaussian normalization condition:

$$\int [\mathcal{D}\delta A^i_\mu] \exp\Big(-\frac{i}{2}\langle \delta A, \delta A\rangle\Big) = 1. \tag{5.31}$$

The measure, defined in this way is manifestly Lorentz invariant. Moreover, it is gauge invariant under (5.29). This is because it is translational invariant under $\delta A^i_\mu(x) \to \delta A^i_\mu(x) + v^i_\mu(x)$, and invariant under global gauge rotations. We may now let $B = \bar{B} + b$ and define

$$Z[\bar{B}] = (Vol(\mathcal{G}))^{-1} \int [\mathcal{D}a^i_\mu] \exp(iS_{inv}[\bar{B} + b]). \tag{5.32}$$

with respect to this induced measure. Here $Vol(\mathcal{G})$ denotes the (infinite) gauge orbit volume that must be divided out, and $\bar{B}$ is a background gauge potential. In order to specify the field coordinates completely one still needs to fix a condition on the b^i_μ of the form,

$$(F \cdot b)^i = (\bar{\nabla}^\mu b_\mu)^i = 0. \tag{5.33}$$

In the geometric approach this is not a gauge "fixing" that breaks the gauge invariance of the theory, but rather a choice of coordinates in a coordinate invariant formulation. The important point is that b^i_μ is gauge invariant, for *any* choice of condition on it, so that gauge invariance is assured by this construction from the start.

The task now is to express the gauge invariant quantity (5.32) in the coordinates (gauge) specified by (5.28) and (5.33). We first compute the Jacobian of the transformation to the new coordinates

$$[\mathcal{D}a^i_\mu] = J\, [\mathcal{D}b^i_\mu][\mathcal{D}\theta^i], \tag{5.34}$$

in the tangent space:

$$\begin{aligned} 1 &= \int [\mathcal{D}a^i_\mu] \exp(-\frac{i}{2}\langle a, a\rangle) \\ &= \int J\, [\mathcal{D}b^i_\mu][\mathcal{D}\theta^i] \exp(-\frac{i}{2}(\langle b, b\rangle + 2\langle \nabla\theta, b\rangle + \langle \nabla\theta, \nabla\theta\rangle)) \end{aligned} \tag{5.35}$$

This integral may be computed by completing the square and using condition (5.33). We find that

$$J = \{\det{}_S(-\nabla^2)\ \det{}_{FV}(\delta^\nu_\mu - \nabla_\mu(\nabla^2)^{-1}\nabla^\nu)\}^{\frac{1}{2}}, \tag{5.36}$$

where the vector determinant is to be evaluated over the space of fields b^i_ν obeying the condition (5.33).

The vector determinant may be converted into a scalar determinant[22] and the Jacobian (5.36) is given finally by:

$$J = \det{}_S^{-\frac{1}{2}}(-\bar{\nabla}^2)\ \det{}_S(-\bar{\nabla} \cdot \nabla) \tag{5.37}$$

The second factor in (5.37) is precisely the Fadeev-Popov Jacobian Δ_{FP} of the gauge fixing method for the specific choice of background gauge, $(F^\mu)^{ij} = (\bar{\nabla}^\mu)^{ij}$, while the first factor is a constant, independent of the field point A_μ, and therefore may be taken out of the path integral, without affecting the result. It is important to retain this factor if one calculates the effective action as a function of the background field, however.

Having determined the correct Jacobian of the transformation to field coordinates, (b_μ, θ), we may now express Z in the form,

$$
\begin{aligned}
Z &= (Vol(\mathcal{G}))^{-1} \int [\mathcal{D}\theta^i] \int J\ [\mathcal{D}b^i_\mu] \exp(iS_{inv}[\bar{B}+b]) \\
&= \det{}_S^{-\frac{1}{2}}(-\bar{\nabla}^2) \int [\mathcal{D}b^i_\mu] \det{}_S(-\bar{\nabla}\cdot\nabla)\Big|_{A=\bar{B}+b} \exp(iS_{inv}[\bar{B}+b]),
\end{aligned}
\tag{5.38}
$$

where the integral is over the gauge invariant field coordinate, b^i_μ, and the gauge volume factor has been cancelled explicitly. It is clear that this form generates the same Feynman rules as the gauge fixed path integral, and that a similar equivalence holds for other choices for F. In the present approach, however, the geometric significance of gauge fixing as simply a choice of coordinates in a coordinate invariant expression is manifest and proofs of gauge invariance are unnecessary. Moreover, the non-perturbative aspects of the path integral and the correct normalization factor $\det{}_S^{-\frac{1}{2}}(-\bar{\nabla}^2)$ are also apparent.

5.3. Gravity

The action of classical general relativity is invariant under general coordinate transformations, $x^\mu \to X^\mu = X^\mu(x)$, the infinitesimal form of which is:

$$x^\mu \to x^\mu + \xi^\mu(x). \tag{5.39}$$

We shall construct the functional measure in the path integral for quantum gravity by requiring that it also be invariant under (5.39), following the same strategy as in the previous examples.[22] Treat an arbitrary spacetime metric $g_{\mu\nu}(x)$ as the coordinate of a point in the function space of all metrics, denoted by $\mathcal{M}$. The infinitesimal one-form, $dg_{\mu\nu}(x) \equiv h_{\mu\nu}(x)$ lies in the cotangent space to $\mathcal{M}$ at the point $g_{\mu\nu}(x)$. We define a quadratic inner product on this cotangent space of one-forms by

$$\langle h, h\rangle_T \equiv \int d^4x \sqrt{-g}\ h_{\mu\nu}(x) G^{\mu\nu\rho\sigma} h_{\rho\sigma}(x). \tag{5.40}$$

The subscript T reminds us that this is an inner product for tensors. Now the scalar $ds^2 = g_{\mu\nu}(x)dx^\mu dx^\nu$ is invariant under the (passive) relabelling of the coordinates of spacetime, (5.39) that leaves the geometric spacetime point unchanged. The corresponding transformation of the metric on the spacetime manifold is

$$h_{\mu\nu}(x) \to h_{\mu\nu}(x) + \nabla_\mu \xi_\nu + \nabla_\nu \xi_\mu. \tag{5.41}$$

This may now be regarded as a relabelling of coordinates on $\mathcal{M}$ which leaves the point, *i.e.* the geometry corresponding to $g_{\mu\nu}(x)$ unchanged. Hence we must require that the measure be invariant under the transformations (5.39) and (5.41) as well. Since $h_{\mu\nu}(x)$ transforms covariantly as a symmetric tensor under (5.39), $G^{\mu\nu\rho\sigma}$ must transform like a contravariant four-tensor. Like $g_{\mu\nu}(x)$, $G^{\mu\nu\rho\sigma}(g)$ has evident symmetry properties: it is symmetric under interchange of its first two indices or its last two indices, as well as interchange of the first two with the last two. Finally, again like $g_{\mu\nu}(x)$, $G^{\mu\nu\rho\sigma}(g)$ must be a purely local function of the coordinates of $\mathcal{M}$. That is, it should contain no derivatives of $g_{\mu\nu}(x)$. The unique ultralocal tensors with these properties are[23]

$$\tfrac{1}{2}(g^{\mu\rho}g^{\nu\sigma}+g^{\mu\sigma}g^{\nu\rho}) \qquad \text{and} \qquad g^{\mu\nu}g^{\rho\sigma}\ . \tag{5.42}$$

If we did not demand ultralocality, an infinite number of tensors involving higher derivatives would appear in this list. Using such tensors in the definition of the inner product and functional measure ultimately would have the effect of defining a different set of dynamical coordinates for the theory. Since we assume that the metric is the fundamental field coordinate, and derivatives of it in the action introduce dynamics, we do not wish to introduce derivatives and spurious dynamics into the essentially *kinematic* definition of the inner product or functional measure. In fact, this is the only principle which justifies an otherwise quite arbitrary distinguishing of the functional measure in the path integral from the action functional.

Restricting the metric on $\mathcal{M}$ to be covariant and ultralocal determines it (up to an overall irrelevant normalization) to be:

$$G^{\mu\nu\rho\sigma}=\tfrac{1}{2}(g^{\mu\rho}g^{\nu\sigma}+g^{\mu\sigma}g^{\nu\rho}+Cg^{\mu\nu}g^{\rho\sigma}), \tag{5.43}$$

where C is an undetermined constant.

Having endowed the function space of metrics $\mathcal{M}$ itself with a metric, we are now in a position to define an invariant volume form on the space. In order to avoid confusion it is useful to introduce the vielbein field, e^m_μ which converts spacetime vector indices to spacetime tangent space indices. If we then define the density,

$$\tilde{h}_{mn}\equiv\sqrt{e}\,e^\mu_m\,e^\nu_n\,h_{\mu\nu}, \tag{5.44}$$

together with the relations,

$$g_{\mu\nu}=e^m_\mu\,e^n_\nu\,\eta_{mn}, \tag{5.45}$$

and

$$e\equiv\det(e^m_\mu)=\sqrt{-g}, \tag{5.46}$$

then the inner product (5.40) may be expressed in terms of the *flat* spacetime metric, $\eta_{mn}=\mathrm{diag}(-1,1,1,1)$ as

$$\langle h,h\rangle_T=\int d^4x\;\tilde{h}_{mn}\tilde{G}^{mnrs}\tilde{h}_{rs}, \tag{5.47}$$

where

$$\tilde{G}^{mnrs} = \frac{1}{2}(\eta^{mr}\eta^{ns} + \eta^{ms}\eta^{nr} + C\eta^{mn}\eta^{rs}) \tag{5.48}$$

is independent of x.

Then by analogy with the invariant volume form on a pseudo-Riemannian manifold, $\prod_m d\tilde{x}^m = \det(e^m_\mu)\prod_\mu dx^\mu$, we would like to define the invariant volume form on the function space of metrics $\mathcal{M}$ by:

$$[\mathcal{D}h_{\mu\nu}] \equiv const. \times \prod_{x, m\leq n} d\tilde{h}_{mn}(x) = const. \times \prod_x e^{\frac{(d-4)(d+1)}{4}} \prod_{\mu\leq\nu} dh_{\mu\nu}(x) \tag{5.49}$$

in d spacetime dimensions. Although useful for some formal manipulations this "definition" leads to ambiguities for continuum functional integrals, because of the ill-defined nature of the product at each spacetime point, which also leaves the normalization of the measure undefined. Instead, as we have seen in the previous examples, it is preferable to define the measure by the Gaussian normalization condition,

$$\int [\mathcal{D}h_{\mu\nu}] \exp(-\tfrac{i}{2}\langle h, h\rangle_T) = 1 \ , \tag{5.50}$$

which is formally satisfied by (5.49). This Gaussian integral is well-defined in the Feynman-Kac sense.

The overall constant in front of the supermetric, (5.43) is irrelevant, since it may always be reabsorbed into the normalization integral, (5.50). The constant C is *not* irrelevant, since it determines the signature of the metric on $\mathcal{M}$. This may be seen by decomposing the arbitrary tangent space tensor into its tracefree and trace parts,

$$h_{\mu\nu} = h^{TF}_{\mu\nu} + \frac{h\, g_{\mu\nu}}{4} \ , \tag{5.51}$$

and operating on this decomposition with the metric G. The $\frac{(d+2)(d-1)}{2}$ tracefree parts of $h_{\mu\nu}$ in d dimensions are mapped onto $h^{TF\mu\nu}$, independent of C. However, on the scalar trace mode G has eigenvalue, $1 + \frac{Cd}{2}$. Hence the signature of G depends on the value of C: for $C > -\frac{d}{2}$, the signature of G in the scalar trace sector is positive, while for $C < -\frac{d}{2}$ the signature is negative. If $C = -\frac{d}{2}$, the metric is non-invertible and becomes a projector onto the tracefree subspace.

For any value of C it is clear that the functional measure defined with reference to the inner product, $\langle h, h\rangle_T$ is invariant under the infinitesimal general coordinate transformation, (5.39). Once we have a coordinate invariant functional measure, we must extract the infinite gauge orbit volume in an invariant way as well. To this end, we introduce a change of coordinates in the tangent space of $\mathcal{M}$ at $g_{\mu\nu}$:

$$h_{\mu\nu} = h^{\perp}_{\mu\nu} + (L\xi)_{\mu\nu} + (2\sigma + \tfrac{1}{2}\nabla_\lambda\xi^\lambda)g_{\mu\nu}, \tag{5.52}$$

where L, (the "conformal Killing form") maps vectors into traceless symmetric tensors,

$$(L\xi)_{\mu\nu} \equiv \nabla_\mu\xi_\nu + \nabla_\nu\xi_\mu - \tfrac{1}{2}(\nabla_\lambda\xi^\lambda)g_{\mu\nu}. \tag{5.53}$$

Thus, $L\xi$ spans all symmetric tensors which are gauge transforms of $h^{TF}_{\mu\nu}$, the traceless part of $h_{\mu\nu}$. The scalar σ is the gauge invariant piece of the trace, and $h^{\perp}_{\mu\nu}$ is the gauge invariant piece of $h^{TF}_{\mu\nu}$. Now $h^{\perp}_{\mu\nu}$ may be chosen to lie in the orthogonal complement to L, with respect to the inner product (5.40), which requires $(L^{\dagger}h^{\perp})_{\mu} = -2\nabla^{\nu}h^{\perp}_{\mu\nu} = 0$. Indeed this is the simplest choice for doing one-loop calculations,[22] which involve only Gaussian integrals, and justifies the notation $h^{\perp}$. However, the choice of orthogonal coordinates on the tangent space of $\mathcal{M}$ is by no means necessary, and $h^{\perp}_{\mu\nu}$ may be required to satisfy an arbitrary coordinate (gauge) condition:

$$(F \cdot h^{\perp})_{\mu} \equiv F^{\nu}h^{\perp}_{\mu\nu} = 0. \tag{5.54}$$

The only condition on F is that the operator $F \circ L$ be locally invertible, so that (5.52) can be solved uniquely for ξ:

$$\xi_{\mu} = (F \circ L)^{-1}_{\mu}{}^{\nu}(F \circ h^{TF})_{\nu}. \tag{5.55}$$

Otherwise, the local coordinate chart, (5.52) is singular at the point $g_{\mu\nu}$.

Following the discussion in Section 5.2 for the case of the non-abelian gauge field, to extract the infinite gauge orbit volume generated by the gauge direction, ξ_{μ} we must find the Jacobian of the transformation to the new field coordinates, $(h^{\perp}_{\mu\nu}, \xi_{\mu}, \sigma)$:

$$[\mathcal{D}h_{\mu\nu}] = J\,[\mathcal{D}h^{\perp}_{\mu\nu}][\mathcal{D}\xi_{\mu}][\mathcal{D}\sigma]. \tag{5.56}$$

This is accomplished by substituting the decomposition, (5.52) into the inner product (5.40), completing the square of the term quadratic in ξ_{μ}, and computing the Gaussian integrals over each of the components respectively:

$$\begin{aligned} 1 &= \int [\mathcal{D}h_{\mu\nu}] \exp(-\frac{i}{2}\langle h, h\rangle_{T}) \\ &= J \int [\mathcal{D}h^{\perp}_{\mu\nu}] \exp(-\frac{i}{2}\langle h^{\perp}, (1-M)h^{\perp}\rangle_{T}) \int [\mathcal{D}\xi_{\mu}] \exp(-\frac{i}{2}\langle \xi', \Delta_{1}\xi'\rangle_{V}) \\ &\qquad \times \int [\mathcal{D}\sigma] \exp(-8i(1+2C)\langle\sigma,\sigma\rangle_{S}), \end{aligned} \tag{5.57}$$

where the vector Laplacian Δ_1 is defined by:

$$(\Delta_1)_{\mu}{}^{\nu} \equiv (L^{\dagger}L)_{\mu}{}^{\nu} = -2\big(\delta_{\mu}{}^{\nu}\nabla^2 + \frac{1}{2}\nabla_{\mu}\nabla^{\nu} + R_{\mu}{}^{\nu}\big), \tag{5.58}$$

the tensor operator M is given by:

$$M_{\mu\nu}{}^{\rho\sigma} \equiv (L(\Delta_1)^{-1}L^{\dagger})_{\mu\nu}{}^{\rho\sigma} = -2\nabla_{\mu}(\Delta_1^{-1})_{\nu}{}^{\rho}\nabla^{\sigma} - 2\nabla_{\nu}(\Delta_1^{-1})_{\mu}{}^{\rho}\nabla^{\sigma} + g_{\mu\nu}\nabla^{\lambda}(\Delta_1^{-1})_{\lambda}{}^{\rho}\nabla^{\sigma}, \tag{5.59}$$

and $\xi'_{\mu} = \xi_{\mu} + (\Delta_1^{-1}L^{\dagger}h^{\perp})_{\mu}$ is the shifted vector obtained by completing the square. The notations $\langle,\rangle_V$ and $\langle,\rangle_S$ denote the covariant inner products on vectors and scalars respectively:

$$\begin{aligned} \langle\xi,\xi\rangle_{V} &= \int d^4x\,\sqrt{-g}\,\xi_{\mu}\,g^{\mu\nu}\,\xi_{\nu}, \\ \langle\sigma,\sigma\rangle_{S} &= \int d^4x\,\sqrt{-g}\,\sigma^2 \end{aligned} \tag{5.60}$$

The remaining tensor and vector Gaussian functional integrals in (5.57) can be simplified[22] to yield finally the result for the Jacobian:

$$J = [\det{}_{\perp T}(1-M)]^{\frac{1}{2}}[\det{}_V\Delta_1]^{\frac{1}{2}} = [\det{}_V(F\circ F^{\dagger})]^{-\frac{1}{2}}\det{}_V(F\circ L), \tag{5.61}$$

in complete analogy with the gauge theory result (5.37). The second factor will be recognized as the Fadeev-Popov determinant for the gauge (5.54) on the tracefree components of $h_{\mu\nu}$, while the first factor is a $h_{\mu\nu}$-independent normalization factor that makes no contribution to the Feynman rules. Notice that the Jacobian factor has been derived by tangent space methods, involving only Gaussian functional integrals, but that this involves no restriction to one-loop order. The result (5.61) is valid to *all* orders of perturbation theory.

With the Jacobian (5.61), we now know how to factor the infinite diffeomorphism gauge group volume out of the covariant quantum measure, (5.56) in a manifestly covariant way. If the action is independent of the vector gauge orbit parameter ξ_μ, integration over ξ_μ would simply yield the infinite volume of the diffeomorphism group, so

$$\begin{aligned}\{Vol(\mathcal{G})\}^{-1}\int[\mathcal{D}h_{\mu\nu}] &= \{Vol(\mathcal{G})\}^{-1}\int[\mathcal{D}\xi_\mu]\int J\,[\mathcal{D}h^{\perp}_{\mu\nu}][\mathcal{D}\sigma] \\ &= \int J\,[\mathcal{D}h^{\perp}_{\mu\nu}][\mathcal{D}\sigma],\end{aligned} \tag{5.62}$$

when integrated over functions independent of ξ_μ.

The final steps in constructing the covariant path integral for quantum gravity involves extending the integration measure defined on the tangent space to a measure on the full metric. The extension of the coordinates on the tangent space (5.52) to coordinates on $\mathcal{M}$ is straightforward. We write:

$$\begin{aligned} g_{\mu\nu}(x) &= \frac{\partial X^\rho}{\partial x^\mu}\frac{\partial X^\sigma}{\partial x^\nu}e^{2\sigma(X)}g^{\perp}_{\rho\sigma}(X), \\ (F\cdot g^{\perp})_\mu &= 0. \end{aligned} \tag{5.63}$$

where σ may be fixed by the requirement that $g^{\perp}$ has constant scalar curvature ($R^{\perp} = const.$):

$$R[g_{\mu\nu}] = 6e^{-3\sigma}\Delta_0^{\perp}e^{\sigma} + e^{2\sigma}R^{\perp} \tag{5.64}$$

and $\Delta_0^{\perp}$ is the scalar Laplace-Beltrami operator with respect to the metric $g^{\perp}$. The vacuum amplitude for quantum gravity may then be written in the succint generally covariant form:

$$\begin{aligned} Z &= \{Vol(\mathcal{G})\}^{-1}\int[\mathcal{D}g_{\mu\nu}]\exp(iS_{inv}[g]) \\ &= \int J\,[\mathcal{D}g^{\perp}_{\mu\nu}][\mathcal{D}\sigma]\exp(iS_{inv}[e^{2\sigma}g^{\perp}]), \\ &= [\det{}_V(F\circ F^{\dagger})]^{-\frac{1}{2}}\int[\mathcal{D}\sigma][\mathcal{D}g^{\perp}_{\mu\nu}]\det{}_V(F\circ L)\Big|_{g=e^{2\sigma}g^{\perp}}\exp(iS_{inv}[e^{2\sigma}g^{\perp}]), \end{aligned} \tag{5.65}$$

where (5.61) has been used.

If we set

$$g^{\perp}_{\mu\nu} = \bar{g}_{\mu\nu} + h^{\perp}_{\mu\nu}, \tag{5.66}$$

the analogy with the gauge theory result (5.38) will be evident. In both cases the $a^{\perp}$ or $h^{\perp}$ independent determinant makes no contribution to the Feynman rules. It arises only because

$$[\mathcal{D}h_{\mu\nu}]\delta(F\cdot h) = [\det_V(F\circ F^{\dagger})]^{-\frac{1}{2}}\,[\mathcal{D}h^{\perp}_{\mu\nu}], \tag{5.67}$$

when both the $[\mathcal{D}h_{\mu\nu}]$ and $[\mathcal{D}h^{\perp}_{\mu\nu}]$ measures are normalized by Gaussian conditions (5.50) in their respective spaces.

6. The Stability of Flat Minkowski Spacetime

In this section I apply the formal analysis of the covariant path integral for quantum gravity to an old question: the stability of flat spacetime to quantum matter fluctuations. For definiteness we shall examine a single massive scalar field coupled to gravity via the (real time) action,

$$S = \frac{1}{16\pi G}\int d^4x\sqrt{-g}R - \frac{1}{2}\int d^4x\sqrt{-g}\left[g^{ab}\nabla_a\Phi\nabla_b\Phi + m^2\Phi^2 + \frac{1}{6}R\Phi^2\right]. \tag{6.1}$$

The choice of $\frac{1}{6}$ for the coefficient of the $R\Phi^2$ term is merely to simplify the algebra in an explicit example. The general case may be treated by the same methods with a bit more effort.

Since the tracefree parts of the metric deformations lead to a positive definite Euclidean action and no instability we restrict ourselves to conformally flat metrics,

$$\begin{aligned} g_{ab} &= \Omega^2(x)\eta_{ab} \equiv e^{2\sigma}\eta_{ab};\\ \eta &= \text{diag }(-+++) \end{aligned}. \tag{6.2}$$

The tracefree parts could be incorporated without any essential difficulty but as they do not change the result, we omit them for simplicity. If we rescale the matter field by the same conformal factor, $\Phi = \Omega^{-1}\phi$ and perform the Gaussian functional integral over the rescaled matter field, we obtain the effective action for the conformal part of the metric:

$$S_{eff}[\Omega] = \frac{3}{8\pi G}\int d^4x(\nabla_a\Omega)^2 - \frac{1}{2}Tr\log\left[-\nabla^2 + m^2\Omega^2\right] + \int d^4x L_c \tag{6.3}$$

where all indices are now raised and lowered with respect to the *flat* spacetime metric η_{ab}. The counterterm L_c is needed to make this one-loop effective action finite and well defined, and can be found by the Schwinger-deWitt proper time technique. It corresponds to renormalization of the cosmological term, the Newtonian constant, and the order R^2 terms in the effective action.

Let us develop the one-loop effective action (6.3) in an expansion in the small conformal deformation σ where $\Omega = 1+\sigma$. It is easiest to work in momentum space and define the usual scalar propagator,

$$G(x,x';m^2) = \langle x|(-\nabla^2+m^2)^{-1}|x'\rangle = \int \frac{d^4k}{(2\pi)^4} e^{ik(x-x')} \frac{1}{k^2+m^2-i\epsilon}. \tag{6.4}$$

The terms zeroth order and first order in σ are cancelled identically by the countert-erm L_c, since they contain no derivatives of σ, and the effective action vanishes for Ω equal to a constant. The vanishing of the tadpole term is a consistency check on the counterterm subtraction, since if it did not cancel identically, flat spacetime would not be a stationary point of the effective action nor a solution of the renormalized one-loop Einstein equations. The terms quadratic in σ are:

$$m^2 \int \frac{d^4q}{(2\pi)^4} \tilde{\sigma}(q)\tilde{\sigma}(-q)[m^2 G_2(q) - \tfrac{1}{2}G(x,x;m^2)] \tag{6.5}$$

where $\tilde{\sigma}(q)$ is the Fourier transform of $\sigma(x)$ and $G_2(q)$ is the Fourier transform of the square of the propagator (6.4),

$$G_2(q) = \int \frac{d^4k}{(2\pi)^4} \frac{1}{k^2+m^2} \frac{1}{(k+q)^2+m^2}. \tag{6.6}$$

The second term in (6.5) is quadratically divergent, but since it has no q^2 dependence, it is removed by the renormalization of the cosmological term. The same cosmological term renormalization converts the logarithmically divergent $G_2(q)$ as given in (6.6) into the convergent expression, $G_2(q) - G_2(0)$. This difference now starts out proportional to q^2 as $q^2 \to 0$. This means that the one-loop matter contribution to the full effective action makes a finite correction to the Einstein action, that is, a finite renormalization of the Newtonian constant. Since the physically measured value of G is determined by the long wavelength limit of the *full* effective action, we must demand that the further (finite) subtraction, $q^2(\frac{dG_2}{dq^2})|_{q^2=0}$ be made in (6.5).

We now obtain the renormalized matter contribution to the effective action at Gaussian order in σ:

$$m^4 \int \frac{d^4q}{(2\pi)^4} \tilde{\sigma}(q)\tilde{\sigma}(-q) \left[G_2(q) - G_2(0) - q^2\left(\frac{dG_2}{dq^2}\right)\Big|_{q^2=0} \right]. \tag{6.7}$$

The twice subtracted function of q^2 in brackets now starts out proportional to q^4 as $q^2 \to 0$, which is to say that both the renormalized cosmological term *and* the Newtonian constant are unchanged from their measured values in flat spacetime, zero and G respectively.

I have deliberately not specified any $i\epsilon$ prescription for the Fourier transforms since that depends on the application. The causal properties are made explicit by introducing the spectral density Källen-Lehmann representation for the retarded function $G_2{}^{(ret)}(q)$,

$$G_2{}^{(ret)}(q) = \int_0^\infty ds\, \rho(s)\, \frac{1}{q^2 + s - i\epsilon q_0}. \tag{6.8}$$

By evaluating the imaginary part of the retarded function $G_2{}^{(ret)}(q)$ from (6.6) in a frame where $\vec{q} = 0$, we arrive at

$$\rho(s) = \int \frac{d^3\vec{k}}{(2\pi)^3 \omega_{\vec{k}}} \delta(s - 4\omega_{\vec{k}}^2) = \frac{1}{16\pi^2}\sqrt{1 - \frac{4m^2}{s}}\,\theta(s - 4m^2)\ . \tag{6.9}$$

Returning now to the effective action (6.7) we observe that

$$G_2^{(F)}(q) - G_2^{(F)}(0) - q^2\Big(\frac{dG_2^{(F)}}{dq^2}\Big)\Big|_{q^2=0} = q^4 \int_{4m^2}^\infty \frac{\rho(s)ds}{s^2(q^2 + s - i\epsilon)} \tag{6.10}$$

so that the entire one-loop effective action in the conformal sector is

$$S_{eff} = \frac{3}{8\pi G}\int \frac{d^4q}{(2\pi)^4} q^2 \tilde{\sigma}(q)\tilde{\sigma}(-q)\left[1 + \frac{8\pi G m^4 q^2}{3}\mathcal{P}\int_{4m^2}^\infty \frac{\rho(s)ds}{s^2(q^2+s)}\right] \tag{6.11}$$

where $\mathcal{P}$ denotes the principal part of the integral. It is worth remarking that this expression for the one-loop effective action represents a non-trivial, non-local polarization of the gravitational vacuum by quantum matter, analogous to the vacuum polarization effects induced by virtual electron-positron pairs in quantum electrodynamics. It is non-local because (6.11) contains arbitrarily high powers of q^2 if expanded as a Taylor series in q^2. This implies that the renormalized one-loop energy momentum tensor of the matter field would contain terms with arbitrarily many derivatives of the metric tensor perturbation, if expressed back in position space. Only in the case $m^2 = 0$ does (6.11) collapse to a single local term. In this case one can easily verify that the effective action (6.11) is just that for a conformally coupled massless field, with the additional local term completely specified by the conformal anomaly, in agreement with the literature.[24]

The second observation is that for $q^2 > 0$, the last factor in (6.11) is non-vanishing. Hence there are no growing or unstable solutions to the one-loop equations of motion following from the σ variation of the effective action (6.11). If $q^2 < 0$, then the factor in brackets may vanish, and additional oscillatory solutions may exist. For the specific case of the conformally coupled massive scalar with spectral function given by (6.9), it is straightforward to evaluate the integral in Eq. (6.11) explicitly. The result is:

$$q^2\mathcal{P}\int_{4m^2}^\infty \frac{\rho(s)ds}{s^2(q^2+s)} = \frac{1}{16\pi^2 q^2}\left(\frac{q^2}{6m^2} + 2 - f\Big(\frac{q^2}{m^2}\Big)\right), \tag{6.12}$$

where

$$f\left(\frac{q^2}{m^2}\right) = \begin{cases} z\ln\frac{z+1}{z-1}, & 0 \le q^2; \\ 2|z|\arctan\frac{1}{|z|}, & -4m^2 \le q^2 \le 0; \\ z\ln\frac{1+z}{1-z}, & q^2 \le -4m^2. \end{cases} \tag{6.13}$$

and

$$z \equiv \sqrt{1+\frac{4m^2}{q^2}}\,. \tag{6.14}$$

From these explicit expressions for the vacuum polarization function it is not difficult to see that (6.12) is strictly positive for *all* real $q^2 \neq 0$. Therefore, the quantity in brackets in (6.11) never vanishes, and there are no additional unphysical solutions to the one-loop linearized Einstein equations following from the effective action (6.11).

In the region of validity of the one-loop semiclassical approximation used flat spacetime is absolutely stable to conformal perturbations of the metric induced by the matter. It is easy to see that the results do not depend on the spin of the matter field or precise values of masses or coupling constants. Indeed, for spin-$\frac{1}{2}$ Dirac fermions one obtains the same form as (6.11) with the spectral density,

$$\rho_{\frac{1}{2}}(s) = \frac{1}{16\pi^2}\sqrt{1-\frac{4m^2}{s}}\left(4+\frac{s}{m^2}\right)\theta(s-4m^2) \tag{6.15}$$

and similar conclusions follow. In the semiclassical limit the positivity of the spectral function $\rho(s)$ is sufficient to guarantee that the flat spacetime is stable.

If you are familiar with the earlier literature on the semiclassical stability of flat spacetime,[25] you will recall that there are modes of the semiclassical theory corresponding to oscillations or exponential growth for q^2 of order of M^2_{Planck}. These are obtained in the present treatment if we recall that actually it is necessary to make one more subtraction of (6.7), corresponding to the renormalization of the order R^2 terms in the effective action. These R^2 terms must then be added to the action with arbitrary renormalized coefficients. Then the conformal sector of the action will indeed have an inverse propagator pole at q^2 of order M^2_{Planck}, corresponding to the new scalar degree of freedom that has been introduced into the theory by the R^2 terms in the Lagrangian. The analysis at low q^2 is unaffected by these terms since in the infrared limit the additional terms are entirely negligible and the theory reduces to the Einstein theory. The additional ultraviolet poles are the analogs of the Landau ghost pole in one-loop quantum electrodynamics, and are to be similarly interpreted as lying outside the range of validity of semiclassical methods. In the case of gravity it is not at all surprising to find peculiar behavior at very short wavelengths or very high frequencies, since the theory is non-renormalizable, and a more fundamental description is required. However, the stability of the vacuum ground state is an issue of the *infrared* behavior of the theory, where semiclassical methods applied to the Einstein action should be wholly adequate, and here there is no instability whatsoever. This is important since flat space is then a good candidate for the ground state of quantum gravity, and in flat space the vacuum cosmological term vanishes. In de Sitter spacetime (with $\Lambda > 0$) the situation is quite different in the infrared,[14,15] and no stable vacuum state exists.

7. The Effective Theory of the Conformal Factor

Having seen how the effective action of the conformal factor σ can be determined to Gaussian order by integration over matter fields in the previous section, we are now ready in this last lecture to consider the effective action of the conformal factor beyond Gaussian order in greater detail. In order to determine the Lagrangian which describes the dynamics of the conformal factor in four dimensions, let us first review the situation in two dimensions.[20] In that case, the general form of the trace anomaly of the energy momentum tensor for matter in a background gravitational field is

$$\begin{aligned} T_a^{a\ (matter)} &= \frac{c_m}{24\pi} R \\ &= \frac{c_m}{24\pi} e^{-2\sigma}(\overline{R} - 2\Box\sigma), \qquad d = 2 \end{aligned} \tag{7.1}$$

in the decomposition $g_{ab} = e^{2\sigma}\bar{g}_{ab}$. The coefficient $c_m = (N_S + N_F)$ for N_S scalar and N_F (Dirac) fermion fields. Considering next the quantization of gravity, the anomaly coefficient receives a contribution of -26 from the corresponding reparameterization ghosts, and $+1$ from the σ field itself. Thus, we should replace c_m by

$$c = N_S + N_F - 25. \tag{7.2}$$

The trace anomaly may be derived from a nonlocal effective action which becomes local in the conformal coordinates $g_{ab} = e^{2\sigma}\bar{g}_{ab}$

$$\begin{aligned} T_a^a &= \frac{1}{\sqrt{-g}}\frac{\delta}{\delta\sigma(x)} S_{anom}, \\ S_{anom} &= -\frac{c}{96\pi}\int d^2x\sqrt{-g}\int d^2x'\sqrt{-g'}\ R(x)\Box^{-1}(x,x')\ R(x'), \\ &= \frac{c}{24\pi}\int d^2x\ \sqrt{-\bar{g}}[-\sigma\Box\sigma + \overline{R}\sigma]. \end{aligned} \tag{7.3}$$

To this induced action we should add the classical action for 2d gravity,

$$\begin{aligned} S_{cl} &= \int d^2x\sqrt{-g}\ (\gamma R - 2\lambda) \\ &= 4\pi\gamma\chi - 2\lambda\int d^2x\ \sqrt{-\bar{g}}e^{2\sigma}, \end{aligned} \tag{7.4}$$

where χ is the Euler number. Notice that the Einstein term alone describes no dynamics, since the scalar curvature is a total derivative. Hence, the anomaly-induced dynamics of the effective 2d gravity theory is very different from the classical theory, even in the infrared, where one's usual prejudice is that quantum gravity or trace anomaly effects are "higher order" and irrelevant.

Let us now consider this same line of reasoning for $d = 4$. In four dimensions the general form of the trace anomaly of the energy-momentum tensor is a linear combination of the four terms[24], R^2, $R_{ab}R^{ab}$, $R_{abcd}R^{abcd}$, and $\Box R$. Each of these corresponds to local counterterms in the dimensionally regulated effective action. Actually, since $\Box R$ is a total divergence in any number of dimensions, adding it to the effective action gives no contribution to the trace of the energy momentum tensor. Hence there can be at most three independent local counterterms possible in dimensional regularization near $d = 4$, and the trace anomaly for matter may be written in the form,[26]

$$T_a^{a\ (matter)} = b(F + \frac{2}{3}\Box R) + b'G + b''\ \Box R, \qquad d = 4 \tag{7.5}$$

where

$$F = R_{abcd}R^{abcd} - 2R_{ab}R^{ab} + \frac{1}{3}R^2 \tag{7.6}$$

is the square of the Weyl tensor in four dimensions, and

$$G = R_{abcd}R^{abcd} - 4R_{ab}R^{ab} + R^2 \tag{7.7}$$

is the Gauss-Bonnet integrand. In four (but only four) dimensions G is a total divergence, and its integral is a topological invariant. The coefficients b and b' have been computed for scalar, Dirac fermion, and vector fields at one-loop order, with the results:

$$\begin{aligned} b &= \frac{1}{120(4\pi)^2}(N_S + 6N_F + 12N_V), \\ b' &= -\frac{1}{360(4\pi)^2}(N_S + 11N_F + 62N_V). \end{aligned} \tag{7.8}$$

If the matter field Lagrangian is invariant under local conformal (Weyl) transformations classically, then the coefficient b'' vanishes at one-loop order.[26] However, in principle it corresponds to an arbitrary parameter, since $\Box R$ is the variation of a local action:

$$\sqrt{-g}\ \Box R = -\frac{1}{6}\ g_{ab}\frac{\delta}{\delta g_{ab}}\int d^4x\ \sqrt{-g}\ R^2, \tag{7.9}$$

which receives divergent contributions in general. Thus we allow for $b'' \neq 0$. In contrast to this term, the first two terms on the right side of (7.5) do not arise from local effective actions in four dimensions. However, in the conformal parameterization of the metric, the anomaly may be derived from a local effective action, just as in the two dimensional case.[27,28] In fact, the quantity $\sqrt{-g}F$ becomes independent of σ (actually vanishing if $\bar{g}_{ab}$ is conformally flat), and the combination $\sqrt{-g}(G - \frac{2}{3}\Box R)$ is particularly simple, being only linear in σ in analogy to Eq. (7.1),

$$e^{4\sigma}(G - \frac{2}{3}\Box R) = 4\ \overline{\Box}^2\sigma + 8\overline{R}^{ab}\overline{\nabla}_a\overline{\nabla}_b\sigma - \frac{8}{3}\overline{R}\overline{\Box}\sigma + \frac{4}{3}(\overline{\nabla}^a\overline{R})(\overline{\nabla}_a\sigma) + \overline{G} - \frac{2}{3}\overline{\Box}\overline{R}. \tag{7.10}$$

Here $\overline{\nabla}$ is the covariant derivative operator with respect to the fiducial metric $\bar{g}_{ab}$, $\overline{\Box} \equiv \overline{\nabla}_a \overline{\nabla}^a$, and all indices are raised and lowered by this fiducial metric tensor. Thus, if we rewrite the trace anomaly (7.5) in the form,

$$\begin{aligned} T_a^{a\ (matter)} &= bF + b'(G - \frac{2}{3}\Box R) + [b'' + \frac{2}{3}(b+b')]\Box R \\ &= \frac{1}{\sqrt{-g}}\frac{\delta}{\delta\sigma(x)} S_{anom}[\sigma], \end{aligned} \tag{7.11}$$

the second term in (7.11) gives a contribution to the effective action which is quadratic in σ, and the full anomaly induced action reads:

$$\begin{aligned} S_{anom}[\sigma] = &\ b \int d^4x\ \sqrt{-\bar{g}}\ \overline{F}\sigma \\ &+ b' \int d^4x\ \sqrt{-\bar{g}}\ \{\sigma[2\overline{\Box}^2 + 4\overline{R}^{ab}\overline{\nabla}_a\overline{\nabla}_b - \frac{4}{3}\overline{R}\overline{\Box}^2 + \frac{2}{3}(\overline{\nabla}^a\overline{R})\overline{\nabla}_a]\sigma + [\overline{G} - \frac{2}{3}\overline{\Box}\overline{R}]\sigma\} \\ &- \frac{1}{12}[b'' + \frac{2}{3}(b+b')] \int d^4x\sqrt{-\bar{g}}[\overline{R} - 6\overline{\Box}\sigma - 6(\overline{\nabla}_a\sigma)(\overline{\nabla}^a\sigma)]^2, \end{aligned} \tag{7.12}$$

where we have used the expression,

$$R = e^{-2\sigma}(\overline{R} - 6\overline{\Box}\sigma - 6(\overline{\nabla}_a\sigma)(\overline{\nabla}^a\sigma)), \tag{7.13}$$

for the Ricci scalar in conformal coordinates. We have omitted a σ independent term in S_{anom} proportional to F. This local action in the conformal parameterization of the metric corresponds to the fully covariant non-local action,

$$\begin{aligned} S_{anom} = &-\frac{1}{4}\int d^4x\sqrt{-g}\ \int d^4x'\sqrt{-g'}\ [bF + b'(G - \frac{2}{3}\Box R)]_x \\ &[2\Box^2 + 4R^{ab}\nabla_a\nabla_b - \frac{4}{3}R\Box^2 + \frac{2}{3}(\nabla^a R)\nabla_a]^{-1}_{xx'}[bF + b'(G - \frac{2}{3}\Box R)]'_x \\ &- \frac{1}{12}[b'' + \frac{2}{3}(b+b')]\int d^4x\sqrt{-g}R^2. \end{aligned} \tag{7.14}$$

The non-local term in (7.14) is the four dimensional analog of the non-local form of the 2d Polyakov action, (7.3). As in two dimensions we add the anomaly induced action (7.12) to the classical Einstein-Hilbert action,

$$\begin{aligned} S_{cl} &= \frac{1}{2\kappa}\int d^4x\sqrt{-g}\ (R - 2\Lambda) \\ &= \frac{1}{2\kappa}\int d^4x\sqrt{-\bar{g}}\ e^{2\sigma}(\overline{R} - 6\overline{\Box}\sigma - 6(\overline{\nabla}_a\sigma)(\overline{\nabla}^a\sigma)) - \frac{\Lambda}{\kappa}\int d^4x\sqrt{-\bar{g}}\ e^{4\sigma}, \end{aligned} \tag{7.15}$$

where $\kappa = 8\pi G$.

In the case that the fiducial metric is conformally flat, *i.e.* $\bar{g}_{ab} = e^{2\bar{\sigma}}\eta_{ab}$, the effective action for σ may be obtained by a translation, $\sigma \to \sigma + \bar{\sigma}$ of the following simple flat space action,

$$S_{eff} = S_{anom} + S_{cl}$$
$$= \int d^4x \left\{2b'(\Box\sigma)^2 - [3b'' + 2(b+b')](\Box\sigma + (\partial_a\sigma)^2)^2 + \frac{3}{\kappa}e^{2\sigma}(\partial_a\sigma)^2 - \frac{\Lambda}{\kappa}e^{4\sigma}\right\}, \tag{7.16}$$

up to a surface term. When $\bar{g}_{ab}$ is conformally flat, the effective action (7.16) may be derived by a completely different method, based on global conformal symmetry, which I shall explain now.

Conformal flatness implies that the fiducial metric has a set of exactly $d+1$ special conformal Killing vectors, $\xi_a^{(i)}$, satisfying:

$$(L\xi^{(i)})_{ab} \equiv \overline{\nabla}_a\xi_b^{(i)} + \overline{\nabla}_b\xi_a^{(i)} - \frac{2}{d}\bar{g}_{ab}\overline{\nabla}^c\xi_c^{(i)} = 0, \qquad i = 1, \ldots, d+1. \tag{7.17}$$

On the other hand, the decomposition, $g_{ab} = e^{2\sigma}\bar{g}_{ab}$ is certainly invariant under the relabelling (Weyl) transformation,

$$\begin{aligned} \bar{g}_{ab}(x) &\to \bar{g}_{ab}(x)e^{2\omega(x)}, \\ \sigma(x) &\to \sigma(x) - \omega(x), \end{aligned} \tag{7.18}$$

for an arbitrary $\omega(x)$. Lastly, any theory of gravity must be invariant under coordinate transformations. The infinitesimal form of a combined coordinate plus relabelling transformation is

$$\begin{aligned} \delta\bar{g}_{ab} &= \overline{\nabla}_a\xi_b + \overline{\nabla}_b\xi_a + 2\omega\bar{g}_{ab}, \\ \delta\sigma &= \xi^a\overline{\nabla}_a\sigma - \omega. \end{aligned}$$

Hence, by choosing

$$\begin{aligned} \xi_a &= \xi_a^{(i)}, \\ \omega &= -\frac{1}{d}\overline{\nabla}^c\xi_c^{(i)}, \end{aligned} \tag{7.19}$$

we find that:

$$\begin{aligned} \delta^{(i)}\bar{g}_{ab} &= (L\xi^{(i)})_{ab} = 0, \\ \delta^{(i)}\sigma &= \xi^{(i)a}\overline{\nabla}_a\sigma + \frac{1}{d}\overline{\nabla}^a\xi_a^{(i)}, \end{aligned} \tag{7.20}$$

must be a symmetry transformation of the σ Lagrangian which leaves the fiducial metric unchanged. In flat space the special conformal Killing vectors are:

$$\begin{aligned} \xi^{(b)a} &= 2x^ax^b - \eta^{ab}x^cx_c, \qquad & i = b = 1, \ldots d \\ \xi^{(i)a} &= x^a, \qquad & i = d+1 \end{aligned} \tag{7.21}$$

The last of these plays a special role since it generates dilations, $x^a \to e^s x^a$. It serves to define the scaling dimension Δ_ϕ of the general field ϕ by:

$$\delta_s\phi \equiv (x\cdot\partial + \Delta_\phi)\phi. \tag{7.22}$$

This definition implies that e^σ has classical scaling dimension equal to unity, (as does $\partial\sigma$) under the global dilation, $\xi^{(d+1)}$. The σ field itself does not have a well-defined scaling dimension, since it transforms inhomogeneously under δ_s:

$$\delta_s \sigma = x \cdot \partial\sigma + 1. \tag{7.23}$$

Since the global conformal symmetry for σ is a remnant of the coordinate invariance of the full theory with total metric g_{ab}, the effective action of the conformal factor must respect this symmetry. Invariance under dilations alone (7.23) determine the 4d action (7.16) up to four derivatives, leaving only the freedom of introducing two different coefficients for the fourth order terms $\Box\sigma(\partial\sigma)^2$ and $(\partial\sigma)^4$. Their ratio is fixed to be 2 to 1 from the invariance under the remaining global conformal transformations (7.21).

It is remarkable that global conformal invariance determines uniquely the form of the effective action for σ. Normally we would expect that this classical scale invariance will be violated in the quantum theory. This is precisely what the trace anomaly of matter fields in a fixed gravitational background geometry (7.1) reflects. Now that we are considering the quantum theory of the geometry itself, and the global conformal transformations are simply particular coordinate transformations, an anomaly in this symmetry would imply a breakdown of general covariance at the quantum level. The invariance of the σ theory under the rigid scale transformations generated by $\xi^{(5)}$ implies that the total trace anomaly of the conformal factor plus matter plus ghosts must vanish. In particular this implies that the beta functions of rigid dilations must vanish identically, corresponding to a critical point of the effective action.

This conclusion will seem less startling if reconsider the 2d case in the covariant conformal parameterization. In the language of string theory, if the target space dimension, $N_S + N_F$ in Eq. (7.2) is different from 26, then the Liouville mode does not decouple, and must be quantized. The fact that all two dimensional metrics are conformally flat guarantees the existence of the residual symmetries (7.20). The conformal group in two dimensions is infinite dimensional, and gives rise to the Virasoro algebra. However, the *global* scale invariance generated by the finite dimensional subgroup, $SO(2,2)$ or $SO(3,1) \simeq SL(2,C)$ is sufficient to enforce the vanishing of the total trace anomaly.

The above construction of the effective theory for σ in four dimensions has led us inevitably to a fourth order action. At linearized order such an action always gives rise to two excitations, one of which has negative metric. It is the role of the ghost determinant (5.61), which is the covariant equivalent of the reparameterization constraints, to enforce unitarity at the quantum level. Recall that the σ field has a negative metric kinetic term already at the level of the classical Einstein action (7.15). In two dimensions as well, the induced Polyakov action contains two derivatives more than the classical action, and σ has negative metric when c_m exceeds 26. Even in that case we know that the theory is unitary.[29] In both two and four dimensions the higher derivative terms in the σ action emerge from the nonlocal terms induced by the trace anomaly (with calculable coefficients). This should be contrasted with local fourth order gravity actions which contain a massive spin-2 ghost, associated with the local F term in the action with arbitrary coefficient. Of course, the four dimensional case is quite different in containing spin-2 excitations which couple to the σ field. For this reason it may not be meaningful to require unitarity of the σ theory alone. Rather, it should probably be viewed as an effective theory from which only limited information can be extracted.

8. Anomalous Scaling and the Infrared Fixed Point

Let us now study the consequences of the quantization of the conformal factor with the four dimensional action (7.16). The trace of the energy momentum tensor of this effective scalar theory must vanish, because of the coordinate invariance of the full theory, as discussed previously. This implies that the beta functions of all couplings must vanish, but does not preclude the possibility of nontrivial anomalous scaling dimensions. In order to allow for this possibility, let us assume that the field $\phi = e^{\sigma}$, which has classical weight equal to unity acquires a scaling dimension, $\Delta_\phi \equiv \alpha$ under the transformation (7.22). Then the transformation of the σ field is modified to:

$$\delta_s \sigma = x \cdot \partial\sigma + \alpha, \tag{8.1}$$

while the rescaled field, $\hat{\sigma} \equiv \frac{\sigma}{\alpha}$ transforms exactly as in (7.22). In order to determine the anomalous scaling dimension α, we study the counterterm structure of the rescaled action,

$$\begin{aligned} S = &-\frac{Q^2}{(4\pi)^2}\int d^4x (\Box\sigma)^2 - \zeta\int d^4x[2\alpha(\partial_a\sigma)^2\Box\sigma + \alpha^2(\partial_a\sigma)^4] \\ &+\gamma\int d^4x\ e^{2\alpha\sigma}(\partial_a\sigma)^2 - \frac{\lambda}{\alpha^2}\int d^4x\ e^{4\alpha\sigma}, \end{aligned} \tag{8.2}$$

where we have introduced the notations,

$$\begin{aligned} \frac{Q^2}{(4\pi)^2} &= 2b + 3b'', \\ \zeta &= 2b + 2b' + 3b'', \\ \gamma &= \frac{3}{\kappa}, \\ \lambda &= \frac{\Lambda}{\kappa}, \end{aligned} \tag{8.3}$$

replaced S by $\alpha^2 S$, and dropped the hat on $\hat{\sigma}$. Simple power counting arguments, combined with the existence of the global symmetry (7.20), show that the fourth order Lagrangian (8.2) is renormalizable in four dimensions in σ perturbation theory. I shall skip over the technical details which may be found in Ref. 28, and simply summarize the results.

Let us first consider the effect of the higher derivative cubic and quartic interaction terms parameterized by the coupling constant ζ, which does not receive any renormalizations from the lower dimension operators in (8.2). Perturbatively in ζ, there are two divergent diagrams at order ζ^2. Because of the propagator involving p^4 these diagrams contain only logarithmic divergences. Using dimensional regularization, the result for the one-loop ζ counterterm is

$$\zeta_{bare} = \mu^{4-d}\left(\zeta_{ren} + \frac{80\pi^2}{4-d}\,\frac{\alpha^2}{Q^4}\,\zeta_{ren}^2\right), \tag{8.4}$$

where ζ_{bare} and ζ_{ren} are the bare and renormalized couplings respectively. From this we find the one-loop beta function for the coupling ζ:

$$\beta_\zeta = \mu\frac{d}{d\mu}\zeta_{ren} = \frac{80\pi^2\alpha^2}{Q^4}\,\zeta_{ren}^2 \geq 0 \tag{8.5}$$

Hence $\zeta_{ren} = 0$ is an infrared stable perturbative fixed point (as in ordinary $\lambda\phi^4$ theory), if $\zeta \geq 0$, the same sign corresponding to boundedness of the Euclidean action. Since we are interested in this effective theory at its IR fixed points, we shall set $\zeta_{ren} = 0$ henceforth.

At this fixed point we see from (8.3) that:

$$Q^2 = -32\pi^2 b', \qquad \zeta = 0. \tag{8.6}$$

Recall that b' was the coefficient of the Gauss-Bonnet term in the trace anomaly (7.5), which is negative definite for all known matter fields: Eqs. (7.8). This corresponds again to a *bounded* Euclidean action, in contrast to the two dimensional case: Eqs. (7.2) and (7.3), or to the Einstein theory.[8] It is worth remarking also that the requirement of a IR fixed point at $\zeta = 0$ fixes the indeterminateness of the b'' term which is usually present in purely local higher derivative theories of gravity.

The remaining theory is a Liouville-like theory, which is actually easy to study in perturbation theory because of the quartic propagator.[28] In fact, it is superrenormalizable(!), and contains only a finite set of divergent diagrams. Let us denote by V_γ the number of vertices of the general graph with the exponential interactions generated by the Einstein term, and by V_λ the vertices generated by the cosmological term. Then, elementary power counting shows that the degree of divergence of the graph is given by:

$$\begin{aligned} D &= 4L - 4I + 2V_\gamma \\ &= 4 - 2V_\gamma - 4V_\lambda, \end{aligned} \tag{8.7}$$

since the number of loops L is related to the number of internal propagators I by the usual topological relation:

$$L = I - V_\gamma - V_\lambda + 1. \tag{8.8}$$

Thus, it is clear that there are only three classes of divergent diagrams:

$$\begin{array}{llll} (i) & V_\lambda = 0, & V_\gamma = 1, & D = 2 \\ (ii) & V_\lambda = 0, & V_\gamma = 2, & D = 0 \\ (iii) & V_\lambda = 1, & V_\gamma = 0, & D = 0. \end{array} \tag{8.9}$$

The logarithmic divergences in the graphs of type (i) give rise to the following beta function for the γ coupling:

$$\beta_\gamma = \mu \frac{d}{d\mu}\gamma_R = \left(2 - 2\alpha + \frac{2\alpha^2}{Q^2}\right)\gamma_R. \tag{8.10}$$

The first two terms correspond to the "classical" scaling dimenension of the γ coupling, for $\alpha \neq 1$, while the α^2 term arises from the one-loop counterterm. The vanishing of this beta function for $\gamma_R \neq 0$ yields a quadratic relation for the anomalous scaling of e^σ:

$$1 - \alpha + \frac{\alpha^2}{Q^2} = 0, \tag{8.11}$$

with the solutions,

$$\alpha_\pm = \frac{1 \pm \sqrt{1 - \frac{4}{Q^2}}}{\frac{2}{Q^2}}. \tag{8.12}$$

This result is completely analogous to the anomalous scaling obtained for the Liouville mode in two dimensions, except that in that case it is the λ beta function which determines the scaling relation[31,32]. The classical scaling $\alpha = 1$ is obtained from the negative branch of the square root, α_- in the limit $Q^2 \to \infty$:

$$\alpha_- \to 1 + \frac{1}{Q^2}, \qquad Q^2 \to \infty \tag{8.13}$$

Unlike two dimensions this limit is obtained by sending the number of matter fields to $+\infty$, rather than $-\infty$. The value $Q^2 = Q^2_{cr} = 4$ corresponds to $c = 1$ in two dimensions, where the theory could exhibit a phase transition or qualitatively new phenomena. However, it seems from (7.8) and (8.6) that the physically relevant case in four dimensions is always $Q^2 > 4$, corresponding to $c < 1$ in two dimensions.

The graphs of type (ii) and (iii) contribute to the λ renormalization and lead to the following beta function for the λ coupling:

$$\beta_\lambda = \mu \frac{d}{d\mu}\lambda_R = \left[4 - 4\alpha + \frac{8\alpha^2}{Q^2}\right]\lambda_R - \frac{8\pi^2\alpha^2}{Q^4}\gamma_R^2\left[1 + \frac{4\alpha^2}{Q^2} + \frac{6\alpha^4}{Q^4}\right]. \tag{8.14}$$

Since α has already been determined by (8.12), setting $\beta_\lambda = 0$ gives a non-trivial relation for the cosmological constant in Planck units:

$$\frac{\lambda}{\gamma^2} = 9\kappa\Lambda = \frac{2\pi^2}{Q^2}\left[1 + \frac{4\alpha^2}{Q^2} + \frac{6\alpha^4}{Q^4}\right], \tag{8.15}$$

which is a function of the only parameter remaining, *viz.* Q^2.

There is an equivalent way of deriving the scaling relations in four dimensions, which is closer in spirit to the methods of conformal field theory in two dimensions, once ζ has been set to zero. This is based on the observation that the operator $e^{p\sigma}$ which has conformal weight p classically, acquires a quantum contribution,

$$\Delta(e^{p\sigma}) = p - \frac{p^2}{2Q^2}, \tag{8.16}$$

which may be computed also by its operator product expansion with the energy momentum tensor derived from the quartic Lagrangian, (7.12):

$$\begin{aligned} T^{ab} = \frac{Q^2}{(4\pi)^2}\Big[& -4\Box\sigma\partial^a\partial^b\sigma - \frac{2}{3}\partial^a\partial^b(\partial\sigma)^2 + 2\Box(\partial^a\sigma\partial^b\sigma) - \frac{2}{3}\partial^a\partial^b\Box\sigma \\ & + \eta^{ab}(-\frac{1}{3}\Box(\partial\sigma)^2 + (\Box\sigma)^2 + \frac{2}{3}\Box^2\sigma)\Big], \end{aligned} \tag{8.17}$$

in the flat space limit. In Wick contracting the σ fields the propagator,

$$G_{flat}(x, x') = -\frac{1}{2Q^2}\ln[\mu^2(x - x')^2], \tag{8.18}$$

of the free quartic action should be used. The anomalous scaling relation, (8.12) may be obtained by requiring that the operator, $\sqrt{-g}R = -6e^{2\alpha\sigma}((\partial\sigma)^2 + \Box\sigma)$ has conformal weight equal to four. Indeed, using Eq. (8.16) and the fact that $\partial\sigma$ has only a classical weight, equal to one, one finds:

$$\Delta(\sqrt{-g}R) = 2\alpha - \frac{2\alpha^2}{Q^2} + 2 = 4. \tag{8.19}$$

This is the necessary condition for the Einstein term to be invariant under the conformal diffeomorphism symmetry (7.20). Once this condition has been imposed, note that one can no longer insist that the cosmological term $e^{4\alpha\sigma}$ have the same conformal weight:

$$\Delta(e^{4\alpha\sigma}) = 4\alpha - \frac{8\alpha^2}{Q^2} = 4(2 - \alpha) \neq 4. \tag{8.20}$$

Instead, as Eq. (8.14) shows, there is a nontrivial mixing between the λ and γ couplings, so that invariance can be enforced only by the relation (8.15).

Since renormalizability and anomalous scaling are properties of the short distance behavior of the Lagrangian, we expect that the results of this section are valid in a general background, although they have been derived with Feynman rules in flat space. Actually, flat space is not a solution of the classical equations of motion following from the Lagrangian, (8.2). Hence, in our flat space calculations, we implicitly assumed the presence of a linear term in σ to cancel tadpole contributions at both the classical and quantum levels.

9. De Sitter Space and the Cosmological Constant

The formal development of the previous sections shows that the effective theory of the conformal factor in four dimensional gravity has some very remarkable properties. It is renormalizable, has a nontrivial, infrared stable fixed point, and anomalous scaling dimensions. Once the fiducial metric $\bar{g}_{ab}$ has been frozen, the analogies with two dimensional gravity are manifest. However, in four dimensions, this truncation of the full theory is serious. Under what circumstances can we expect the σ theory to give a reliable approximation to nature?

Certainly, a necessary condition for neglecting the transverse graviton modes of the metric is that we are discussing phenomena at energies far below the Planck scale: we can apply the σ theory only to the infrared aspects of metric fluctuations. However, quantum fluctuations of the metric seem to be negligible at all accessible energy scales below the Planck scale. At such scales we would expect the classical Einstein theory to provide an excellent phenemenological description.

The only situation where this might not be the case is when the classical background spacetime described by $\bar{g}_{ab}$ has associated infrared divergences. Such is known to be the case in de Sitter spacetime, which is of particular interest both because it is the maximally symmetric spacetime which solves Einstein's equations with a positive cosmological term, and because of its role in inflationary cosmological models. Massive matter is redshifted away exponentially rapidly in a background with a cosmological term, becoming essentially noninteracting, cold, and decoupled in a few expansion times. Hence, de Sitter space is approached rapidly in the classical picture, and treating the remaining massless fields as conformally invariant would seem to be valid. Then the conditions needed for the applicability of the effective theory of the conformal factor have been satisfied. Let us attempt to apply the σ theory to this case.

The first observation we can make is that if the expectation value of the Ricci scalar is different from zero, then the global conformal symmetry discussed in the previous sections must be spontaneously broken. This will be clear from the expression for the conformal weight for the operator R, using (8.16):

$$\Delta(R) = -2\alpha - \frac{2\alpha^2}{Q^2} + 2 = 4(1-\alpha). \tag{9.1}$$

In the semiclassical limit, when the anomaly induced fluctuations are suppressed by $\frac{1}{Q^2}$ for large Q^2, $\alpha \to 1$ and the weight of R is zero, *i.e.* it transforms like a scalar under global conformal transformations. If α differs from unity, then this is no longer the case. This should not be viewed with alarm if we recall that in two dimensions physical matter field operators become "dressed" by their interactions with the Liouville mode, and acquire modified scaling dimensions.[31] In four dimensions the Ricci scalar acquires an anomalous scaling from the self interactions of the conformal part of gravity itself. This implies that $\langle R\rangle$ becomes an order parameter for the spontaneous breaking of global conformal symmetry, in sharp contrast to the classical situation in which $\langle R\rangle$ can take on any value consistent with the symmetry. Therefore, the cosmological "constant" problem reduces to the question of whether this symmetry remains spontaneously broken, or is restored in the quantum theory.

To address this question, let us consider the σ theory at the $\zeta = 0$ fixed point, in a maximally symmetric space. The effective Lagrangian becomes:

$$L_{eff} = -\frac{Q^2}{(4\pi)^2}\Big[\sigma\Box(\Box - \frac{\overline{R}}{6})\sigma + \frac{\overline{R}^2}{12}\Big] + \gamma e^{2\sigma}[(\overline{\nabla}\sigma)^2 + \frac{\overline{R}}{6}] - \lambda e^{4\sigma}. \qquad (9.2)$$

At the fixed point, the theory behaves like a "free" theory with an anomalous dimension. This can be seen by shifting σ by a constant σ_0, under which the partition function scales according to:

$$Z(\gamma) = Z(\gamma e^{2\alpha\sigma_0})\, e^{-Q^2\chi\sigma_0}, \qquad (9.3)$$

where χ is the Euler number ($\chi = 2$ for the sphere), and we have used the translational invariance of the measure. The partition function depends only on γ, because of the relation (8.14). From (9.3), we see that even this dependence is trivial at the fixed point, and we can scale it away by taking an appropriate limit ($\sigma_0 \to -\infty$). This is completely analogous to string theory, restricted to the worldsheet toplogy of the sphere. Hence, for calculating correlation functions we may replace the effective Lagrangian (9.2) by the free quartic one, with $\gamma = \lambda = 0$.

The propagator corresponding to the quartic term, $\Box(\Box - \frac{\overline{R}}{6})$ is dominated by the $\Box^2$ term at short distances, and by the $-\frac{\overline{R}}{6}\Box$ term at large distances. However, in de Sitter space,[15]

$$\Box^{-1}_{xx'} = -\frac{1}{8\pi^2}\left[\frac{2}{s^2(x,x')} - H^2\ln\Big(\frac{H^2 s^2(x,x')}{4}\Big)\right], \qquad \overline{R} = 12H^2, \qquad (9.4)$$

where $s(x,x')$ is the geodesic distance between the two points. Hence, the quartic propagator has the *same* logarithmic behavior in both limits:

$$\langle\sigma(x)\sigma(x')\rangle = G_\sigma(x,x') \to -\frac{1}{2Q^2}\begin{cases}\ln\mu^2 s^2, & \text{as } s\to 0,\\ \ln H^2 s^2, & \text{as } s\to\infty,\end{cases} \qquad (9.5)$$

with the ultraviolet scale μ replaced by the infrared horizon scale H. This means that the fixed point calculations we performed in the previous section, using the short distance behavior of the theory in dimensional regularization or the operator product expansion, in flat space, can be carried out in the infrared with the same results, in de Sitter space. This is not surprising, as it is known from critical phenomena that there is a close interplay between ultraviolet and infrared behavior in systems with conformal symmetry. Notice that unlike a purely local R^2 action whose effects are unimportant in the infrared, the effects of the σ - propagator arising from the anomaly induced effective action (9.2) grow logarithmically at large distances, and dominate the physics.

Consider now the correlation function of Ricci scalars, $\langle R(x)R(x')\rangle$ at two different points. Using the expression, (7.13) with σ replaced by $\alpha\sigma$, and the free action which describes the theory at its critical point, we find:

$$\begin{aligned}\langle R(x)R(x')\rangle &\to \overline{R}^2\langle e^{-2\alpha\sigma(x)}e^{-2\alpha\sigma(x')}\rangle \\ &\to \overline{R}^2 e^{4\alpha^2 G_\sigma(x,x')} \\ &= \overline{R}^2\left|\frac{Hs(x,x')}{2}\right|^{-\frac{4\alpha^2}{Q^2}}, \qquad |s(x,x')|\to\infty,\end{aligned} \tag{9.6}$$

where only the dominant infrared behavior has been retained.

The result (9.6) states that the *effective* cosmological "constant" goes to zero with a definite power law behavior for large distances. In other words, there is screening in the infrared of the effective value of vacuum energy at larger and larger scales. The value of the power is *universal*, depending only on Q^2 which counts the effective number of massless degrees of freedom. In particular, it depends neither on the classical value of the background curvature $\overline{R}$, nor on the Planck scale. This is essential for a scale invariant, phenemenologically acceptable solution of the cosmological constant problem.

We may compare this case to that of spontaneous breaking of a continuous symmetry in 1 + 1 dimensions.[30] Consider a complex scalar field with a tree-level potential energy function of $|\phi|^2$ corresponding to symmetry breaking. When the field is quantized, the corresponding massless Goldstone boson has a propagator which grows logarithmically at large distances. This infrared divergence implies instability of the spontaneously broken vacuum to quantum fluctuations. Because of this instability of the ordered state, the system becomes disordered, the $U(1)$ symmetry is restored at the quantum level and the Mermin-Wagner-Coleman theorem is satisfied. Locally, there are regions of broken symmetry in which the classical description remains valid. However, as we consider regions of larger and larger size, the local classical vacuum executes a random walk in the phase angle, and averages to zero. The quantitative description of this phenemenon is given by the power law fall-off of the correlation function, $\langle\phi(x)\phi^\dagger(x')\rangle$. Introducing the nonlinear polar field decomposition, $\phi = \rho\exp(i\theta)$, and neglecting the fluctuations of the massive ρ field in the infrared, one finds that the angular field θ may be treated as a free field with the propagator,

$$\langle\theta(x)\theta(x')\rangle = -\frac{1}{4\pi\rho^2}\ln(\mu^2(x-x')^2)\,, \tag{9.7}$$

so that the model is now equivalent to the x-y model in two dimensions. In this infrared scaling limit, the correlation function for ϕ has a power law behavior analogous to (9.6):

$$\begin{aligned}\langle\phi(x)\phi^\dagger(x')\rangle &\to \rho^2\langle e^{i\theta(x)}e^{-i\theta(x')}\rangle \\ &\to \rho^2\exp\langle\theta(x)\theta(x')\rangle \\ &= \rho^2\,|\mu(x-x')|^{-\frac{1}{2\pi\rho^2}}, \qquad |x-x'|\to\infty.\end{aligned} \tag{9.8}$$

In this case the power law is non-universal, as it depends on the classical value of the background field ρ.

10. Conclusions and Open Problems

To summarize, I believe that the effective theory of the conformal factor presented here provides a useful, novel framework for studying the infrared behavior of gravity in four dimensions and addressing the cosmological constant problem. The anomalous scaling of the conformal factor may be the key to understanding why $\langle R \rangle = 0$ in the observed universe. However, many unanswered questions remain.

First of all, there are some important formal elements missing. The σ action is fourth order in derivatives. One must understand if a mechanism similar to that in string theory[29] applies to render the theory unitary nonetheless, or if not, how the effective action (7.12) is to be interpreted in the infrared. A related issue is the question of how to incorporate properly the transverse-traceless graviton modes into the framework.

Secondly, there are questions of confrontation with observational cosmology. The infrared fixed point can at most describe the conformally invariant vacuum state of gravity. How is that fixed point approached in the real universe? Can a realistic inflationary universe cosmology be developed with these ideas? How should correlation functions computed in this framework be compared to observations of the cosmic microwave background or the large scale distribution of matter in the universe? Is there a residual cosmological energy density from this mechanism which behaves like dark matter and closes the universe?

If the theoretical framework proposed in these lectures withstands an ongoing thorough analysis, then it should have quite novel effects on the dynamics of the universe in its early inflationary epoch(s) and dramatic predictive consequences for observational cosmology, large scale structure and the missing matter puzzles.

Acknowledgement

I would like to thank Drs. O. Bertolami, M. C. Bento, J. Mourao for organizing a very enjoyable, successful First Iberian Meeting on Gravity, and hope that it will indeed be the first of many more to come.

References

1 E. Loh and E. Spillar, *Astrophys. J. Lett.* **307** (1986) L1;
E. Loh, *Phys. Rev. Lett.* **57** (1986) 2865.

2 H. B. G. Casimir, *Proc. Kon. Ned. Acad. Wetenschap.* **51** (1948) 793;
M. J.Sparnaay, *Physica* **24** (1958) 751;
T. M. Boyer, *Ann. Phys.* **56** (1970) 474.

3 S. Weinberg, *Rev. Mod. Phys.* **61** (1989) 1.

4 S. W. Hawking, *Phys. Rev.* **D18** (1978) 1747;
Recent Developments in Gravitation, M. Levy and S. Deser eds. (Plenum, New York, 1979);
General Relativity: An Einstein Centennial Survey, S. W. Hawking and W. Israel eds. (Cambridge University Press, Cambridge, 1979).

5 E. Baum, *Phys. Lett.* **B133** (1983) 185.
S. W. Hawking, *Phys. Lett.* **B134** (1984) 403;
A. Linde, *Phys. Lett.* **B200** (1988) 272.

6 T. Banks, *Nucl. Phys.* **B309** (1988) 493.
S. Coleman, *Nucl. Phys.* **B310** (1988) 643.

7 I. Klebanov, L. Susskind, and T. Banks, *Nucl. Phys.* **B317** (1989) 665;
W. Fischler and L. Susskind, *Phys. Lett.* **B217** (1989) 48.
J. Preskill, *Nucl. Phys.* **B323** (1989) 141.

8 G. W. Gibbons, S. W. Hawking and M. J. Perry, *Nucl. Phys.* **B138** (1978) 141.

9 Voltaire, *Candide*, translated by J. Butt (Penguin Books, Harmondsworth, Middlesex, 1947).

10 J. J. Van der Bij, H. Van Dam and Y. J. Ng, *Physica* **A116** (1982) 307;
W. Buchmüller and N. Dragon, *Phys. Lett.* **B207** (1988) 292;*Phys. Lett.* **B223**, (1989) 313.

11 S. Weinberg, *Phys. Rev. Lett.* **59** (1987) 2607.

12 J. Schwinger, *Phys. Rev.* **82** (1951) 664.

13 D. W. Sciama, P. Candelas and D. Deutsch, *Adv. Phys.* **30** (1981) 327.
B. L. Hu, *Phys. Lett.* **A90** (1982) 375; *Phys. Lett.* **A97** (1983) 368;

L. Parker, *Phys. Rev. Lett.* **50** (1983) 1009.
E. Calzetta and B. L. Hu, *Phys. Rev.* **37** (1988) 2878;

14 N. P. Myhrvold, *Phys. Rev.* **D28** (1983) 2439;
E. Mottola, *Phys. Rev.* **D31** (1985) 754; *Phys. Rev.* **D33** (1986) 1616, 2136;
P. O. Mazur and E. Mottola, *Nucl. Phys.* **B278** (1986) 694;

15 I. Antoniadis and E. Mottola, *Jour. Math. Phys.* **32** (1991) 1037.

16 F. Cooper and E. Mottola, *Phys. Rev.* **D40** (1989) 456.
F. Cooper, J. M. Eisenberg, Y. Kluger, E. Mottola, and B. Svetitsky, *Phys. Rev. Lett.* **67** (1991) 2427.

17 I. M. Gel'fand and A. M. Yaglom, *Jour. Math. Phys.* **1** (1960) 1;
R. P. Feynman and A. R. Hibbs, *Quantum Mechanics and Path Integrals*, (McGraw-Hill, New York, 1965).

18 E. S. Fradkin and G. A. Vilkovisky, *Phys. Lett.* **B55** (1975) 2;
I. A. Batalin and G. A. Vilkovisky, *Phys. Lett.* **B69** (1977) 309; *Nucl. Phys.* **B234** (1984) 106.

19 O. Babelon and C.-M. Vaillet, *Phys. Lett.* **B85** (1979) 246;

20 A. M. Polyakov, *Phys. Lett.* **B103** (1981) 207, 211; *Gauge Fields and Strings*, (Harwood Academic, Chur, 1987);

21 L. C. L. Botelho, *Phys. Rev.* **D38** (1988) 2464;
P. Ellicott, G. Kunstatter and D.J. Toms, *Mod. Phys. Lett.* **A4** (1989) 2397;

22 P. O. Mazur and E. Mottola, *Nucl. Phys.* **B341** (1990) 187;
S. K. Blau, Z. Bern, and E. Mottola, *Phys. Rev.* **D43** (1991) 1212.

23 B. S. DeWitt, *General Relativity: An Einstein Centenary Survey*, S. W. Hawking and W. Israel editors, (Cambridge Univ. Press, Cambridge, 1979)

24 S. Deser, M. J. Duff, and C. J. Isham, *Nucl. Phys.* **B111** (1976) 45;
N. D. Birrell and P. C. W. Davies, *Quantum Fields in Curved Space*, (Cambridge Univ. Press, Cambridge, 1982).

25 G. T. Horowitz, *Phys. Rev.* **D21** (1980) 1445;
J. B. Hartle and G. T. Horowitz, *Phys. Rev.* **D24** (1981) 257;
B. Biran, R. Brout and E. Gunzig, *Phys. Lett.* **B125** (1983) 399.

26 M. J. Duff, *Nucl. Phys.* **B125** (1977) 334.

27 R. J. Riegert, *Phys. Lett.* **B134** (1984) 56;
E. S. Fradkin and A. A. Tseytlin, *Phys. Lett.* **B134** (1984) 187;
S. Odintsov and I. Shapiro, *Class. Quant. Grav.* **3** (1991).

28 I. Antoniadis and E. Mottola, *Phys. Rev.* **D45** (1992) 2013.

29 R. C. Myers, *Phys. Lett.* **B199** (1987) 371;
I. Antoniadis, C. Bachas, J. Ellis and D. V. Nanopoulos, *Nucl. Phys.* **B328** (1989) 117.

30 N. D. Mermin and Wagner, *Phys. Rev. Lett.* **17** (1966) 1133;
V. I. Berezinskii, *Soviet Physics, JETP* **32** (1971) 493;
S. Coleman, *Comm. Math. Phys.* **31** (1973) 259;
S. K. Ma and R. Rajaraman, *Phys. Rev.* **11** (1975) 1701.

31 A. M. Polyakov, *Mod. Phys. Lett.* **A2** (1987) 899;
V. G. Knizhnik, A. M. Polyakov, and A. B. Zamolodchikov, *Mod. Phys. Lett.* **A3** (1988) 819.

32 F. David, *Mod. Phys. Lett.* **A3** (1988) 1651;
J. Distler and H. Kawai, *Nucl. Phys.* **B321** (1989) 509.

CLASSICAL AND QUANTUM GRAVITY FROM STRING THEO

G.Veneziano
Theory Division, Cern, Geneva, Switzerland

FOREWORD

These three lectures will be devoted to String Theory as a candidate consistent theory of classical and quantum gravity. The selection of the topics has not been an easy one. Eventually, I decided for the following breakup of the lectures:

I. **Strings, Fields, Particles, Forces.**

II. **An example of Classical String Gravity: Large-distance String Cosmology.**

III. **An example of Quantum String Gravity: Planckian-energy String Collisions.**

By making this choice I have neglected many important topics such as Conformal Field theory for its own sake, $2D$-black holes, non-critical, non-perturbative strings, matrix models, supersymmetry and its breakdown in string theory, string-inspired phenomenology, etc. The selection was motivated by my own feelings about which aspects of string theory are enough generic and physically interesting to serve as truly illustrative examples for the subject of these lectures.

I. STRINGS, FIELDS, PARTICLES, FORCES

Outline:

1. Introduction, history.
2. Classical and Quantum Strings in flat space-time.
3. Origin of Fundamental Constants in Quantum Field and Quantum String Theory.

1. Introduction, history

Tremendous progress in understanding the fundamental forces of Nature started right at the beginning of the century. It has come to a sort of completion in the early 70s with the advent of the Standard Model(SM).

The SM is our state-of-the-art theory of three of the fundamental forces of Nature: Thus, in the SM, electromagnetic interactions are described, at low energy, by the usual $U(1)$ gauge theory but get unified at higher energies with the weak interactions within an $SU(2) \times U(1)$ gauge group. Finally, strong interactions are added in through an extra factor $SU(3)$, to yield the full SM gauge group: $SU(3) \times SU(2) \times U(1)$. As any theory, the SM too has its own pluses and minuses:

A) SM Pluses

These are many, not the least one the SM seems to explain the data (and maybe I should stop there)! But, also, from the theorist's view point, it reduces *three* Fundamental Forces to *a single* physical principle :

Gauge Invariance

It so happens that, precisely in D=3+1 dimensions of space-time, gauge theories are just "renormalizable". Let me explain in simple terms what is meant by this and why it is so important:

The classical Lagrangian of the standard model has the schematic form:

$$L_{class} = F^2_{\mu\nu} + \bar{\psi}\gamma D\psi + L_{Higgs} \tag{1.1}$$

However, experimentally, what one measures is the full, quantum-corrected, so-called effective Lagrangian:

$$L_{eff} = L_{class} + L_{quantum} = Z_1 F^2_{\mu\nu} + Z_2\bar{\psi}\gamma D\psi + Z_3 \ldots + L_{finite} \tag{1.2}$$

A renormalizable theory is defined by the property that, although (as opposed to a finite theory) it does have quantum infinities, these can all be lumped into a *finite* number of constants Z_i which multiply precisely the terms appearing in L_{class}. Thus, even if someone had given us the parameters appearing in L_{class}, there would still be no hope to compute those appearing in the *measured* L_{eff}. Rather, one fixes the "renormalized" parameters (masses, couplings) from the experiments while everything else is determined, in principle, by the theory.

To summarize:

In a renormalizable theory L_{eff} is a *finite* function of a *finite* number of *arbitrary* parameters.

The above statement shows, at the same time, the strength and weakness of renormalizable theories: it is good to have a *finite* number of arbitrary parameters; yet it would have been even better if they were not *arbitrary*.

B) SM Minuses

1. There are "many" (15 or so) arbitrary parameters:
α, G_F, $sin^2\,\theta_W$, $\Lambda_{QCD}\ldots$ but also m_q, m_l, m_H, $sin\theta_C\ \ldots$
2. The Higgs sector is "ugly" with
$m^2_H = m^2_H(class.) + O(\Lambda^2)$,
where Λ is the ultraviolet cut-off, so that a huge fine tuning is needed in order to avoid that $m_H \gg m_W$.

3. Some sectors of the theory may be sick and/or trivial (e.g. the $U(1)$ and Higgs sectors).

What is meant by this is the so-called Landau pole phenomenon by which, in non-asymptotically free theories, the effective coupling becomes infinite at a finite energy scale which is to be interpreted as an ultraviolet cut-off. Alternatively, if the cut-off is sent to infinity, the effective coupling goes to zero at any finite scale (triviality). This is what, in 1955, led L. Landau to say:

"The Lagrangian is dead. It should be buried, with all due honour, of course."

Today we know of some possible remedies to the above problems: they go under the name of supersymmetry, Grand Unified Theoris (GUT's), technicolour and all possible combinations thereof. Yet none of this appears todate to solve another fundamental problem with the SM:

The framework of Quantum Field Theory does not work for the 4th fundamental force, gravity.

Why is it so? It has to do again with quantum corrections to L_{class}:

Let us compare the correction due to the emission and reabsorption of a photon by an electron with those of a graviton. Let us call x the distance between the point of emission and that of absorption. An elementary calculation shows that the corresponding probability amplitudes (to be integrated over x) have the approximate form:

$$dP(x) \sim \alpha \frac{dx}{x} \rightarrow P \sim \alpha ln \frac{l}{\epsilon} \tag{1.3}$$

for the photon and

$$dP(x) \sim G_N \hbar \frac{dx}{x^3} \rightarrow P \sim \frac{\lambda_P^2}{\epsilon^2} \tag{1.4}$$

for the graviton, where α is the fine structure contant, $\lambda_P = \sqrt{G_N \hbar}$ is the Planck length, or about $10^{-33} cm.$, l is the typical scale involved in the process and we have introduced an ultraviolet (short-distance) cut-off ϵ.

We see that the divergence as $\epsilon \to 0$ is much more violent for the graviton than for the photon, but this per-se would not be a disaster (after all we have already seen the quadratically divergent renormalization of the Higgs mass in the SM). What is lethal though is that "a singularity may hide another singularity". A careful computation reveals indeed that there is also a logarithmically divergent contribution for the graviton case which, unlike the quadratic one, cannot be reabsorbed into a redefinition of the electron mass or of the Newton constant. In formulae we find:

$$L_{eff} = L_{class} + L_{quantum} = (16\pi G_N)^{-1}R + (ln\epsilon)R^2 + \ldots \qquad (1.5)$$

and it gets worse and worse as we go to higher and higher loops! At each order infinities appear multiplying more and more complicated terms none of which existed in L_{class}. Conventional quantum gravity is a *non – renormalizable* theory, i.e. one in which infinities multiply an *infinite* number of (theoretically unpredictable) terms. Put differently:

No predictivity is left for non-renormalazable theories Unfortunately, standard quantum gravity is one of them.

We now leave the SM for a while to talk about an apparently unrelated subject: Strings.

2. Classical and Quantum String Theory in flat space-time

Let us recall the form of the action for a system of points and for the string:

$$\begin{aligned} S_{points} &= \sum_i m_i c \int d\ell_i \; + \text{ interactions} \\ S_{string} &= T \int d\sum \end{aligned} \qquad (2.1)$$

The first is proportional to the sum of the length of the paths described by the points each one weighted by a different constant, its mass. Interactions modify the action and are quite arbitrary. The second is given by a single term proportional to the area swept by the string. A single constant appears in place of the masses (the string tension T) and, most remarkably, interactions will come out naturally and uniquely.

The area is swept in space-time and therefore space and time have to be measured in the same units. A constant c (later to be recognized as the speed of light), is implicitely multiplying time intervals. Having c in our list, we proceed without writing it explicitely each time.

In order for the action to have its usual dimensions, T has to have dimensions energy/length as appropriate for a tension. However, as long as we work at the classical level, T is completely irrelevant (for points, analogously, only mass ratios are important). This is because, classically, only the stationary points of S do matter and these are invariant under a rescaling of the action.

Still classically, free points move along straight lines, while the string has a rich variety of motions all of which, however, satisfy the constraint:

$$M^2 \geq 2\pi T J \tag{2.2}$$

The mass-squared of the string is bound from below by a multiple of its angular momentum, the equality sign being reached for a rigid, rotating stick whose end points (for an open string) move at the speed of light.

At the quantum level the difference between points and strings becomes even bigger. Of course we know that angular momentum becomes quantized in units of $\hbar$:

$$J = n\hbar \tag{2.3}$$

Not surprisingly because of (2.2), also $M^2(2\pi T)^{-1}$ becomes quantized in units of $\hbar$:

$$M^2(2\pi T)^{-1} = m\hbar \tag{2.4}$$

Afterall the string is a collection of harmonic oscillators!

What becomes of the classical inequality (2.2)? Here the first surprise comes. Instead of a naive $m \geq n$ one finds:

$$m \geq n - a_0 \text{ with } a_0 = +1(+2) \text{ for the open (closed) string} \tag{2.5}$$

The origin of a_0 is that of a zero point energy (or normal-ordering constant) analogous to the famous 1/2 of the harmonic oscillator levels $(n + 1/2)\hbar\omega$. This normal ordering constant is of paramount importance for string theory in that it allows the existence of classically forbidden, massless spin 1 and spin 2 states (for open and closed strings, respectively). These objects are believed to exist in Nature, in the form of Gauge bosons and Gravitons respectively, and to mediate all known forces, including gravity. That's a good start indeed for *Quantum* String Theory!

Quantization is responsible for another important property of strings: strings acquire a typical, better a *minimal*, size. Recall again that position and momentum uncertainty behave rather symmetrically for the harmonic oscillator, scaling both like the square root of the Planck constant. The same is true for the string where we get:

$$\Delta x \simeq (\hbar/T)^{1/2} \quad ; \quad \Delta p \simeq (\hbar T)^{1/2} \tag{2.6}$$

So far we have introduced three fundamental constants:

$$c \quad , \quad T \quad , \quad \hbar$$

Classically, there was no $\hbar$, of course, and T was irrelevant. Quantum mechanically it is only a combination of $\hbar$ and T that appears. Since the quantum theory depends on S only through the pure number, $S/\hbar$, the combination that survives is obviously:

$$T/\hbar = \lambda_s^{-2} \tag{2.7}$$

which defines a fundamental length, λ_s. It looks funny, at first sight, that we managed to get rid of $\hbar$ and T in favour of a single quantity, λ_s. Actually, in order to do so, we have implicitely changed units of energy. The natural unit of energy in string theory is *length*, with T providing the *conversion* factor to and from c.g.s. (or normal particle theory) units:

$$E \to \overline{E} \equiv E/T \quad \text{which is a length} \tag{2.8}$$

In other words, using the string tension, I can give the energy of a string by giving its length: energy is length and length is energy in string theory! If natural string units are consistently used, only two dimensionful constants are needed for the relativistic and quantum nature of the theory [1].

It looks like no achievent at all to have replaced c and $\hbar$ by c and $\lambda_s$through a change of units. But, actually, we have gained a lot:

In String Theory, the Planck constant λ_s is also the short-distance cut-off.

This is so because, as we have seen, strings acquire, through quantization, a finite, minimal size $O(\lambda_s)$. The finite size induces in turn a cut-off (in the way of a form-factor) on large virtual momenta. Incidentally, this is what makes sense of Quantum String Gravity. A typical one-loop (quantum gravity) correction normalized to the tree (classical) value has the form given in eq. (1.4). With the string cut-off $(\hbar T)^{1/2}$ on the virtual momentum one gets a finite correction of order λ_s^2/λ_P^2: Quantum gravity corrections are (typically) small if the string length parameter λ_s is somewhat larger than the Planck length λ_P!

3. Fundamental constants in Quantum Field and String Theory

a) Quantum Field Theory

As mentioned in the introduction the basic object of QFT is the so-called effective Lagrangian (or its space-time integral, the effective action S_{eff}) of eq. (1.2). Let us discuss some of its basic properties:

1.S_{eff} possesses all the symmetries of the quantum theory. In general these are fewer than those of the classical theory because of anomalies).

2. The stationary points of S_{eff} correspond to the possible ground states of the theory (at the quantum level, of course). A most famous example is the so-called Coleman-Weinberg mechanism [2] in which, typically, the classical action has a symmetric ground state (say $\phi = 0$) while the effective action has a symmetry breaking ground state (say $\phi \neq 0$)

3. Ground state expectation values are obtained by setting to 0 the first derivatives of S_{eff}. Higher derivatives provide other physical quantities such as masses, coupling constants, scattering and decay amplitudes. Thus we can say that it is S_{eff} that is being tested at LEP!

4.S_{eff} depends on its arguments (the renormalized fields) via a finite number of (finite) parameters λ_j:

$$S_{eff} = S_{eff}(\phi_i; \lambda_j) \tag{3.1}$$

which are nothing but the uncalculable fundamental constants that a renormalizable field theory has to live with. These are QFT's Fundamental Constants!

5. Last, but not least, there must be a recipe for computing S_{eff} from the S_{class}. It can be expressed in a deceptively simple form (which took about 30 years to be defined):

$$\exp(-S_{eff}(\phi_i^c; \lambda_j)/\hbar) \Leftrightarrow \int D\phi_i(x) \exp(-\tilde{S}(\phi_i, \phi_i^c; \lambda_j)/\hbar) \tag{3.2}$$

Here ϕ^c is called the "classical" field as opposed to ϕ, the quantum field on which one "integrates". $\tilde{S}$ is a functional that we need not specify. At the end of the calculation one will get a (generally non-local) functional of ϕ^c which also depends on the fundamental constants λ_i. Then ϕ^c is simply renamed ϕ.

b) Quantum String Theory

1. String physics is also encoded in some S_{eff}. However, the recipe to compute S_{eff} is now

$$\exp(-S_{eff}(\phi_i^c)/\hbar) \Leftrightarrow \int DX^\mu(\sigma, \tau) \exp(-\tilde{S}(X^\mu; \phi_i^c)/\hbar) \tag{3.3}$$

indicating some major differences w.r.t. QFT.

2. There is only *one* set of $\phi's$ and it's clearly the analog of ϕ^c (it is not integrated over). Instead, the role of the quantum fields is now

played by the string coordinate $X^\mu(\sigma,\tau)$ (and, in canonical quantization, by its conjugate momentum $P_\mu(\sigma,\tau)$). In other words we are "back", in QST, to *first* quantization! Recall, incidentally, how things went for relativistic quantum particles. There were problems with the relativistic generalizations of Schroedinger's equation and thus people preferred to take them as *classical* field equations to be quantized again (hence the name of second quantization). Somehow, all this is not needed (and might even be wrong) in string theory!

3. Finally, $\tilde{S}$ contains no arbitrary parameters since it is possible to absorb them in the redefinition of the classical fields. Furthermore, the ultraviolet finiteness of loop calculations guarantees that no "renormalization" is needed in order to compute S_{eff} through (3.3) and therefore that no new arbitrary parameters are generated from the loops. We conclude that, in string theory:

S_{eff} does not contain arbitrary constants

But then where do the Constants of Nature come from in QST? The (obvious?) answer is [3]:

The fundamental constants of QST are vacuum expectation values!

We shall illustrate this point with some examples concentrating our attention on two $\phi's$ which are of capital importance in string theory:

The so-called "Metric" : $G_{\mu\nu}$

The so-called "Dilaton" : ϕ

These two fields are defined by the way they enter $\tilde{S}$ of eq. (3.3) [5]:

$$\tilde{S}/\hbar = -1/2 \int d^2\xi g^{1/2}\{g^{\alpha\beta}\partial_\alpha X^\mu \partial_\beta X^\nu G_{\mu\nu}(X) + {}^{(2)}R(\xi)\phi(X)/4\pi\} \tag{3.4}$$

Here, as usual, X^μ is the string coordinate, $g_{\alpha\beta}$ is the metric on the world sheet whose coordinates σ and τ are collectively denoted by ξ. ${}^{(2)}R(\xi)$ is the curvature scalar on the world sheet which satisfies the well known formula:

$$(2-2g) = \int d^2\xi g^{1/2} \, {}^{(2)}R(\xi)/4\pi \tag{3.5}$$

where g is the genus of the two-dimensional Riemann surface described by the metric $g_{\alpha\beta}$ (thus $g = 0$ for a sphere, $g = 1$ for a torus, $g = 2$ for a sphere with two handles etc). In string theory, the sum over g implicit in the integral over all metrics in eq. (3.3), corresponds to the sum over loops of ordinary Quantum Field Theory.

Note that, in order for (3.4) to define a dimensionless quantity, we have to attribute to $G_{\mu\nu}$ dimensions of $(length)^{-2}$ once we call $(length)$ the dimension of X. By contrast, ϕ is dimensionless.

What is the general structure of S_{eff}? First of all, it can be written as sum of contributions from different Riemann surfaces (string-loop expansion):

$$S_{eff} = \sum S^g_{eff} = S^0_{eff} + S^1_{eff} + S^2_{eff} + \ldots$$

Next, if we look at the lowest order ($g = 0$) term, we find:

$$S^{g=0}_{eff} = \int d^D x \sqrt{-detG_{\mu\nu}} e^{2\phi} [\frac{D-10}{3} + R(G) + 4\partial_\mu\phi\partial^\mu\phi + F^2_{\mu\nu} + \ldots] \tag{3.6}$$

where D is the number of components of X (i.e. the number of dimensions of space-time) and, as promised, no Fundamental Constant (not even c, or λ_s of our Sect.2) appears! The dots indicate here terms with more than two derivatives acting on the fields.

We have now to put to zero the first derivatives of S_{eff} i.e.

$$\delta\Gamma/\delta G_{\mu\nu} = \delta\Gamma/\delta\phi = \ldots = 0 \tag{3.7}$$

If we look for a slowly varying solution of these equations we are entitled to neglect the dots in (3.7). More doubiously, we shall also neglect the higher genus contributions a point we shall come back to. We then find that a solution of this kind exists only in $D = 10$ and has the form:

$$G_{\mu\nu} = \lambda_s^{-2}\eta_{\mu\nu} \quad , \quad \eta_{\mu\nu} = \text{diag}\,(-c^2, +1, +1, \ldots + 1)$$

$$\phi = constant \tag{3.8}$$

We have finally recovered our fundamental constants c and λ_s as VEV's. They came from our desire of finding a flat-metric vacuum and from the unconventional dimensions of $G_{\mu\nu}$. Precisely the same dimensions for G have been advocated for different reasons in [5] .

It can be argued that the role of Einstein's metric $g_{\mu\nu}^{Einst}$ is played, in string theory, by $\lambda_s^2 G_{\mu\nu}$ so that

$$\lambda_s^2 (ds^2)_{GR} = G_{\mu\nu} dx^\mu dx^\nu \equiv (ds^2)_{String} \tag{3.9}$$

String distance larger or smaller than 1 defines, in string theory what we mean by large or short distance.

Who determines the actual values of c and λ_s? Let us distinguish two cases:

a) If the dimension we are looking at is non-compact (the case we believe to be true for our 3+1 dimensional world) the actual value of these constants is irrelevant. What the theory should determine, in this case, are the ratios of c and λ_s to some physical speed and length that <u>we</u> call cm./sec. and cm. Thus, in principle, the theory allows to compute c, λ_s in centimetres and seconds.

b) If a dimension is instead compact (those in excess of 4 should, physically), say a circle of radius R, the ratio

$$\rho = R/\lambda_s \tag{3.10}$$

i.e. how many Planck constants is the dimension large <u>is</u> relevant. Actually, a very non-trivial property of (closed) string theories says that ρ and $1/\rho$ define the same theory [6].

The case $\rho = 1$ is of course special: it is the fixed point of the "duality" transformation. One finds that, at this special value, the gauge symmetry generated from compactification (a la Kaluza-Klein) is enhanced. It has been argued [1,7] that this special value (a radius equal to the fundamental Planck constant) will represent the preferred value for the extra dimensions (thus stabilizing the conventional KK picture) and [8] some minimal value of the Hubble parameter of the Universe at the big bang (equivalently a maximal temperature) thus avoiding the initial singularity (see lecture II for further details).

Let us now suppose that, in the early universe, the metric was not really flat (because of high matter density or of an effective cosmological constant, see my previous lecture) but just conformally flat i.e.

$$dx^\mu dx^\nu \lambda_s^2 G_{\mu\nu} \;=\; g_{\mu\nu} dx^\mu dx^\nu \;=\; -dt^2 + a^2(t) d\vec{x}^2 \qquad (3.11)$$

The conventional interpretation of $a(t)$ is that it represents the scale factor of the universe. The distance of any two points expands (for a an increasing function of t) by a factor $a(t)/a(0)$ as we go from time 0 to time t. Quantum String Theory affords an amusing reinterpretation of this expansion. Because of the way the metric enters the action, we may reinterpret a varying scale factor $a(t)$ as a varying λ_s

$$\lambda_s \to \lambda_s(t) \;=\; \lambda_s / a(t) \qquad (3.12)$$

For an increasing $a(t)$, the Planck constant of string theory is becoming smaller with time... and the Universe is becoming more and more classical! The Universe expands because our rods, made of atoms, shrink as a consequence of (3.12). The light that reaches us today from a distant star was emitted a long time ago i.e. when the Planck constant was larger by a factor $a(now)/a(t_{emission})$. All other fundamental constants being equal, its wavelength was larger precisely by the same factor: the red shift formula is thus perfectly recovered. Incidentally, a similar description of the red shift was given by Dirac in a different context, that of his Large Number Hypothesis [9].

After having discussed the dimensionful constants we now turn to the dimensionless ones, typically the fine structure constant(s) α_i where i labels the various gauge interactions.

We find again, and not surprisingly, that these are again related to the Vacuum Expectation Values (VEV) of certain fields. The most fundamental dimensionless coupling is related, in string theory, to the dilaton field ϕ appearing in eq (3.4). The reason why ϕ is related to a coupling constant was first pointed out by Witten [10]. It is related to eq. (3.5) which implies that a definite dependence of S^g_{eff} of eq. (3.6) from ϕ :

$$S^g_{eff} \simeq \exp(2\phi(1-g)) \tag{3.13}$$

But this means a factor $(\alpha_{SL})^g$ per loop where:

$$\alpha_{SL} = \exp(-2\phi) \tag{3.14}$$

Thus, indeed, α_{SL} is the loop-expansion parameter, the analog of the fine structure constant in QED. The identification can also be seen by noting that, for $g = 0$, we get from (3.7) a factor $1/\alpha_{SL}$ sitting in front of the tree level action. The latter has the form of a gauge-plus-gravity action where the fields are already rescaled so that the appropriate powers of the gauge and Newton couplings multiply simply the action. The above identification yields the two fundamental relations of (closed or heterotic) string theory:

$$\begin{aligned} &\alpha_{GUT} = \alpha_{SL}\lambda_s^{D-4} \\ &\lambda_P^{D-2} \equiv G_N\hbar = \alpha_{SL}\lambda_s^{D-2} \end{aligned} \tag{3.15}$$

where α_{GUT} is the grand unification coupling at the grand unification scale and D is the number of space-time dimensions. The string loop expansion parameter (the dilaton VEV) thus gives the ratio between the Planck length and the fundamental length of the theory! This means, unfortunately, that the string scale cannot be a lot larger than the Planck scale and that, consequently, the new physics implied by the string will only be felt at energies of the order of the Planck mass.

From (3.15) we immediately get the relation:

$$G_N\hbar = \alpha_{GUT}\lambda_s^2 \tag{3.16}$$

Compactification of string theory from D=26 or D=10 to D=4 affects the relations (3.15) whithout changing (3.16). The physical meaning of the latter equation is striking. It says that:

Heavy, charged Strings have identical gauge and gravitational interactions

a property I am tempted to call GUT^2. Of course neutral and heavy, or light and charged (e.g. the electron) strings are the exceptions confirming this rule.

This unification of all forces works fine at tree level. What about loops? Obviously α_{GUT} controls gauge loops. For gravitational loops let us go back to the estimate (1.4) and insert eq. (3.16). We find, not without satisfaction, that quantum gravity corrections are also controlled by α_{GUT}. Thus string unification seems to persist beyond tree level. At a closer look [7] this turns out to be only "logarithmically true" i.e. modulo running. Only the short distance couplings are unified while the low energy can be substantially different (even for heavy strings).

Finally I shall mention two important aspects of string unification. Because of the UV finiteness of the theory, the tree level, unified couplings are not bare, unobservable quantities: they are physical and observable, in principle, by a short distance experiment. As a corollary, non-asymptotically free theories make perfect sense in string theory. Their couplings, being finite at tree level, will just decrease towards lower energies as a result of loops (screening). Landau poles and similar deseases of field theories will be pushed to scales where the field theory approximation is certainly invalid.

Conclusions

My main conclusions are briefly summarized below:

1. Quantum Strings have few (maybe too few!) fundamental constants.
2. The dimensionful constants are "easy" : c and λ_s for relativity and quantum mechanics, respectively.
3. The dimensionless constants are harder to pin down. Among these:
   a) the overall string-fine-structure-constant enters in a most elegant way, controlling the importance of geometric interactions.

b) others (Yukawa and Higgs couplings, for instance) come from the details of compactification.

4. All of them are vacuum parameters.

Thus the great open problem in string theory is:

What determines the string vacuum?

Had we been lucky we would have found that eqs. (3.7) already determine a unique solution which looks like the real world. Unfortunately, this is not (and by far) the case at least order by order in the expansion (3.6). In perturbation theory there are too many degenerate, inequivalent vacua corresponding to different numbers of uncompactified dimensions, different gauge symmetries, different numbers of generations etc. Hopefully, non-perturbative phenomena of the kind known from supersymmetric gauge theories will resolve this huge degeneracy.

In any event, String Theory makes a new, interesting connection between arbitrariness in the fundamental constants and the existence of massless scalars, whose exchange could affect the basic tests of General Relativity. Thus, far from being of Academic value only, the question of Fundamental Constants in String Theory becomes, probably, one of survival of the theory itself!

REFERENCES TO LECTURE I

1. G. Veneziano, Europhysics Lett. **2** (1986) 133.

2. S. Coleman and E. Weinberg, Phys. Rev. D7 (1973) 1888.

3. G. Veneziano, in "The Challenging Questions, Erice 1989 (A. Zichichi editor, Plenum Press, N.Y. and London) p. 171.

4. A. M. Polyakov, Phys. Lett. **103B** (1981) 207, 211; see also M. Ademollo et al., Nuovo Cim. **21A** (1974) 77; E.S. Fradkin and A.A. Tseytlin, Phys. Lett. **158B** (1985) 316.

5. V. de Alfaro, S. Fubini and G. Furlan, Nuovo Cim. **A50** (1979) 523; ibid. **B57** (1980) 227.

6. K. Kikkawa and M. Yamasaki, Phys. Lett. **149B** (1984) 357; N. Sakai and I. Senda, Prog. Theor. Phys. **75** (1986) 692.

7. T.R. Taylor and G. Veneziano, Phys. Lett. **212B** (1988) 147.

8. R. Brandemberger and C. Vafa, Nucl. Phys. **B316** (1988) 391.

9. P. A. M. Dirac, Nature **139** (1937) 323.

10. E. Witten, Phys. Lett. **149B** (1984) 351.

II. AN EXAMPLE OF CLASSICAL STRING GRAVITY: LARGE DISTANCE STRING COSMOLOGY

Outline:

1. Standard, inflationary, and string cosmology.
2. R and Scale-Factor duality in string theory.
3. SFD as subgroup of O(d,d).
4. Examples of string cosmology.
5. Observable tracks of the pre-big-bang.

1. Standard, inflationary and string cosmology

I shall start this talk by recalling some very general concepts pertaining to the standard and to the inflationary cosmological models. This will allow me to point out where string cosmology, even in the approximate sense in which I shall define it, differs drastically from its predecessors.

Let us first look at the basic ingredients and assumptions of the standard cosmological model (SCM). The ingredients are:

a. The gravitational field, identified with the space-time metric $g_{\mu\nu}$, whose dynamics is described by the Einstein-Hilbert action

$$S_g = (16\pi G)^{-1} \int g^{1/2}(2\Lambda - R)$$

where, for the sake of generality, we have included a cosmological constant Λ.

b. Matter, generically described by a matter action S_m, which provides the source for gravity, i.e. the right-hand side of Einstein's equation:

$$R_{\mu\nu} - 1/2 g_{\mu\nu} R + g_{\mu\nu}\Lambda = -8\pi G T_{\mu\nu}$$

where:

$$T_{\mu\nu} = -2g^{-1/2}\frac{\delta S_m}{\delta g^{\mu\nu}}$$

The assumptions are:
a. Homogeneity and isotropy of the metric.
b. A vanishing cosmological constant.
c. A perfect-fluid ansatz for the sources:

$$T_0^0 = \rho \ , \ T_i^j = -\delta_{ij} p$$

Combining all this we get the usual Einstein-Friedman equations:

$$H^2 \equiv (\frac{\dot{a}}{a})^2 = \frac{8\pi G}{3}\rho \ , \ \frac{\ddot{a}}{a} = -\frac{4\pi G}{3}(\rho + 3p)$$

where we have considered, for simplicity, the case of a spacially flat Universe with line element:

$$ds^2 = -dt^2 + a^2(t)dx^2$$

The above equations imply the energy conservation equation:

$$\dot{\rho} + 3H(\rho + p) = 0$$

In order to get unique solutions to the above equations (up to integration constants) we have to provide an equation of state, usually written as a fixed relation between ρ and p. Thus, for instance, $p = \frac{\rho}{3}$ gives the radiation-dominated cosmology ($a \sim t^{1/2}$), while $p = 0$ gives the matter-dominated ($a \sim t^{2/3}$) expansion of the present era. Note that, since usual matter has non-negative pressure, standard cosmologies are always decelerating ($\ddot{a} < 0$). Many of the problems of the SCM follow directly from this simple fact.

The SCM has many merits. Among others, it predicts the red-shift law, the Cosmic Microwave Background (CMB), the abundance of light elements, and so on. It also has, however, several shortcomings. Some of these are phenomenological: they go under the names of the homogeneity, the horizon, the flatness, the entropy etc. problem. A major theoretical problem is instead the prediction, which looks inescapable in the SCM, of an initial singularity.

The celebrated remedy for these problems is called inflation [1]. It comes in many brands called Old, New, Extended, Eternal, Chaotic etc. All of these have a new ingredient beyond those of the SCM, a scalar field, the so-called inflaton σ, whose potential $V(\sigma)$ provides energy and pressure during an early era of vacuum dominance. Since, from Lorentz invariance of the vacuum,

$$T_{\mu\nu} \sim g_{\mu\nu},$$

the inflaton provides a negative pressure equal in magnitude to the energy density. Such an equation of state yields an exponential (hence accelerated) expansion:

$$a \sim \exp Ht\ , H^2 = \frac{8\pi G}{3}\rho_{vac},$$

which, eventually, should give way to the standard (decelerated) radiation-dominated era of the SCM.

Inflation appears to be today the best solution to the above-mentioned problems of the SCM. However, even inflation has its own difficulties: a lot of fine tuning is needed on the parameters of $V(\sigma)$ in order to provide on one hand enough inflation and, at the same time, a "gracious" exit to the radiation dominated era in which $V(\sigma) = 0$. Another problem is that of initial conditions which, again, have to be chosen very carefully. Finally, although the initial singularity problem is certainly alleviated in inflationary scenarios, geodetic completeness seems to be still somewhat problematic [2].

Let us now define, by comparison, what we shall refer to here as "*string cosmology*". It is nothing but the string-theory analogue of the usual Einstein-Friedman (EF) equations of the SCM. The string-cosmology equations are thus obtained by varying a total action consisting of two terms:

i) the low-energy effective action describing the massless background fields of string theory (the metric G, the "torsion" B and the dilaton ϕ):

$$S_{eff} = -\frac{1}{16\pi G}\int dx\sqrt{|g|}e^{-\phi}(R + \partial_\mu\phi\partial^\mu\phi - \frac{dB^2}{12} + V)$$

(here dB is the antisymmetric tensor field strength and V is a constant which is vanishing in a critical number of space-time dimensions),

ii) a phenomenological source term that accounts for possible additional bulk string matter at the classical level.

This definition of string cosmology is probably insufficient for a truly "stringy" description of the Universe in the high curvature regime (near the Planck scale), where higher-order string corrections cannot be neglected. These are probably essential in order to solve the initial-singularity problem (see below).

We hope, nonetheless, that our provisional definition of string cosmology does represent a good starting point. In particular, new, stringy symmetries show up already in this approximation: these might survive (possibly after some modification) the higher order effects.

2. R and scale factor duality in string theory

In order to introduce the new symmetries of string cosmology it is convenient to recall first a better known string symmetry, R-duality.

This symmetry applies to the case of closed strings propagating in a space with (some) compact dimensions, e.g. on the product of ordinary Minkowski space-time M_4, and a fifth dimension taken to be a flat circle R_1. R-duality follows from the observation that a closed string moving on such a space can wind around the compact dimension any number of times. Such a winding string is described by:

$$X^5(\sigma,\tau) = q^5 + 2\alpha' p_5 \tau + L_5 \sigma + \dots$$

where L_5 is quantized in units of R in order to ensure that $X^5(2\pi)$ and $X^5(0)$ are identical points. Quantum mechanically, also the string momentum p_5 is quantized, this time in units of $\hbar/R$. The Hamiltonian of the system is given by:

$$H \sim \dot{X}\dot{X} + X'X' \sim n^2 \frac{\lambda^4}{R^2} + m^2 R^2 + \dots,$$

$$\lambda^2 \equiv 2\alpha' \hbar,$$

and contains both momentum (Kaluza-Klein) and winding excitations. If we now compare the above Hamiltonian with the one obtained by replacing R with $\tilde{R} = \lambda^2/R$, we realize that the two Hamiltonians have the same spectrum with the rôles of momentum and winding interchanged. R-duality is just the statement [3] that the closed string theories defined on these two manifolds are completely equivalent.

If one introduces a (momentarily constant) space-time metric G_{55} on the circle, it is easy to convert the R-duality transformation into a G_{55}-duality transformation. Indeed, since only the proper size $G_{55}R^2$ matters, one can fix R to be just λ and define duality as $G_{55} \rightarrow G_{55}^{-1}$ [4].

In doing this we have done absolutely nothing. However, this reformulation of R-duality lends itself better to generalizations to the case in which G_{55} is not constant e.g. to the case of a time-dependent G_{55}. We shall thus ask the question:

Is there an R-duality-like symmetry in string cosmology?

Amazingly, the answer to this question is yes! The equations of String cosmology, as defined in the previous section, quite irrespectively of the type of strings one considers and quite irrespectively of compactification, do possess duality-like symmetries. Unlike in the case of strict R-duality, however, these are not symmetries connecting equivalent theories but rather transformations that generate, from a given solution to the string cosmology equations, new, generally inequivalent solutions.

The simplest example of this kind is what I have termed [5] scale-factor duality (SFD) * How does SFD act? Consider, as before, a homogeneous, isotropic, spatially flat Universe, described by the FRW line element in $d+1$ space-time dimensions:

$$ds^2 = -dt^2 + a^2(t)dx^2$$

* A formally identical symmetry was considered independently by Tseytlin [6] and developed by Tseytlin and Vafa [7]. Physically, however, their transformation is very different from ours in that it only works for closed strings in compactified target space and does not actually relate expanding to contracting cosmologies.

Let us also set $B = 0$ for the time being and simply add a time-dependent dilaton $\phi(t)$. Under SFD:

$$a(t) \rightarrow \tilde{a}(t) = a^{-1},$$

$$\phi \rightarrow \tilde{\phi} \equiv \phi - d \ln a.$$

Two observations are relevant here:

1. Recalling that the effective Newton constant in string theory is given by

$$G \sim e^{\phi},$$

it is clear that, under SFD, a constant G is transformed into a time-dependent G. This is why ordinary (Einstein-Friedman) cosmology does not possess SFD invariance.

2. The effective action of string theory resembles that of Brans-Dicke (BD) [8], with a special value of the BD parameter, $\omega_{BD} = -1$. This special value appears to be picked up by SFD and it is dangerously small (if the dilaton remains massless)!

This is the way the backgrounds transform under SFD. What about the sources? It is quite simple to answer that question if a convenient notation is used.

Let $Z_k = (X^0, X', P \equiv \dot{X}a^2)$ denote a solution of the equations of motion for the kth string in the cosmological background with scale factor a. The solution in the SFD related background $\tilde{a}$ is simply given by $\tilde{Z}_k = (X^0, \tilde{X}' = P, \tilde{P} \equiv \dot{\tilde{X}}\tilde{a}^2 = X')$. Consider now a perfect fluid made out of string sources and the one obtained from the dual sources. It is not hard to find the relation between the energy and the pressure of the two fluids. One finds:

$$\overline{\rho} \equiv \rho\sqrt{|g|} = \tilde{\overline{\rho}} = \tilde{\rho}\sqrt{|\tilde{g}|}$$

$$\overline{p} \equiv p\sqrt{|g|} = -\tilde{\overline{p}} = -\tilde{p}\sqrt{|\tilde{g}|}$$

and, indeed, $\overline{\rho}$ ($\overline{p}$) acts as a source for SFD-even (odd) string cosmology equations.

The peculiar behaviour of pressure under SFD can be easily understood from the study of string propagation in fastly-expanding geometries, the SFD transform of a rapidly contracting Universe (say the radiation-dominated era played backward). In the latter case one obtains the usual relativistic equation of state $p = \rho/3$ (in three spatial dimensions) while in the former strings stretch and become long and slow. Since pressure gets opposite contributions from the $\dot{X}$ and X' terms in the action, the equation of state now becomes $p = -\rho/3$.

I refer to the original literature [9] for more details on the study of these interesting regimes. Incidentally, an important insight that one gains from those considerations concerns the identification of the physical metric in string theory, i.e. of what gives distances as measured by stringy rods. One finds very simply that the rôle of the physical metric is played by the original σ-model metric $G_{\mu\nu}$ and not by a Weyl-rescaled version of it which contains the dilaton. This is a crucial point when discussing, for instance, the possibility of inflation in string theory.

3. SFD as a subgroup of $O(d,d)$

I shall now discuss some extensions of SFD. The first one is almost trivial [5]. Starting from a homogeneous, spacially flat, anisotropic cosmology:

$$ds^2 = -dt^2 + a_i^2(t)dx_i^2 \ ,$$

we can generalize SFD to the Z_2^d group generated by

$$a_i \to \tilde{a}_i = a_i^{-1} \quad , \quad \phi \to \tilde{\phi} = \phi - 2\ln a_i \ , \ i = 1, 2, \ldots d.$$

A similar transformation applies, of course, to the sources.

The second extension of SFD [10,11] is much less trivial and necessitates the most general spatially flat, homogeneous metric and torsion. In this case SFD can actually be extended to a full continuous, non-compact group, $O(d,d;R)$ which acts in a complicated non-linear way on G and B but linearly on the $2d$-vector Z and on the $2d \times 2d$ matrix:

$$M = \begin{pmatrix} G^{-1} & -G^{-1}B \\ BG^{-1} & G - BG^{-1}B \end{pmatrix}$$

One finds :

$$M \to \Omega^T M \Omega,$$

$$Z \to \Omega^{-1} Z, \; \Omega \in O(d, d; R)$$

Actually, it is possible to formulate all the equations of string cosmology in a way which is manifestly $O(d, d)$ covariant [12]. The sources are represented by an energy density ρ and by a generalized stress tensor T which itself becomes an antisymmetric $2d \times 2d$ matrix.

Solutions to the string cosmology equations, thus formulated, can be found but, in analogy with ordinary cosmology, their uniqueness requires an equation of state i.e. a relation between T and ρ. Since, after all, we do start with a specific model for the sources, it should be possible, in principle, to derive such an equation of state, which will automatically be $O(d, d)$ covariant.

Unfortunately, it is not easy to find simple, yet consistent equations of state. The simplest ones, e.g. the perfect fluid ansatz, can be easily shown to be incompatible with $O(d, d)$. Work is in progress to circumvent this difficulty. For the time being, I shall limit myself to a few illustrative examples of solutions in which the equation of state is given from the start.

4. Examples of string cosmologies

a) Vacuum cosmologies

Let us start from the case of "vacuum" solutions, i.e. with solutions holding in the absence of sources ($\rho = p = 0$). The most general solution of this kind can be given in terms of quadratures [10] and is known even if one adds an arbitrary $O(d, d)$-preserving dilaton potential $V = V(\Phi)$ where $\Phi \equiv \phi - 1/2 \ln \det G$. It is given implicitly by:

$$M(t) = exp(-A\eta\tau)M(0)$$

$$t = \int^{\Phi} dx(\frac{1}{8}e^{2x}Tr(A\eta)^2 + V(x)), \; d\tau = e^{\Phi}dt$$

where the matrix A stands for a matrix of "Constants of the Motion" which follows from $O(d,d)$ symmetry and is given by:

$$e^{-\Phi}M\eta\dot{M} = A, \; A^T = -A, \; M\eta A = -A\eta M$$

where η is the $O(d,d)$ metric in off-diagonal form [10]. As particular examples of this general solution, by setting $V = const.$ and by taking a special form for A, we recover the "Kasner-like"[13] solutions found in refs. [14,5]:

$$B = 0, \; G_{ij} = a_i^2(t)\delta_{ij},$$

$$a_i(t) = a_i(\pm\infty)(\tanh(\pm\xi/2))^{\alpha_i}, \; \sum \alpha_i^2 = 1$$

$$\Phi = -\ln\sinh(\pm\xi); \; \xi \equiv \sqrt{\frac{D - D_c}{3\alpha'}}t$$

We note, en passant, that, for $D = 2$ and after the replacement $t \rightarrow ix$, the above solutions go over to the $2D$ black hole solutions recently considered by many authors [15]. The analogue of SFD can be defined for black holes [16] and $O(d,d)$-like transformations can also be used to generate new solutions [11,17].

b) Radiation-like solution and its pals

We now turn to particularly simple solutions which occur in the presence of string sources. A nice result [7,12] in this context is the existence of "self-consistent" radiation-like cosmologies. We have seen that, in one particular regime, the equation of state of stringy matter approaches that of radiation:

$$p = \frac{\rho}{d}$$

Inserting this relation into the equations of string cosmology, one finds that a particular solution is indeed given by the conventional radiation-dominated cosmology, which, luckily, leads to the equation of state one has started from. In formulae:

$$B = 0, \; \phi = const, \; a \sim t^{\frac{2}{d+1}}$$

and, of course,

$$Ta = const, \ \rho \sim t^{-2}$$

The Z_2^d symmetry of string cosmology allows one to construct immediately many more (generally anisotropic) solutions of the form:

$$B = 0, \ a_i \sim t^{\frac{2}{d+1}} (i = 1, \dots n), \ a_j \sim t^{-\frac{2}{d+1}} (j = 1, \dots m)$$

$$\phi = const. - \frac{4m}{d+1} \ln t, \ T_i^j = (p, \dots p, q, \dots q), \ p = -q = \frac{\rho}{n+m}$$

Of course, in order for these to be actual solutions, one has to transform the sources as well, or, if we prefer, their equation of state. Z_2^d symmetry guarantees that the new equations of state are again self-consistent.

Analogously, one can reach many more solutions with torsion (i.e. a non-trivial dB) by acting on the above solutions with the full $O(d,d)$ group. This time the self-consistent fluid can no longer satisfy the equation of state of a perfect fluid and exhibits bulk and shear viscosity. Needless to say, all these solutions would have been extremely hard to find without the use of the symmetries of string cosmology.

One could ask now the following question: given the fact that a radiation-dominated era has existed and that, in string theory, there are many other cosmologies simply related to it by a symmetry, is it possible that any of those new cosmologies might have played a rôle at some stage in the evolution of the Universe? The most daring of these possibilities is that the Universe has evolved in a self-dual way, in a sense that we shall now explain.

c) SFD-symmetric cosmologies.

Again, analogy with R-duality can be helpful here. In that case, it is generally expected that, if a duality-invariant potential $V(R)$ is generated by some non-perturbative effects, this will tend to freeze the compactification radii at their self-dual value $R = \lambda$.

Similarly, can we find cosmologies which, being fixed points of SFD, might be dynamically preferred once the appropriate non-perturbative dynamics is taken into account? In answering this question we face an immediate problem: SFD has only trivial fixed points, since:

$$a(t) = a^{-1}(t) \Rightarrow a(t) = 1$$

Fortunately, we can combine SFD and time reversal (which is a symmetry of both ordinary and string cosmology):

$$SFD \times T \;:\; a(t) \to a^{-1}(-t)$$

The above transformation, unlike straight SFD, has non-trivial, physically interesting fixed points. Indeed $a(t) = a^{-1}(-t)$ implies:

$$H(-t) = H(t) \Rightarrow \; expansion \; at \;\; t < 0$$

$$\dot{H}(-t) = -\dot{H}(t) \Rightarrow \; growing \; H \; at \;\; t < 0$$

The latter result automatically implies pre-big-bang inflation. It is not hard to write down self-dual cosmologies which interpolate smoothly between an inflationary pre-big-bang phase and a decelerated post-big-bang cosmology. For instance:

$$a(t) = \sqrt{\frac{t + \sqrt{\lambda^2 + t^2}}{\lambda}}$$

is one such example. The problem is to show that such kinds of scale factors naturally emerge from solving some form of improved string cosmology equations, e.g. with higher orders in α' and/or e^{ϕ} included.

Note that the above scale factor also avoids the singularity at $t = 0$. Should something like this happen in string cosmology? Arguments can be given for an affirmative answer in the sense that curvatures in excess of λ^{-2} are either unattainable or not seen by test strings.

Such arguments are based on the study of Planckian-energy (gedanken) string collisions at small impact parameters [18,19] and of string scattering in highly inhomogeneous external gravitational fields [20]. Using the folk theorem that high energy and early Universe physics are closely related, one infers that string theory must have its way around the initial singularity problem. The question is, which way?

One possibility is that, near $t = 0$, the torsion field B plays a major rôle. There are examples [21] of singular cosmologies which become non-singular after an $O(d, d)$ transformation: necessarily, this brings in a non-trivial torsion. Alternatively one can play with a dilaton potential [12].

My own belief is that the mechanism through which curvature singularities should be avoided in string theory is rooted in its short-distance behaviour, and should therefore be inaccessible to the low-energy approximation in which we have been working so far.

5. Observable tracks of the pre-big-bang

As a final topic, I should like to mention that, far from being just a speculation, a pre-big-bang era of the type suggested by string theory makes physical predictions that could be tested, in principle, by near-future gravitational-wave detectors.

The most direct predictions concern a background of relic gravitons that should exist today throughout space very much like the well known electromagnetic microwave background. Since the gravitons should decouple from everything else just after the Planck era, their spectrum should have remained unchanged since then, up to the standard red-shift.

Even in ordinary (de Sitter) inflation one predicts [22] a spectrum of relic gravitons (or of gravitational waves if we prefer). In the pre-big-bang scenario the shape of the spectrum is quite different though (it is not of the flat Harrison-Zeldovich type). An interesting consequence of this fact is that the bounds [23] on the maximal value of the Hubble parameter which hold for ordinary inflation ($H_{max} < 10^{-5} M_P$) can be avoided in our scenario, so that Planckian (or string) scale curvatures are attainable near $t = 0$ making the peculiarities of string cosmology felt.

We refer to the original literature [24,12] for details and just mention that a more indirect and less unambiguous -but also easier-way to test the pre-big-bang idea would be through the distorsion induced by gravitational radiation on the usual (electromagnetic) CMB [25]. This is precisely what COBE is measuring [26]. The problem would be, of course, to disentangle our contribution to the distorsion from many other possible effects: it seems that, at best, only some upper limit on the amount of distorsion provided by relic gravitons will be attainable.

REFERENCES TO LECTURE II

1. For a review on Inflation see, e.g., M.S. Turner in "Intersection between elementary particle physics and cosmology" (T. Piran and S. Weinberg eds. WSPC, Singapore 1986) Vol.1, p.100.
2. A.Vilenkin, "Did the universe have a beginning?" Caltech prepr. CALT-68-1772 (1992).
3. K. Kikkawa and M. Yamasaki, Phys. Lett. 149B (1984) 357; N. Sakai and I. Senda, Prog. Theor. Phys. 75 (1986) 692.
4. A. Giveon, E. Rabinovici and G. Veneziano, Nucl. Phys. B322 (1989) 167; A. Shapere and F. Wilczek, Nucl. Phys. B320 (1989) 669.
5. G.Veneziano, Phys.Lett.B265(1991)287.
6. A.A.Tseytlin, Mod.Phys.Lett.A6(1991)1721; A.A.Tseytlin, "String cosmology and dilaton", these proceedings.
7. A.A.Tseytlin and C.Vafa, Nucl.Phys.B372(1992)443; A.A.Tseytlin, Class. Quantum Grav.9(1992)979.
8. C. Brans and R. Dicke, Phys. Rev. 124 (1961) 965.
9. H.J.De Vega and N.Sanchez, Phys.Lett.B197(1987)320; N.Sanchez and G.Veneziano, Nucl.Phys.B333(1990)253; M.Gasperini, N.Sanchez and G.Veneziano, Int.J.Theor.Phys.A6 (1991)3853; Nucl.Phys.B364(1991)365; M.Gasperini, Phys.Lett.B258(1991)70; M.Gasperini, Gen.Rel.Grav.24(1992)219;

Nguyen S.Han and G.Veneziano, Mod.Phys.Lett.6(1991)1993; G.Veneziano, Helvetica Physica Acta 64 (1991) 877.

10. K.A.Meissner and G.Veneziano, Phys.Lett.B267(1991)33; Mod.Phys.Lett.A6(1991)3397.
11. A.Sen, Phys.Lett.B271(1991)295; S.F.Hassan and A.Sen, Nucl.Phys.B375(1992)103.
12. M.Gasperini and G.Veneziano, Phys.Lett.B277(1992)256; "Pre-Big-Bang in String Cosmology", preprint CERN-TH. 6572/92 (1992).
13. E. Kasner, Amer. J. Math. 43 (1921) 217.
14. M. Mueller, Nucl. Phys. B337 (1990) 37.
15. K. Bardakci, A. Forge and E. Rabinovici, Nucl. Phys. B344 (1990) 344; S. Elitzur, A. Forge and E. Rabinovici, Nucl. Phys. B359 (1991) 581; I. Bars and D. Nemeschansky, Nucl. Phys. B348 (1991) 89; G. Mandal, A.M. Sengupta and S.R. Wadia, Mod. Phys. Lett. A6 (1991) 1685; E. Witten, Phys. Rev. D44 (1991) 314.
16. A.Giveon, Mod.Phys.Lett.A6(1991)2834.
17. A. Giveon and M. Rocek, Nucl. Phys. B380 (1992) 128; J. Horne, G. Horowitz and A. Steif, Phys. Rev. Lett. 68 (1992) 568; J. Panvel, Phys. Lett. B284 (1992) 50; J. Maharana and J. H. Schwarz, "Non-compact symmetries in string theory", Caltech preprint CALT-68-1790 (1992); A. Kumar, "Gauged string actions and $O(d,d)$ transformations", preprint CERN/TH.6530/92 (1992); S. Kar, S. Pratik Khastgir and S. Sengupta, "Four dimensional stringy black membrane", Bhubaneswar preprint IP/BBSR/92-35; S. Mohapatra, "Four dimensional 2-brane solution in chirally gauged Wess-Zumino-Witten models", Tata preprint, TIFR/TH/92-28; S. Kar, A. Kumar and G. Sengupta, "Hidden isometry and a chiral gauged WZW model", Bhubaneswar preprint IP/BBSR/92-67.

18. D. Amati, M. Ciafaloni and G. Veneziano, Phys. Lett.B216(1989) 41;
G. Veneziano, Proc. Superstring '89 Workshop, Texas A. and M. University, R. Arnowitt et al. eds. (WSPC, Singapore, 1990) p. 86.
19. D.J. Gross and P.F. Mende, Nucl. Phys. B303(1988) 407;
P.F. Mende and H. Ooguri, Nucl. Phys. B339(1990)641.
20. M. Fabbrichesi and G. Veneziano, Phys. Lett. B233(1989)135.
21. M.Gasperini, J.Maharana and G.Veneziano, Phys.Lett. B272(1991) 277; Phys. Lett.B296(1992)51;
A. Giveon and A. Pasquinucci, " On cosmological string backgrounds with toroidal isometries", Princeton preprint, IASSNS-HEP-92/55.
22. L.P.Grishchuk, Sov.Phys.Usp.31(1988)940;
L.P.Grishchuk and M.Solokhin, Phys.Rev.D43(1991)2566;
V.Sahni, Phys.Rev.D42(1990)453.
23. V.A.Rubakov, M.V.Sazhin and A.V.Veryaskin, Phys.Lett. 115B(1982)189;
R.Fabbri and M.D.Pollock, Phys.Lett. 125B(1983)445;
L.F.Abbot and M.B.Wise. Nucl.Phys.B244(1984)54.
24. M.Gasperini and M.Giovannini, Phys.Lett.B282(1992)36; "Dilaton contributions to the cosmic gravitational wave background", Turin Univ. Preprint (July 1992).
25. L.Krauss and M.White, "Grand unification, gravitational waves and the CMB anisotropy",Yale preprint YCTP-P15-92;
T. Souradeep and V.Sahni, "Density perturbations, gravity waves and the cosmic microwave background", preprint IUCAA (Pune, India), July 1992;
F.Lucchin, S.Matarrese and S.Mollerach, "The gravitational wave contribution to CMB anisotropies and the amplitude of mass fluctuations from COBE results", Fermilab-Pub-92/185-A (July 1992).
26. G.Smoot et al., COBE collaboration preprint 92-04 (to appear in Ap.J.Lett.).

III. AN EXAMPLE OF QUANTUM STRING GRAVITY: PLANCKIAN-ENERGY STRING COLLISIONS

Outline:

1. Introduction, motivations.
2. Scales in Planckian-energy Collisions.
3. Results via the ACV method.
4. Comparison with the GMO approach.
5. Other considerations, conclusions.

1. Introduction, motivations

In the previous lecture we have seen an example of Classical String Gravity. In order to proceed to the Quantum case, it turns out to be useful to resort to some Gedanken Experiment (GE). Why?

This century has seen some of the greatest revolutions of scientific thought, in particular the birth and development of Relativity and Quantum Mechanics. In both instances a new constant of Nature was discovered first, almost accidentally, and then led, through discussion of some (GE), to a new physical principle:

c $(< \infty)$ led to the impossibility of defining an absolute time and, eventually, to the Equivalence Principle of General Relativity;

$\hbar(> 0)$ led to the impossibility of exact particle localization in phase space and, eventually, to the uncertainty principle and to commutation relations among observables.

Ideally, we would like to repeat this successful game for String Theory. We have seen in Lecture I that, at the quantum level, String Theory possesses another fundamental constant, $\lambda_s \equiv \sqrt{\alpha' \hbar}$. Encouraged by history, we would like to consider some stringy GE which could reveal the underlying principle upon which the magic of strings lies. This is very ambitious, of course. We shall be happy to be able to scratch just the surface of the problem later in the talk.

It turns out that Planckian-energy string collisions are quite ideal for the purpose. There are several reasons for that:

a) We wish to test Quantum String Gravity and energy is the source of gravity. Thus, high energy means a space-time metric which is sensibly different from flat.

b) Substantial quantum-string modifications are expected at short distances and large curvatures. For this we need high energies. We shall see later under which conditions this is also sufficient.

c) Last but not least the S-Matrix is unambiguously defined in String Theory (off shell concepts are much more subtle).

Before proceeding to the discussion of the GE we recall from Lecture I that a fundamental dimensionless constant of string theory is the string-loop-expansion parameter

$$\alpha_{SL} = \exp(\langle\phi\rangle)$$

In our case case, although we work at the critical number of dimensions in order to be in the flat Minkowski vacuum, we shall compactify a number of dimensions to very tiny circles, leaving only d of them macroscopic (note that, unlike in Lecture II, d now includes time).

It is known that in this case the effective coupling coupling constant differs from α_{SL} by a factor related to the compactification volume. We shall denote this new coupling by α_d and play with the fact that, in the supersymmetric case:

i) α_d is undetermined in perturbation theory owing to a flat direction in the dilaton's potential;

ii) α_d gives the ratio between the fundamental length of string theory λ_s and the Planck scale λ_P via:

$$\lambda_P^{d-2} = (\hbar G) = \alpha_d \lambda_s^{d-2}, \quad G = \text{Newton Constant}$$

To summarize, QSG has a fundamental scale, the ultra-violet cut-off λ_s, and a dimensionless coupling α_d. The latter allows us to take the ratio of the string and Planck lengths at will. We shall make crucial use of this freedom in the rest of this talk.

2. Scales in Planckian-energy Collisions

A typical process to be investigated is shown in Fig. 1. Two massless particles (e.g., two gravitons) come in with centre-of-mass energy $(s)^{1/2}$ of order M_P and with total angular momentum J. For large J we can define a semi-classical impact parameter $b = 2J(s)^{-1/2}$, corresponding to the transverse distance between the colliding gravitons. One would like to know what happens after the collision as a function of s, b and of the fundamental constants of the theory, α_d and λ_s.

As it turns out, this is not, at very high energy, an easy problem. The reason is that, even for small values of α_d, perturbation theory is non-trivial for sufficiently high energies. This is actually a good thing since the tree-level results for, say, the elastic amplitude are at variance with expectations. Indeed:

i) At fixed b (or t) the amplitude becomes too large at high energy, e.g., in $d = 4$:

$$A(B, s) \simeq G\frac{s}{\hbar} log\ b$$

which eventually exceeds the unitarity bound [1]

ii) at fixed scattering angle θ the amplitude is exponentially damped:

$$A(\theta, s) \simeq exp(\frac{-\alpha' s}{\hbar} f(\theta))$$

which is too small to be consistent with the Classical General Relativity (CGR) expectation of a deflection angle $\theta = O(GE/b)$.

Obviously, if string theory is to reproduce CGR in some limit, loops should manage to shift some of the cross section from the fixed t region on to the fixed θ region.

Two (somewhat complementary) approaches have been proposed to this aim:

a) The Gross-Mende-Ooguri approach [2,3] consists in computing, genus by genus (i.e., at each loop order), the fixed angle limit and to attempt then a resummation which, hopefully, gives a much bigger sum.

b) The Amati-Ciafaloni-Veneziano (ACV) approach computes fixed-b amplitudes and then resums them to find a much smaller (and hopefully unitary) result.

The kinematical regions from which the two methods start are shown in Fig.2. With some luck, one may hope to find intermediate regions (like zero angle but large t), where both methods could retain some validity and should thus be compared. This we shall do at the end.

3. Results via the ACV method

The results of our group [5-7] can de described best with the help of Fig.3. Indeed, the outcome of the collision turns out to depend crucially on the number d of non-compactified (i.e., macroscopic) dimensions and on the relative ratio of three lengths: the fundamental string scale λ_s, the impact parameter b and the classical radius $R = R(E)$ associated, in CGR, with the centre-of-mass collision energy $E = \sqrt{s}$:

$$\mathrm{R}(E) \simeq (GE)^{\frac{1}{d-3}}$$

It is thus convenient to consider, as in Fig. 3, the $R - b$ plane and to mark on it the lines $R = \lambda_s, b = \lambda_s$ and $R = b$. We could have also marked on the figure the Planck length, given by 1, and assumed to be always smaller (possibly much smaller) than λ_s.

Most of this talk will consist of describing (<u>not</u> deriving) what we have understood about the physics of the collision in these various kinematical regions. In a nutshell, the main parameter that determines the kind of physics exposed in the process is the ratio R/λ_s:

i) if $R > \lambda_s$, CGR effects tend to screen short-distance physics.

ii) if $R < \lambda_s$, string-size (minimal length) effects tend to modify sustantially, and for the better, CGR at short distances.

Paradoxically, it is the latter regime which turns out to be the easier one to analyze. We shall discuss, in succession, case i) and case ii), beginning in each case from large impact parameters (top of Fig. 3) and then coming down as much as our present ability allows.

A) $R > \lambda_s$. At $b \gg R$ the leading contributions can be shown to consist (at loop order, or genus, g) of the exchange of $(g+1)$ Reggeized gravitons between the two fast-moving external particles. These contributions can be neatly resummed [4] in an operator-eikonal formula reading:

$$a(s,b) \equiv \int dq\; e^{-iq\cdot b}\frac{1}{4s}A(s,t\approx -q^2) =$$

$$= \langle 0|\frac{exp(2i\hat{\delta})-1}{2i}|0\rangle$$

$$\hat{\delta} \equiv \frac{1}{4\pi^2}\int_0^{2\pi} d\sigma_u d\sigma_d :\; a_{tree}(s, X_u(\sigma_u) - X_d(\sigma_d))$$

where $X_{u,d}(\sigma)$ are mutually commuting string position operators (at $\tau = 0$) for the upper and lower strings (moving fast in opposite direction). When b is extremely large one can further replace these operators by their centre-of-mass parts and obtain in this way a more conventional (field-theoretic) eikonal formula (also obtained, independently, in Refs.[8,9].

It is easy to go from the impact parameter back to momentum transfer and to find that, at a given scattering angle $\theta \ll 1$, the integral over b is dominated by a real saddle point given by:

$$b_s^{d-3} = \frac{8\pi GE\Gamma(d/2-1)}{\pi^{d/2-1}\cdot\theta} \tag{1}$$

The above relation between b_{saddle} and θ is precisely the one expected in CGR, once Einstein's famous 1916 formula is generalized to the case of two relativistic particles in any number d of space-time dimensions. It also coincides with the deflection suffered [10] by the geodesic of one graviton (seen as the test-body) in the Aichelburg-Sexl (AS) metric [11] produced by the other graviton (seen as the source).

Furthermore, the existence of a real saddle in b implies an amplitude which is no longer exponentially suppressed, at fixed, small θ. The real saddle is absent at any finite order (fixed genus as

$E \to \infty$) but is recovered after the infinite resummation contained in the eikonal. In other words, CGR is recovered as an infinite genus effect!

As we move down to smaller impact parameters (but still with $b >> R$) we encounter another phenomenon [4] (dashed-dotted line in Fig.3) due to the fact that $X_{u,d}$ can no longer be replaced by their centre-of-mass operators. This leads, physically, to the opening of inelastic, "diffractive" channels. The S-matrix is still unitary in this "diffractive-two-body" Hilbert space. Corrections of this sort can be termed quantum-stringy and typically involve the small parameter λ_s/b. They are already relevant at largish impact parameters because they effectively add an imaginary part (absorption) to the purely real leading eikonal.

Quantum string corrections compete, in this regime, with what we may call CGR corrections: the latter are controlled by the expansion parameter:

$$(R/b)^{2(d-3)} \simeq \frac{G^2 s}{b^{2(d-3)}}$$

They originate diagramatically from loops (corners of moduli space) in which the exchanged gravitons reinteract (at tree level) before connecting the fast external lines. The leading correction comes at two-loop level through the so-called H-diagram and was studied in detail in Ref. [6] (see also [12])

The analysis is made technically hard by the appearance of infrared singularities, which can be fortunately handled [6] by Bloch-Nordsieck-type techniques. The final outcome shows a breakdown of the external metric approximation already at this level: the behaviour of one graviton not only feels now the energy of the other graviton: the quantum numbers of the system as a whole are important.

The above-mentioned technical difficulties make it impossible to penetrate, at present, the region $R \leq b$, where the expansion parameter becomes $O(1)$. This is unfortunate since, precisely at $b \leq R$,

$$Gs \simeq R(s)^{1/2} > b(s)^{1/2} = J$$

The above condition coincides with the condition that the whole system be in a collapsed state, a kind of (Kerr) black hole. Following classical work by Hawking [13], we would expect to observe a very soft spectrum coming out of the collision process, that of Hawking's thermal radiation with a temperature $O(\hbar/R)$. The higher E, the higher is R and the lower should be the temperature (energy) of the spectrum. This is not what we see at lowest order in the classical corrections but should come out (another miracle!) as the result of a resummation.

An easier way to tackle this kind of problems would consist of studying the scattering of a massless string on a very massive one which is already, in itself, a candidate black hole. Processes of this kind have been recently analyzed [14] at lowest orders and, again, no sign of black holes is seen at that level. Certainly, multiloop technologies should be developed in order to extract the relevant asymptotic limits and to be able to attempt a resummation.

In conclusion, at $R > \lambda_s$, it is very difficult to enter at present the region $b < R$ (and a fortiori $b < \lambda_s$, i.e. the right bottom corner of Fig. 3). If and when we do so, we should find lots of interesting strong gravity and black hole physics lying there. However, genuine string physics will probably remain screened behind the horizons provided by CGR.

B) $R < \lambda_s$. We turn now to the left part of our Fig. 3, the case of large, but not extreme energies or of very small Newton constant ("large" strings in Planck-length units). The basic reason why this regime can be analyzed more easily than the previous one is the suppression of classical corrections at $R < \lambda_s$.

Indeed, the above estimates for the magnitude of the corrections is only valid, in string theory, at $b \gg \lambda_s$. When $b < \lambda_s$, string-size effects modify substantially the correction so as to make it no longer singular in the limit $b \to 0$. A rough account of this stringy effect is given by the replacement:

$$\frac{R^2}{b^2} \to \frac{R^2}{(b^2 + \lambda_s^2)}$$

Thus, for $R \ll \lambda_s$, the string-size-modified classical corrections always remain very small, presumably preventing their resummation to yield dramatic effects. Let us describe what happens in this case as we move down in b.

At $b \gg \lambda_s (> R)$ we find quasi-elastic scattering with deflection angles still given by (1). There are small string-size "diffractive" corrections of the type described earlier as well as even smaller classical corrections $O(R/b)$. As we move down to $b \simeq \lambda_s$, the two strings "graze" each other and the deflection angle attains a maximal value [5]:

$$\theta = \theta_M \simeq (R/\lambda_s)^{d-3} \tag{2}$$

This is illustrated in Fig.3 by the turnover of the lines of constant θ around $b = \lambda_s$. As we move down to $b = R$ or even towards $b = 0$ (bottom left of Fig. 3) nothing spectacular seems to occur (the singular line $b = R$ does not continue below λ_s). The deflection angle starts decreasing proportionally to b, while the inelasticity of the process keeps increasing, suppressing more and more the elastic channel.

The linear dependence of θ on b is characteristic [10] of geodesics in a Generalized Aichelburg-Sexl (GAS) metric, the one of an homogeneous massless "beam" of spacial extent $O(\lambda_s)$, in agreement with the naive picture that a stringy graviton has a smeared energy distribution as opposed to a point-like graviton.

We can say that, in the case at hand, string-size effects have managed to tame the otherwise divergent CGR phenomena. Black holes of radius $R < \lambda_s$ (hence with curvatures at the horizon and temperatures larger than λ_s^{-2} and λ_s^{-1}, respectively) do not appear to be formed in our gedanken collisions. Most probably they do not exist in string theory!

A related observation concerns the scale $\Delta x \simeq b$ which is tested in a GE at a given energy and scattering angle. Let us enumerate all possible cases:

i) For $R > \lambda_s$, we have seen that a scale $O(R/\theta)$ is probed. This shows, incidentally, that, at over-Planckian energies, increasing the energy (at fixed θ) does not improve the spacial resolution. In any case, the scale tested is larger than the string scale.

ii) For $R < \lambda_s$ we have to distinguish two subcases: subitema) $\theta < \theta_M$. As quite evident from Fig. 3, there are two real saddles in this case, one at $b > \lambda_s$ and one at $b < \lambda_s$. As it turns out, the former dominates over the latter (it has smaller absorption), thus also in this case distances larger than λ_s are effectively probed.

subitemb) $\theta > \theta_M$. The two real saddles disappear in favour of a complex saddle (in agreement with the fixed genus case, see below). Even then, $|b| > \lambda_s$.

We conclude that in no case have we been able to explore scales smaller than λ_s. We can actually give to this result a more elegant form, that of an enlarged Uncertainty Principle (UP) obeyed by strings and reading [15] :

$$\Delta x > \frac{\hbar}{\Delta p} + \alpha' \Delta p$$

The first term on the r.h.s. is the usual one while the second represents the fact that a momentum spread Δp, if carried by stringy probes, will necessarily correspond to a spacial extent of the probe $O(\alpha' \Delta p)$.

The way the generalized UP reflects itself in our GE is rather subtle. At genus g the overall momentum transfer is shared among $(1+g)$ strings each carrying a $\Delta p_{ind.} \simeq \frac{\Delta p}{(1+g)}$. As a result, the usual UP term gets enhanced by a factor $(1+g)$ (the "hardness" of the process is really $\Delta p_{ind.}$.), while the string correction is divided by the same factor (each exchanged string is only $\alpha' \Delta p_{ind.}$ long). Amusingly, this genus effect introduces extra $\hbar$-dependence in the final result converting the usual UP term into the *classical* impact parameter and the classical-string-size term into a *quantum* effect. In any event, the $(x + 1/x)$ form of Δx provides immediately the already described lower bound.

3. Comparison with the GMO approach

In a couple of interesting papers Gross and Mende [2] managed to generalize to any g the fixed angle result obtaining:

$$A^{(g)} \sim exp\left(-\frac{\alpha' s}{\hbar} \cdot \frac{f(\theta)}{(1+g)}\right)$$

A salient feature of this formula is that the exponential damping, while present at any fixed g, gets smaller and smaller with increasing g, suggesting that the actual behaviour of the sum depends on the high-genus limit. Gross and Periwal have argued [16] that, unfortunately, the sum over g not only diverges but is even not Borel summable, at least for vacuum graphs in the bosonic string.

Remarkably, however, Mende and Ooguri [3] found that this is not the case for the scattering amplitude, the object we are interested in. These authors were able to perform the Borel summation and keep their (typically saddle) approximations under control only in a limited kinematical regime, i.e.,

$$1 \ll \frac{\alpha' s}{\hbar} \ll log^3(\alpha_d^{-2})(\ll \alpha_d^{-2}) \tag{3}$$

Because of the last inequality this regime obviously corresponds to the case $R < \lambda_s$, fortunately the case we know more about. Furthermore, GMO always work at fixed angle, which, owing to (3), is much larger than the maximal angle (2). In conclusion, the MO regime overlaps slightly (see shaded region in Fig. 2) with one of the regimes encountered in the ACV analysis, interestingly enough with the one in which there was no real saddle!

The quantitative comparison is quite amazing: both the elastic amplitude and the scale which is explored agree beyond expectation. In particular, the amplitude falls off like $exp(-(tY)^{1/2}f(\alpha_d))$ where the only discrepancy lies in the meaning of Y (log $\alpha' s$ in ACV, $log(s/t)$ in MO). This discrepancy (tiny because of (3)) can be traced back to the tree level, where it appears in the transition between the fixed t and the fixed angle regimes. Otherwise the agreement looks to me spectacular, given the fact that the two approaches start very far away from each other and use very different techniques and approximations. The reason for the agreement could be due, a posteriori, to the fact that, for $1 \gg \theta \gg \theta_M, \Delta p_{ind.}$ is neither too large nor too small, so that both approaches retain some validity.

If the smooth joining of the ACV and MO results are confirmed and possibly extended further out in energies, the combination of the two methods could eventually cover all regions for the case $R < \lambda_s$. In the opposite case, $R > \lambda_s$, the ACV approach encounters, as we explained, technical difficulties while the MO method appears to be doomed by its (present?) inability to reproduce CGR expectations at large impact parameters.

4. Other considerations, conclusions

Another often-quoted argument for a minimal scale is target space duality [17], a stringy symmetry of closed strings, stemming (see Lecture II) from the interchange of compactified momenta and windings (X' and P). The enlarged UP does exhibit some sort of duality (the exchange of $\hbar/\Delta p$ with $\alpha'\Delta p$) which is again a kind of exchange between P and X', but in non-compact space.

It is natural to ask at this point: since we have been unable to explore scales shorter than λ_s, does that mean that they are unphysical or irrelevant? A possibility is that short distance degrees of freedom are like gauge degrees of freedom and thus unphysical. This point of view can be given the form of an enlarged Equivalence Principle (EP) in the following way [18]:

We know (see Lecure I) that string physics is encoded in the partition function in presence of a background. If only a space-time metric is included, this is

$$Z(g_{\mu\nu}) = \int dX^{\mu} ... exp\left(-\frac{1}{4\pi\alpha'}\int d\sigma d\tau \partial X^{\mu}\bar{\partial}X^{\nu} g_{\mu\nu}(X)\right)$$

$Z(g_{\mu\nu})$ is believed to be invariant under usual general co-ordinate transformations (i.e., the usual EP). The Extended EP would claim $Z(g_{\mu\nu})$ to be also invariant under changes of $g_{\mu\nu}$ which leave unchanged averages over a string-size volume. In this sense, degrees of freedom living on scales shorter than the string scale can be gauged away (e.g., by making $g_{\mu\nu}$ smooth over such scales). The comparison of point and string motion in GAS metrics endowed with inhomogeneities below the string scale [19] lends further support to this idea.

This point of view also fits well with some results by Karliner, Klebanov and Susskind [20] who reconstruct continuum string theory starting from a lattice of finite spacing and with those of Konishi, Paffuti and Provero [21] who have discretized the string action and have found that, unlike in the point particle case, the spacial resolution does not improve below λ_s even if the lattice is made more and more fine.

Finally, the thermodynamics of strings above Hagedorn's temperature indicates [22] that the free energy, instead of exhibiting the typical field theory T^d behaviour, grows just like T^2 as if, again, the number of physical degrees of freedom was reduced at high temperatures (energies).

Possibly, in quantum string theory, the very concept of a space-time continuum in which strings move has to be abandoned below a certain scale. What, if any, replaces it is the big open question. Perhaps a sort of phase transition to a more symmetric phase will occur in which the fundamental length itself disappears. Obviously, our considerations, being limited to calculations in flat space-time, are not sufficient to answer such questions.

REFERENCES TO LECTURE III

1. M. Soldate, *Phys. Lett.* **B186** (1987) 129; *Nucl. Phys.* **B303** (1988) 407.
2. D.J. Gross and P.F. Mende, *Phys. Lett.* **B197** (1987) 129; *Nucl. Phys.* **B303** (1988) 407.
3. P.F. Mende and H. Ooguri, *Nucl. Phys.* **B339** (1990) 641.
4. D. Amati, M. Ciafaloni and G. Veneziano, *Phys. Lett.* **B197** (1987) 81; *Int. Journ. Mod. Phys.* **3A** (1988) 1615.
5. D. Amati, M. Ciafaloni and G. Veneziano, *Phys. Lett.* **B216** (1989) 41.
6. D. Amati, M. Ciafaloni and G. Veneziano, *Nucl. Phys.* **B347** (1990) 550.
7. G. Veneziano, Proc. of Superstring '89 Workshop, Texas A&M University, R. Arnowitt et al., Editors, (World Scientific, Singapore, 1990), p. 86.
8. G. 't Hooft, *Phys. Lett.* **B198** (1987) 61.
9. I. Muzinich and M. Soldate, *Phys. Rev.* **D37** (1988) 353.
10. V. Ferrari, P. Pendenza and G. Veneziano, *Gen. Rel. Grav.* **20** (1988) 1185.
11. P.C. Aichelburg and R.V. Sexl, *Gen. Rel. Grav.* **2** (1971) 303.
12. E. Gava, R. Iengo and C.-J. Zhu, *Nucl. Phys.* **B323** (1989) 585; G. Cristofano, M. Fabbrichesi and K. Roland, *Phys. Lett.* **B236** (1990) 159.
13. S.W. Hawking, *Phys. Rev.* **D13** (1976) 191.
14. R. Iengo and K. Lechner, *Nucl. Phys.* **B335** (1990) 221;
G. Cristofano, M. Fabbrichesi and K. Roland, Nordita preprint 90/6P (1990).
15. G. Veneziano, Invited Talk at the Annual Meeting of the Italian Physical Society (Naples, Oct. 1987);
D. Amati, M. Ciafaloni and G. Veneziano, Ref. [5];
D.J. Gross, Proc. XXIV Int. Conference on HEP (Munich, Aug. 1988), R. Kotthaus and J.H. Kühn, Editors, (Springer Verlag, Berlin), p. 310.

16. D.J. Gross and V. Periwal, *Phys. Rev. Lett.* **60** (1988) 2105.
17. K. Kikkawa and M. Yamasaki, *Phys. Lett.* **149B** (1984) 357; N. Sakai and I. Senda, *Progr. Theor. Phys.* **75** (1986) 692; see also A. Giveon, E. Rabinovici and G. Veneziano, *Nucl. Phys.* **B322** (1989) 167.
18. G. Veneziano, Proc. of the 5th Marcel Grossmann Meeting, D.G. Blair and M.J. Buckingham, Editors, (World Scientific, Singapore, 1989), p. 173.
19. M. Fabbrichesi and G. Veneziano, *Phys. Lett.* **233B** (1989) 135.
20. M. Karliner, I. Klebanov and L. Susskind, *Int. Journ. Mod. Phys.* **A3** (1988) 1981.
21. K. Konishi, G. Paffuti and P. Provero, *Phys. Lett.* **234B** (1990) 276.
22. J.J. Atick and E. Witten, *Nucl. Phys.* **B310** (1988) 291; see also Ya.I. Kogan, *JETP Lett.* **45** (1987) 709; B. Sathiaplan, *Phys. Rev.* **D35** (1987) 3277.

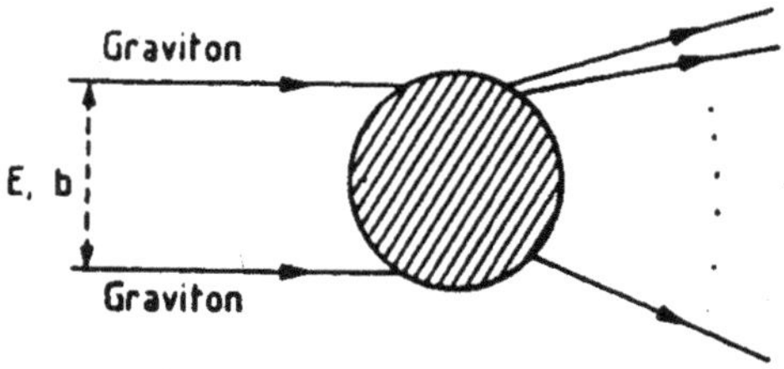

Fig.1 The gedanken experiment discussed in this talk.

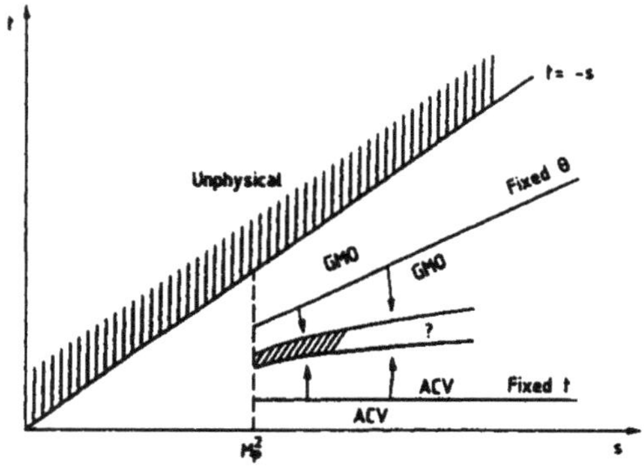

Fig.2 $s - t$ plane for elastic amplitude and the kinematical regions studied by GMO and ACV. The shaded region is where a comparison can be presently made.

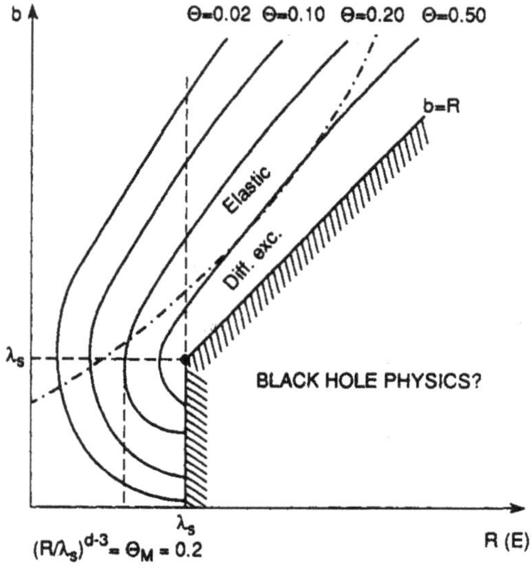

Fig.3 $R - b$ "phase diagram" of the our gedanken experiment. The various curves are explained in the text.

BLACK HOLES AND THERMODYNAMICS

ROBERT M. WALD
Enrico Fermi Institute and Department of Physics
University of Chicago, 5640 S. Ellis Ave.
Chicago, IL 60637, USA

ABSTRACT

A review is given of the relationship between laws of black hole physics and the laws of thermodynamics.

One of the most remarkable developments to arise from the theory of black holes in classical general relativity was the discovery of a close mathematical analogy between certain laws of "black hole mechanics" and the ordinary laws of thermodynamics[1]. In this analogy, the mass of the black hole plays the role of the energy of a thermodymamic system, its surface gravity plays the role of temperature, and the area of its event horizon plays the role of entropy. This analogy was put on a firm physical foundation by Hawking's discovery[2] of the fact that the gravitational field of a black hole will have an important effect upon quantum fields, resulting in the emission of particles by the black hole in a manner identical to that of a black body. The quantum effects also play a key role in making viable the "generalized second law of thermodynamics", which states that the sum of the total entropy of matter outside of black holes plus 1/4 the area of black holes never decreases with time.

My first lecture reviewed the basic properties of black holes and the derivation of the classical laws of black hole mechanics. One recent development has taken place on this topic: It has been recognized that the firat law of black hole mechanics arises in a very simple and direct way from the Hamiltonian formulation of general relativity[3,4]. This helps explain why forms of this law continue to hold for a variety of types of matter fields outside the black hole (such as Yang-Mills fields) and why this law holds in other theories (such as the two-dimensional theory treated in Frolov's lectures).

My second lecture reviewed the nature of quantum field theory in curved spacetime and the derivation of the effect of particle creation by black holes. As I reviewed recently elsewhere[5], the theory of quantum fields in a fixed, curved spacetime can be put on a rigorous mathematical footing, and particle creation by black holes is an unambiguous prediction of this theory. However, some issues remain open concerning the "back-reaction" of the quantum field upon the gravitational field and the final state thereby resulting from black hole "evaporation". These issues remain active areas of research.

My final lecture discussed the generalized second law and the evidence for its validity. In particular, on account of subtle quantum effects, one cannot violate the generalized second law by slowly lowering a box of matter containing high entropy to near the horizon of a black hole before dropping it in[6].

All of the topics of these lectures (except for the new derivation of the first law[3,4]) were treated extensively in my recent Erice lectures[7], to which I refer the reader for further details.

References

1. J.M. Bardeen, B. Carter, and S.W. Hawking, *Commun. Math. Phys.* **31** (1973) 161.
2. S. W. Hawking, *Commun. Math. Phys.* **43** (1975) 199.
3. D. Sudarsky and R.M. Wald, *Phys. Rev.* **D46** (1992) 1453.
4. R.M. Wald, "The First Law of Black Hole Mechanics", to appear in the Misner Festschrift, ed. by B.L. Hu and T.A. Jacobson.
5. R.M. Wald, "Quantum Field Theory in Curved Spacetime", to appear in the Proceedings of the 1992 Les Houches School on Gravitation and Quantization, ed. by B. Julia.
6. W.G. Unruh and R.M. Wald, *Phys. Rev.* **D25** (1982) 942.
7. R.M. Wald, "Black Holes and Thermodynamics", in *Black Hole Physics*, ed. by V. De Sabbata and Z. Zhang, (Kluwer Academic Publishers, Dordrecht, 1992).

THE CONFRONTATION BETWEEN GENERAL RELATIVITY AND EXPERIMENT

Clifford M. Will
McDonnell Center for the Space Sciences, Department of Physics,
Washington University, St. Louis, Missouri 63130

ABSTRACT

In these lectures, the status of experimental tests of general relativity and of theoretical frameworks for analysing them were reviewed. Einstein's equivalence principle is well supported by experiments such as the Eötvös experiment, tests of special relativity, and the gravitational redshift experiment. Tests of general relativity have reached high precision, including the light deflection, the Shapiro time delay, the perihelion advance of Mercury, and the Nordtvedt effect in lunar motion. Gravitational wave damping has been detected to half a percent using the binary pulsar, and new binary pulsar systems promise further improvements. The status of the "fifth force" was discussed, along with the frontiers of experimental relativity, including proposals for testing relativistic gravity with advanced technology and spacecraft.

1. Summary

Because the basic material contained in these lectures has been recently summarized elsewhere, it will not be repeated here. Instead, the reader is referred to the following publications:

"The Confrontation between General Relativity and Experiment: A 1990 Update", in *Gravitation: A Banff Summer Institute*, eds. R. Mann and P. Wesson (World Scientific, Singapore, 1991), p. 439.

"The Confrontation between General Relativity and Experiment: A 1992 Update", *International Journal of Modern Physics D* **1**, 13 (1992).

Theory and Experiment in Gravitational Physics, 2nd Edition (Cambridge University Press, Cambridge, 1992).

COMMUNICATIONS

Classical Gravity

Symmetric Fields and Harmonic Analysis

D. KAPETANAKIS[1]
Physik Department
Technische Universität München
D-8046 Garching
and
G. ZOUPANOS [2]
Physics Deptartment
Nat. Technical University
157 73 Zografou, Athens, Greece

Abstract

We review the results of the Coset Space Dimensional Reduction (CSDR) scheme applied on Gauge Theories defined in higher dimensions and we compare them with those obtained from harmonic analysis over coset spaces.

1. Introduction

Unification of all observed interactions is fascinating physicists for many decades. In fact one of the most ambitious unification scenarios was proposed seventy years ago by Kaluza and Klein [1]. It was suggesting the way to obtain electromagnetism coupled to gravity starting from pure gravity in five dimensions. In the recent years a revival of interest in the Kaluza-Klein scheme started when it became clear that non-abelian gauge groups emerge starting from pure gravity in further higher dimensions. Then assuming that the total spacetime manifold can be written as a direct product $M^D = M^4 \times B$, where B is a compact Riemannian space with a non-abelian isometry group S, one obtains after dimensional reduction of the theory gravity coupled to a gauge theory based on a gauge group that contains S and scalars in four dimensions. The obvious advantages of this scenario is the unification of all interactions and the natural emergence of the gauge symmetries.

[1]Partially supported by an C.E.C. fellowship (ERB4001GT910195).
[2]Partially supported by a C.E.C. project (SC1-CT91-0729)

However, there exist also some serious drawbacks in the Kaluza-Klein scenario, as for instance the lack of chiral fermions in four dimensions [2] which rules it out from being a candidate for a realistic theory describing observed interactions. Another problem is that there exist no classical ground state corresponding to the assumed direct product structure of M^D. The reason is that the D-dimensional vacuum Einstein equations lead to vanishing Ricci tensor, which should be non-zero for a compact space with a non-abelian isometry group.

At least the above mentioned serious obstacles can surppased by adding suitable matter fields and in particular Yang-Mills fields to the gravity action in higher dimensions. Then one can obtain a classical ground state of the required form [3] as well as chiral fermions in four dimensions [4]. Thus introducing Yang-Mills fields in higher dimensions one obtains a viable framework, which is not as ambitious as the original Kaluza-Klein proposal but certainly much more promising [5]. One should also remember that some other popular schemes such as supergravity and superstring theories contain naturally an Einstein-Yang-Mills theory in higher dimensions. Next one has to consider the dimensional reduction of the higher dimensional theory to four dimensions.

2. Symmetric fields

Let us start with a D-dimensional manifold M^D assumed to be of the direct product form $M^4 \times B$, where M^4 is a four-dimensional manifold (corresponding to space-time) while B is a compact one. We shall be interested in the case where B admits a transformation group T, acting trivially on M^4. We write the coordinates of M^D as $X^M = (x^\mu, y^\alpha)$, where x^μ denote the coordinates of M^4 and y^α, $\alpha = 1, \dots, d$ the coordinates of B. Thus T is considered to act on M^D as:

$$q(X)^M = (x^\mu, q(y)^\alpha), \quad q \in T. \tag{1}$$

The action of T on M^D induces naturally a transformation of the tensor fields.

Let us now define the notion of symmetric fields on M^D. A field V is symmetric under the action of T if it obeys [6]:

$$V(q(X)) = \Omega(q,y)V(X), \quad \forall q \in T, \quad \forall X \in M^D, \tag{2}$$

where $\Omega(q,y)$ is a tangent space rotation depending on the $SO(d)$ representation content of the field $V(X)$. Since T acts trivially on M^4, $\Omega(q,y)$ does not depend on x^μ. The symmetry equation (2) determines $\Omega(q,y)$ and leads to certain constraints which the symmetric fields have to obey. In the following we shall consider the case when T is a continuous transformation group acting transitively on B.

However, we should stress that since we are primarily interested in gauge theories, the statement that the symmetric fields have to obey eq. (2) is too strong. We can merely require that the transformation of a field under the group T to be a gauge transformation [7, 8, 9]. This requirement leads to a wider class of fields that are allowed. Therefore in the presence of a gauge group G, a field is symmetric if instead of eq. (2)) it satisfies:

$$V(q(X)) = D(W(q,y))\Omega(q,y)V(X)\,, \tag{3}$$

where $D(W(q,y))$ is a gauge transformation in the representation of the field $V(X)$. In this case the symmetry equation (3) mixes non-trivially the gauge group structure and the geometry of the space B.

Let us now examine the implications of eqs. (2) and (3) on the fields in the theory. The consequences are more transparent if we consider the infinitesimal case. So for scalar ϕ, vector A_α and spinor ψ fields on B, respectively, we demand:

$$\begin{aligned} \delta_A\phi &= \xi_A^\alpha\partial_\alpha\phi = D(W_A)\phi\,, & (4)\\ \delta_A A_\alpha &= \xi_A^\beta\partial_\beta A_\alpha + \partial_\alpha\xi_A^\beta A_\beta = \partial_\alpha W_A - [W_A, A_\alpha]\,, & (5)\\ \delta_A\psi &= \xi_A^\alpha\partial_\alpha\psi - \frac{1}{2}G_{Abc}\Sigma^{bc}\psi = D(W_A)\psi\,. & (6) \end{aligned}$$

We will assume that W_A depends only on the B coordinates y.

Since the commutator of two transformations satisfies $[\delta_A, \delta_B] = f_{AB}{}^C\delta_C$, we also have the additional consistency relation on the gauge transformation W_A:

$$\xi_A^\alpha\partial_\alpha W_B - \xi_B^\alpha\partial_\alpha W_A - [W_A, W_B] = f_{AB}{}^C W_C\,, \tag{7}$$

which corresponds to the equivalence of the interpretation of δ_A as S as well as gauge transformation. Let us see how the W_A's will change when the fields are transformed under the gauge group. Consider the gauge transformation of the scalar field $\phi \longrightarrow \phi' = D(g)\phi$. Then the W_A needed to compensate an S-transformation acting on ϕ' will be:

$$W_A' = gW_A g^{-1} + \xi_A^\alpha(\partial_\alpha g)\cdot g^{-1}\,. \tag{8}$$

The requirement that W_A transforms according to eq. (8, under gauge transformations, ensures that the constraints (5)–(6) remain invariant under general coordinate and gauge transformations. Furthermore to simplify the calculation somewhat we can use the gauge freedom (8) to select

$$W_a(y^\alpha = 0) = 0\,. \tag{9}$$

Under these assumptions eq. (7) yields:

$$\partial_a W_b - \partial_b W_a = f_{ab}{}^i W_i \,, \tag{10}$$

$$\partial_a W_i = 0 \,, \tag{11}$$

$$[W_i, W_j] = -f_{ij}{}^k W_k \,. \tag{12}$$

From eq. (12) we see that the W_i's are constants over the coset. Defining

$$J_i \equiv -W_i \tag{13}$$

eq. (12) implies that the J_i's form the algebra of R. Since the W's live by definition in the Lie algebra of the gauge group G, eq. (12) makes sense only if R is embedded in G. In that case the J_i's are the generators of an R-subgroup R_G of G. The usefulness of eq. (11) will become transparent when we will analyse the potential of the theory.

We proceed by analysing the constraints on the fields in the theory [7]. Consider first a T-symmetric scalar field, ϕ on M^D, singlet under the action of the gauge group G i.e. ϕ obeys: $\phi(x, q(y)) = \phi(x, y)$, which in turn means that a T-invariant singlet scalar field is independent of the coordinates of the coset space S/R.

A gauge field A_M on M^D splits into A_μ on M^4 and A_α on S/R; A_μ behaves as a scalar under S-transformations and lies in the adjoint representation of G. From eq. (5) we obtain at $y = 0$:

$$\partial_a A_\mu = 0 \,, \tag{14}$$

$$[J_i, A_\mu] = 0 \,. \tag{15}$$

Eq. (15) indicates that the four-dimensional gauge field is completely independent from the coset space coordinates. Furthermore the gauge group in four dimensions is dictated by eq. (15). Since A_μ commutes with J_i's, which are the generators of R_G in G, the surviving gauge symmetry H is that subgroup of G which commutes with R. In other words the gauge group H in four dimensions is the centralizer of R in G, i.e.

$$H = C_G(R_G) \,. \tag{16}$$

The remaining components of the higher-dimensional gauge field A_α become vectors under the coset space transformations. As they will be the scalar fields in the resulting four-dimensional theory we write $\Phi_a = e^\alpha_a A_\alpha$. The constraint (6) at $y = 0$ implies:

$$\partial_a \phi_b - \partial_b W_a = \frac{1}{2} f_{ab}{}^c \phi_c \,, \tag{17}$$

$$[J_i, \phi_a] = f_{ia}{}^c \phi_c \,. \tag{18}$$

Eq. (18) will be useful when we shall calculate the potential of the theory. From eq. (18) we see that ϕ_a's act as intertwining operators connecting the induced representations of R in G and in S. Indeed we have already shown that J_i form an R-subalgebra of G and also $(f_i)_a{}^b$ can be regarded as the R-generator within S. This implies, exploiting Schur's lemma that the transformation properties of the fields $\phi_a(x,y)$ under H can be found if we express the adjoint irreducible representation of G in terms of $R_G \times H$:

$$\begin{aligned} G &\supset R_G \times H \\ \mathrm{adj}G &= (\mathrm{adj}R, 1) + (1, \mathrm{adj}H) + \textstyle\sum(r_i, h_i)\,. \end{aligned} \tag{19}$$

and S under R:

$$\begin{aligned} S &\supset R \\ \mathrm{adj}S &= \mathrm{adj}R + \textstyle\sum s_i\,. \end{aligned} \tag{20}$$

Then for every pair r_i, s_i where r_i and s_i are identical irreducible representations of R, there remains a Higgs multiplet transforming under the representation h_i of H. All other scalar fields vanish. Note that in the case of a vector field singlet under the gauge group G ($J_i = 0$) eq. (18) implies that ϕ_b will be zero unless it is a singlet under R. Therefore in this case there will remain in the theory n scalars, where n is the number of singlets in v$=\sum s_i$.

The analysis of the constraints imposed on spinors is analogous to the cases discussed before [10]. Eq. (6) at $y = 0$ implies:

$$\partial_a \psi = 0\,, \tag{21}$$

$$T_i \psi = J_i \psi\,, \tag{22}$$

where $T_i = -\frac{1}{2} f_{iab} \Sigma^{ab}$ and the J_i's are in the representation of G corresponding to that of ψ. Eq. (22) states that the spinors are independent of the coset coordinates, while eq. (22) has the same form as eq. (18). In this case ψ acts as an intertwining operator connecting induced representations of R in $SO(d)$ and in G. Similarly to vectors, in order to obtain the H representation content of the four-dimensional fermions we have to decompose the initial fermion representation F (of G) under $R_G \times H$:

$$F = \sum (t_i, h_i) \tag{23}$$

and the spinor of $SO(d)$ under R:

$$\sigma = \sum \sigma_j\,. \tag{24}$$

Then for each pair t_i and σ_i, where t_i and σ_i are identical irreducible representations, there is an h_i multiplet of spinor fields in the four-dimensional theory. In analogy with the previous case we can draw also the conclusion that in the case of a spinor field, singlet under the gauge group G, we will have n surviving spinor fields, independent of the coset space coordinates, with n the number of singlets in the decomposition of the spinor of $SO(d)$ under R. In order, however to obtain chiral fermions in the effective theory we have to impose further requirements. We first impose the Weyl condition in D dimensions. In D=4n+2 dimensions, the decomposition of the left-handed, say spinor under $SO(4)\times SO(d)$ is $\sigma_D = (2,1,\sigma_d)+(1,2,\overline{\sigma}_d)$. So we have the decompositions $\sigma_d = \sum \sigma_k$, $\overline{\sigma}_d = \sum \overline{\sigma}_k$. Then for a vectorlike representation F for the fermions. In this case each term (t_i, h_i) in eq.(23) will be either selfconjugate or it will have a partner $(\overline{t}_i, \overline{h}_i)$. According to the rule described in eqs.(23) and (24) if we consider σ_d we will have at four dimensions left-handed fermions transforming according to $f_L = \sum h_k^L$. It is important to notice that since σ_d is non-selfconjugate f_L is non-selfconjugate too. Similarly from $\overline{\sigma}_d$ we will obtain the right-handed representation $f_R = \sum \overline{h}_k^R$ but as we have assumed that F is vectorlike $\overline{h}_k^R \sim h_k^L$. Therefore there will appear two sets of Weyl fermions with the same quantum numbers under H. This is already a chiral theory, but still one can go further and try to impose the Majorana condition in order to eliminate the doubling of the fermionic spectrum. We should remark now that if we had started with F complex we would have again a chiral theory since in this case $\overline{h}_k^R$ is different from h_k^L (σ_d non-selfconjugate). Nevertheless since starting with F vectorlike is a more appealing situation we proceed by considering the Majorana condition. The Majorana condition can be imposed in D=2,3,4+8n dimensions and is given by $\psi = C(\overline{\psi})^T$ where C is the D-dimensional charge conjugation matrix. Majorana and Weyl conditions are compatible in D=4n+2 dimensions. Then in our case if we start with Weyl–Majorana spinors in D=4n+2 dimensions we force f_R to be the charge conjugate of f_L, thus arriving in a theory with fermions only in f_L. Furthermore if F is to be real then we have to have D=2+8n while for F pseudoreal D=6+8n.

The above analysis can be in a straightforward way generalized to tensors and Rarita- Schwinger fields.

3. The four-dimensional Lagrangian

Consider the action of a Yang–Mills theory with gauge group G, coupled to fermions defined on a manifold M^D:

$$A = \int d^D x \sqrt{-g}[-\frac{1}{8}\,\mathrm{Tr}(F_{MN}F_{K\Lambda})g^{MK}g^{N\Lambda} + \frac{i}{2}\overline{\psi}\Gamma^M D_M \psi], \tag{25}$$

where $D_M = \partial_M - \theta_M - A_M$ with $\theta_M = \frac{1}{2}\theta_{MN\Lambda}\Sigma^{N\Lambda}$ the spin-connection of M^D, $F_{MN} = \partial_M A_N - \partial_N A_M - [A_M, A_N]$. Then, if we compactify M^D to $M^4 \times S/R$ and assume that the fields in the theory are symmetric in the sense described in sec. 2, the action of the theory becomes independent of the coset space coordinates. It follows that we can integrate over the coordinates of the compact space S/R, obtaining a theory in four dimensions. The fields, however, have to obey certain constraints which mix non-trivially the gauge group structure and the geometry of the coset space S/R. The solution of these constraints gives the four-dimensional gauge group as well as the field content of the resulting theory.

Since M^D compactifies to $M^4 \times S/R$, it is natural to assume that the metric acquires the form:

$$g^{MN} = \begin{pmatrix} \eta^{\mu\nu} & 0 \\ 0 & -g^{ab} \end{pmatrix}, \tag{26}$$

where $\eta^{\mu\nu} = \mathrm{diag}(1, -1, -1, -1)$ and g^{ab} is the coset space metric. Obviously the F_{MN} will split into $F_{\mu\nu}$ the usual field tensor of the gauge field A_μ, $F_{\mu a}$ and F_{ab}. For $F_{\mu a}$ part of the tensor, using eq. (15) we find:

$$F_{\mu a} = \partial_\mu \phi_a - [A_\mu, \phi_a] = D_\mu \phi_a, \tag{27}$$

while using eqs. (11) and (18) for the F_{ab} part we obtain:

$$F_{ab} = f_{ab}{}^{C}\phi_C - [\phi_a, \phi_b], \tag{28}$$

where we have defined for convenience $\phi_i \equiv J_i$.

To consider the fermionic part of the Lagrangian we split the D-dimensional Γ-algebra in accordance with eq. (26). The Γ-algebra satisfies:

$$\{\Gamma^M, \Gamma^N\} = 2g^{MN}, \tag{29}$$

and can be rewritten as:

$$\{\Gamma^\mu, \Gamma^\nu\} = 2\eta^{\mu\nu}, \qquad \{\Gamma^a, \Gamma^b\} = -2g^{ab}, \tag{30}$$

with

$$\Gamma^\mu = \gamma^\mu \otimes I_d \quad \text{and} \quad \Gamma^a = \gamma_5 \otimes \gamma^a, \tag{31}$$

where γ^μ are the usual four-dimensional Dirac matrices and γ^a's satisfy the algebra given in eq. (30). Putting together the above equations we finally obtain the four-dimensional action (for the gauge group generators we assume the normalization $\mathrm{Tr}(G_t G_s) = 2\delta_{ts}$):

$$A = C \int d^4x \left(-\frac{1}{4}F^t_{\mu\nu}F^{t\mu\nu} + \frac{1}{2}(D_\mu\phi_a)^t(D^\mu\phi^a)^t + V(\phi) + \frac{i}{2}\overline{\psi}\Gamma^\mu D_\mu\psi - \frac{i}{2}\overline{\psi}\Gamma^a D_a\psi\right), \tag{32}$$

where $D_\mu = \partial_\mu - A_\mu$ and $D_a = \partial_a - \theta_a - \phi_a$ with $\theta_a = \frac{1}{2}\theta_{abc}\Sigma^{bc}$ the connection of the coset space, $V(\phi)$ is given by:

$$V(\phi) = -\frac{1}{4}g^{ac}g^{bd}Tr(f_{ab}{}^C\phi_C - [\phi_a, \phi_b])(f_{cd}{}^D\phi_D - [\phi_c, \phi_d]) \tag{33}$$

and C is the volume of the coset space.

We see that the gauge theory defined on M^D yields a Yang–Mills theory in four dimensions which is of the canonical form if we rescale the scalar fields. In addition to the usual kinetic terms the theory contains also the potential given in eq. (33), which usually leads to spontaneous breakdown of the gauge theory. The last term in eq. (32) is a Yukawa term of the fermions with the Higgs fields and explicit mass terms of geometrical origin.

4. Harmonic Analysis and Masses

In case the extra dimensions have physical significance, it is reasonable to derive the four-dimensional theory by identifying the massless fields and their interactions. This can be done by applying the techniques of harmonic expansion [11, 12], when the D-dimensional fields are expanded in 'normal modes' or harmonics of the internal space B. It is instructive to compare the CSDR approach to the results obtained from harmonic analysis, i.e. taking the extra dimensions more seriously, but ignoring fluctuations of the metric. The comparison shows which of the fields obtained via CSDR are massless. Moreover we find that the gauge fields obtained from CSDR are the only massless gauge fields after the first stage of the symmetry breaking, at least in the case of symmetric coset spaces with rank S = rank R. The hope is that the latter has wider validity.

In order to perform the harmonic expansion of the fields on $M^4 \times S/R$ we have to find a suitable ground state configuration around which to expand [13, 14]. Such a configuration should satisfy the equations of motion on $M^4 \times S/R$:

$$D^\nu F_{\mu\nu} + D^\alpha F_{\mu\alpha} = 0\,, \tag{34}$$

$$D^\mu F_{\alpha\mu} + D^\beta F_{\alpha\beta} = 0\,. \tag{35}$$

Since the symmetry of the vacuum $M^4 \times S/R$ is the Poincaré group times S, we demand that the background configuration A^B_M should possess these symmetries. Therefore A^B_μ has to vanish, thus eqs. (35) are identically satisfied, while A^B_α has to be S-symmetric. Now finding a solution of the Yang–Mills eqs. (35) on the internal space is less trivial. Clearly the problem of solving eqs. (35) for our case (when A^B_α is S-symmetric) amounts to classifying the stationary points of the quartic Higgs potential. Although this is an algebraic problem, it is by no means trivial as discussed in detail before. Let us recall here that the Yang–Mills field strength $F_{\alpha\beta}$ is:

$$F_{\alpha\beta} = e^a_\alpha e^b_\beta (f_{ab}{}^i J_i + f_{ab}{}^c \phi_c - [\phi_a, \phi_b]) . \tag{36}$$

As one can easily prove for symmetric coset spaces, $\phi_a = 0$ corresponds to a local maximum of $V(\phi)$ and therefore

$$A^B_\alpha = e^i_\alpha J_i \tag{37}$$

should satisfy the Yang–Mills eqs. (35). That this is indeed so is well known [13]. For the background configuration (37) $F^B_{\alpha\beta}$ is found to be:

$$F^B_{\alpha\beta} = e^a_\alpha e^b_\beta f_{ab}{}^i J_i . \tag{38}$$

Now for the covariant divergence of $F^B_{\alpha\beta}$ one obtains:

$$D^\beta F_{\alpha\beta} = e^a_\alpha \frac{1}{2} f_{ab}{}^c f_{ci}{}^b J^i . \tag{39}$$

For symmetric spaces eq. (39) vanishes identically since then $f_{ab}{}^c = 0$. Note also that

$$f_{ab}{}^c f_{ci}{}^b \propto \gamma_{ai} , \tag{40}$$

where the γ_{ai}'s are the corresponding components of the Cartan–Killing metric

$$\gamma_{AB} = \mathrm{tr}(T_A T_B) = f_{AD}{}^C f_{CB}{}^D \propto \delta_{AB} , \tag{41}$$

(assuming that the T_A's are in the adjoint representation). Therefore if $\mathcal{R}$ is *regular* subalgebra of S, $\gamma_{ai} = f_{ab}{}^c f_{ci}{}^b$ vanishes. This includes the most relevant case for our purposes, i.e. when rank R = rank S.

The background field A^B_α is also S-symmetric in the broader sense described in section 2., i.e. including a suitable gauge transformation. Indeed we obtain that:

$$\delta_A A^B_\alpha = -(\partial_\alpha \Omega^i_A) J_i + [\Omega^i_A J_i, A^B_\alpha] , \tag{42}$$

and therefore the gauge transformation needed to compensate this S-transformation is given by:

$$W_A = -\Omega^i_A J_i \,. \tag{43}$$

Note that the gauge transformation (43) satisfies eqs. (9)–(13).

Let us now discuss the harmonic expansion of a field on a coset space. It is well known that the matrix elements $D^{(m)}_{pq}$ of the inequivalent unitary irreducible representations of S, indexed by (m), serve as a complete orthonormal basis into which any function defined on S can be expanded (this is known as the Peter–Weyl theorem in group theory [11]). Since S is the isometry group of S/R, we can still use, in the case of the coset space, the matrix elements of the unitary irreducible representations of S as an orthonormal basis for the harmonic analysis, although these now have to be properly restricted [12]. The harmonic expansions of the fields A_μ, A_a and ψ can be written as:

$$A_\mu(x,y) = \sum_m \sum_{pq} \sqrt{\frac{d_m}{d_{\mathcal{D}}}}\, D^{(m)}_{pq}(L(y))\, a_\mu{}^{(m)}_{pq}(x)\,, \tag{44}$$

$$A_a(x,y) = A^B_a + \sum_m \sum_{pq} \sqrt{\frac{d_m}{d_{\mathcal{D}}}}\, D^{(m)}_{pq}(L(y))\, \phi_a{}^{(m)}_{pq}(x) = A^B_a + \overline{A}_a\,, \tag{45}$$

$$\psi(x,y) = \sum_m \sqrt{\frac{d_m}{d_{\mathcal{D}}}}\, D^{(m)}(L(y))\, \psi^{(m)}(x)\,, \tag{46}$$

where $L(y)$ is a representative element $L(y)$ of each R-equivalence class. As we will see only those irreducible representation of S contribute to the harmonic expansion of a field on S/R in the representation $\mathcal{D}$ of R, which contain $\mathcal{D}$ in their decomposition under R. Furthermore, all the relevant $D^{(m)}$'s contribute to the harmonic expansion only with those of their rows which correspond to the particular R-representation $\mathcal{D}$. In order to determine the relevant representations of S which contribute to the harmonic expansions (45)–(46), we should demand that under a general S-transformation, which now includes the gauge transformation given in eq. (43), the coefficients in these expansions should transform according to some irreducible representation of S [15]. We find:

$$D^{(m)}_{ps}(Q_i)\phi_a{}^{(m)}_{pq} + f_{iab}\phi_b{}^{(m)}_{sq} - [J_i, \phi_a{}^{(m)}_{sq}] = 0\,, \tag{47}$$

$$D^{(m)}_{ps}(Q_i)a_\mu{}^{(m)}_{pq} - [J_i, a_\mu{}^{(m)}_{sq}] = 0\,, \tag{48}$$

$$D^{(m)}(Q_i)\psi^{(m)} + T_i\psi^{(m)} - J_i\psi^{(m)} = 0\,. \tag{49}$$

Eqs. (48)–(49) specify which representations of S are present in the harmonic expansions of the corresponding fields given in eqs. (45)–(46). From the solution of eq. (48) we find that the relevant representations of S for the harmonic expansion of A_a are those which include in their restriction to R the representations of R appearing in the cross product of r_i and s_i given in eqs. (19) and (20). Similarly, for the harmonic expansion of A_μ one should include only those representations of R which appear in eq. (19). Finally, for the fermions, the relevant S-representations are those which include, when restricted to R, the representations appearing in the product of r_i and σ_i given in eqs. (23) and (24). All these S representations will contribute only with their rows which correspond to the R representation. Furthermore, comparing eqs. (48)–(49) with eqs. (6), (18) and (22), we find that the S-symmetric fields are the y-independent ones, i.e. the first terms in the expansions (45)–(46).

Let us next calculate the masses of all the four-dimensional fields appearing in the harmonic expansions of the bosonic fields [15, 16]. We obtain:

$$\begin{aligned}\nabla_b\nabla^b\, D^{(m)}_{pq}(L(y))\; a_{\mu}{}^{(m)}_{pq}(x) &= D^{(m)}_{pq}(Q_bQ_bL(y))\; a_{\mu}{}^{(m)}_{pq}(x) \\ &= [C_S(D^{(m)}) - C_R(\mathcal{D})]D^{(m)}_{pq}(L(y))\; a_{\mu}{}^{(m)}_{pq}(x)\,,\end{aligned} \tag{50}$$

where $\mathcal{D}$ is the representation of R under which the fields a_μ are transforming according to eq. (49) and $C_S(D^{(m)})$, $C_R(\mathcal{D})$ are the quadratic Casimirs of S and R in the representations $D^{(m)}$ and $\mathcal{D}$ respectively. We readily see that since R is a subgroup of S then $C_S(D^{(m)}) \leq C_R(\mathcal{D}) \leq 0$ (with equality holding only for the trivial representation). Therefore the symmetric fields are the only massless fields. The mass of the ϕ_a's can also be calculated and we find:

$$-\,C_S(D^{(m)})\;\phi_{a}{}^{(m)}_{pq}(x) + [[\phi_{a}{}^{(m)}_{pq}(x), J_i], J_i]\,, \tag{51}$$

which clearly shows the presence of tachyons. These tachyons belong to the trivial representation of S. This fact corresponds to the spontaneous symmetry breaking in the case of a symmetric coset space.

Let us next consider the case of fermions [15, 16, 17, 18], As we saw in section 2, after the dimensional reduction the fermionic Lagrangian is split into two terms. The first term is just the kinetic term for the fermions while the second one will give us the mass term. Then all one has to do is to calculate the square of the mass operator $i\Gamma^a\nabla_a$ acting on the spinor fields:

$$M^2\psi = -\Gamma^a\Gamma^b\nabla_a\nabla_b\psi = (\nabla^a\nabla_a - \Sigma^{ab}[\nabla_a, \nabla_b])\psi\,. \tag{52}$$

Finally we obtain:

$$M^2 = C_S(D^{(m)}) - C_R(\mathcal{D}) + C_R(\sigma)\,, \tag{53}$$

where σ is the spinorial representation of $SO(d)$. It is not difficult to check that the symmetric fields are massless.

When the coset space is non-symmetric, some partial results concerning the masses of the fermion fields are already known in the literature [19, 20]. In this case we have:

$$L_M = -\frac{i}{2}\overline{\psi}\Gamma^a\nabla_a\psi + \overline{\psi}V\psi\,, \tag{54}$$

where

$$V = \frac{i}{4}\Gamma^a G_{abc}\Sigma^{bc}\,. \tag{55}$$

We now restrict ourselves to six-dimensional cosets. Let us examine the second term appearing in eq. (54) [18, 19]. One can easily show that V anticommutes with the six-dimensional helicity operator. Furthermore one can show that the operator V commutes with the $T_i = -\frac{1}{2}f_{ibc}\Sigma^{bc}$ (T_i as we have seen generate an R-subalgebra of $SO(6)$). In turn we can draw the conclusion – exploiting Schur's lemma – that the non-vanishing elements of V are only those which appear in the decomposition of both $SO(6)$ irreducible representations 4 and $\overline{4}$ i.e. the singlets. Since this term is a pure geometric term we arrive at the conclusion that the singlets in 4 and $\overline{4}$, will acquire large geometrical masses, a fact that has interesting phenomenological implications. For example in supersymmetric theories defined in higher dimensions, it means that the gauginos surviving in four dimensions obtain, after dimensional reduction, superheavy masses, i.e. supersymmetry is broken at the compactification scale.

After the spontaneous symmetry breaking, some of the scalar fields will acquire v.e.v.'s and give masses to fermions comparable to the compactification scale. In general the fermion mass matrix, after spontaneous symmetry breaking, is model dependent. However there exist a class of models in which the square fermion mass matrix can be expressed in terms of Casimirs [21]. This is the class of models that satisfy $S \subset G$ resulting in $V(\phi)\mid_{\phi=\phi_{min}}= 0$ or $F_{ab} = 0$, which is analogous to what happens in Kaluza–Klein theories. Thus for a symmetric coset space we find:

$$M^2\psi = (\phi^a\phi_a + 2\phi^i\phi_i)\psi = (C_S + C_R)\psi\,, \tag{56}$$

where C_S and C_R are the quadratic Casimirs of S and R, respectively, having non-negative eigenvalues for all representations.

In this way a very interesting class of models is ruled out because the four-dimensional fermions acquire superheavy masses. Of course, if we do not consider the extra dimensions seriously, this drawback could be avoided.

5. Conclusions

The CSDR scheme is a very interesting framework for studying the dimensional reduction of higher dimensional theories and examining the details of the resulting theories in four dimensions. Nevertheless the CSDR scheme technically is a consistent truncation of the original theory. Here we have presented in some detail the CSDR scheme and we have paved the way for a serious comparison between the full and the truncated theory. Our conclusion is that the fields obtained using the CSDR approach are the first terms in the complete expansions of harmonic analysis. It is very interesting that in the wide class of coset spaces that have been examined here the fields obtained via CSDR are massless after the first stage of the symmetry breaking. It is obviously of great interest to examine under which conditions these fields are the only massless states.

Acknowledgements

It is a pleasure to thank the organizing committee for the invitation and the warm hospitality.

References

[1] T. Kaluza, *Sitz. Preuss Akad. Wiss.* **K1** (1921) 966, O. Klein, *Z. Phys.* **37** (1926) 895.

[2] E. Witten, 'Fermion quantum numbers in Kaluza–Klein theory' in *Shelter Island II: Proceedings of the 1983 Shelter Island Conference on Quantum Field Theory and the Fundamental Problems of Physics*, eds. R. Jackiw et al. (MIT Press, Cambridge, Mass. 1985)

[3] Z. Horvath, L. Palla, E. Cremmer and J. Scherk, *Nucl. Phys.* **B127** (1977) 57.

[4] N. S. Manton, *Nucl. Phys.* **B193** (1981) 502; C. Chapline and R. Slansky, *Nucl. Phys.* **B209** (1982) 461; C. Wetterich, *Nucl. Phys.* **B222** (1983) 20.

[5] R. Coquereaux and A. Jadczyk, *Riemannian Geometry, Fiber Bundles, Kaluza–Klein Theories and all that...*, (World Scientific, Singapore, 1988); Y.A. Kubyshin, J.M. Mourão, G. Rudolph and I.P. Volobujev, *Dimensional Reduction of Gauge Theories, Spontaneous Compactification and Model Building*, (Springer Notes in Physics Vol. 349, Heidelberg 1989); D. Kapetanakis and G. Zoupanos, *Phys. Rep.* **C 219** 1.

[6] N.S. Manton, *Ann. Phys.* **167** (1986) 328.

[7] P. Forgács and N.S. Manton, *Commun. Math. Phys.* **72** (1980) 15.

[8] E. Witten, *Phys. Rev. Lett.* **38** (1977) 121.

[9] R. Jackiw, *Acta Phys. Austriaca Suppl.* **XXII** (1980) 383.

[10] N. S. Manton, *Nucl. Phys.* **B193** (1981) 502

[11] N.J. Vilenkin, *Special functions and the Theory of Group Representations*, Translations of Mathematical Monographs, Vol. 22 (American Mathematical Society, Providence, Rhode Island, 1968).

[12] A. Salam and J. Strathdee, *Ann. Phys.* **141** (1982) 316.

[13] S. Randjbar-Daemi and E. Percacci, *Phys. Lett.* **117B** (1982) 41.

[14] K. Pilch and A.N. Schellekens, *Nucl. Phys.* **B256** (1985) 109.

[15] L. Palla, *Z. Phys.* **C24** (1984) 195.

[16] A.N. Schellekens, *Nucl. Phys.* **B248** (1984) 706, **B250** (1985) 252.

[17] K. Pilch and A.N. Schellekens, *J. Math. Phys.* **25** (1984) 3455.

[18] P. Forgács, Z. and L. Palla, *Z. Phys.* **C30** (1986) 261, *Acta Phys. Hung.* **62** (1987) 177.

[19] D. Kapetanakis and G. Zoupanos, *Phys. Lett.* **B249** (1990) 73.

[20] Y.S. Wu and A. Zee, *J. Math. Phys.* **25** (1984) 2696.

[21] K.J. Barnes, P. Forgács, M. Surridge and G. Zoupanos *Z. Phys* **C33** (1987) 427.

PHYSICAL PROPERTIES OF AN EXACT SPHERICALLY SYMMETRIC SOLUTION WITH SHEAR IN GENERAL RELATIVITY

HENNING KNUTSEN
Høgskolesenteret i Rogaland, Box 2557, Ullandhaug, N-4004 Stavanger, Norway

ABSTRACT

The properties of an exact spherically symmetric perfect fluid solution obtained in non-comoving coordinates are examined. This solution contains shear. We discuss the nature of the pressure p and the density ρ, and also give the kinematic quantities for this fluid. It is not possible to match this solution with an exterior vacuum Schwarzschild solution. Further, the dominant energy condition produces imaginary values for the sound speed.

1 Introduction and basic equations

An exact solution of Einstein's field equations which appear in simple form in non-comoving coordinates may show a bewildering appearance when transformed to a comoving system. Hence, the late McVittie and Wiltshire[1] thought it worthwhile to examine the possibility of solving Einstein's equations in terms of non-comoving coordinates. However, they obtained their solution by ancillary mathematical assumptions and abstained from a detailed physical analysis of their models. Moreover, very few of the non-static solutions found in the literature[2] contain shear. In this paper we examine the properties of a particular model given by McVittie and Wiltshire, and we show that this model contains shear.

The spherically symmetric metric is given by

$$\mathrm{d}s^2 = \mathrm{e}^{2\lambda}\mathrm{d}\eta^2 - \mathrm{e}^{2\mu}\left[\mathrm{d}\xi^2 + f^2\left(\mathrm{d}\theta^2 + \sin^2\theta\mathrm{d}\phi^2\right)\right], \tag{1}$$

where the functions λ and μ depend on both the time coordinate η and the spatial coordinate ξ, while f is dependent on ξ alone.

Assuming the material to be a perfect fluid, and using Einstein's field equations, the pressure isotropy condition takes the following form:

$$\mathrm{e}^{2\lambda}\left(G_1{}^4\right)^2 + \mathrm{e}^{2\mu}\left(G_2{}^2 - G_1{}^1\right)\left(G_2{}^2 - G_4{}^4\right) = 0, \tag{2}$$

where $G_a{}^b$ is the Einstein tensor.

One particular solution obtained by McVittie and Wiltshire is given by:

$$\mathrm{d}s^2 = R^2\left[\mathrm{d}\eta^2 - \frac{\mathrm{d}\xi^2}{\xi^2} - \left(\mathrm{d}\theta^2 + \sin^2\theta\mathrm{d}\phi^2\right)\right], \tag{3}$$

where $R^2 = \exp\left[A\xi^{\frac{1}{L}}\exp\left(\frac{\varepsilon}{L}\eta\right) + \frac{1+2L^2}{2L}\ln\xi + \frac{1-2L^2}{2L}\varepsilon\eta\right]$, $\varepsilon = \pm 1$, L, A constants.

2 Pressure and density

Using Einstein's equations *in reverse* we obtain:

$$8\pi p = G_2{}^2 = \frac{1}{2R^2}(2X+1), \quad 8\pi\rho = G_2{}^2 - G_1{}^1 - G_4{}^4 = \frac{1}{2R^2}(1-6X), \tag{4}$$

where $X = A\xi^{1/L}e^{\epsilon\eta/L}$. We find that the weak, strong and dominant energy conditions[3] are fulfilled iff $X \le \frac{1}{6}$, $X \le \frac{1}{2}$ and $X \le 0$ respectively.

Taking u^4 to be positive we find that the velocity four-vectors are given by:

$$u^4 \equiv \frac{d\eta}{ds} = \frac{-\sqrt{2}\,(2X - 2L^2 - 1)}{4LR\sqrt{1-2X}}, \quad u^1 \equiv \frac{d\xi}{ds} = \frac{-\epsilon\xi\sqrt{2}\,(2X + 2L^2 - 1)}{4LR\sqrt{1-2X}}. \tag{5}$$

These expression may be checked using the following relation obtained from the equations of motion:

$$\rho\frac{\partial}{\partial\xi}(u_4) + \frac{\partial}{\partial\xi}(pu_4) = \rho\frac{\partial}{\partial\eta}(u_1) + \frac{\partial}{\partial\eta}(pu_1). \tag{6}$$

3 Speed of sound and ratio of gradients

Denoting the radial coordinate and the time coordinate for an orthogonal comoving system by r and t respectively, the speed of sound is given by:

$$v^2_{\text{sound}} = \frac{\frac{\partial p}{\partial t}}{\frac{\partial \rho}{\partial t}} = \frac{\frac{\partial p}{\partial \eta}\frac{\partial r}{\partial \xi} - \frac{\partial p}{\partial \xi}\frac{\partial r}{\partial \eta}}{\frac{\partial \rho}{\partial \eta}\frac{\partial r}{\partial \xi} - \frac{\partial \rho}{\partial \xi}\frac{\partial r}{\partial \eta}} = \frac{X}{2-3X}. \tag{7}$$

The speed of sound is thus less than the speed of light in a vacuum iff $X \le \frac{1}{2}$ or $X \ge \frac{2}{3}$. Moreover, the sound speed is real iff $0 \le X \le \frac{2}{3}$.

The ratio of the pressure and the density gradients with respect to comoving radial coordinates is given by:

$$\frac{\frac{\partial p}{\partial r}}{\frac{\partial \rho}{\partial r}} = \frac{e^{2\lambda}u^4\frac{\partial p}{\partial \xi} + e^{2\mu}u^1\frac{\partial p}{\partial \eta}}{e^{2\lambda}u^4\frac{\partial \rho}{\partial \xi} + e^{2\mu}u^1\frac{\partial \rho}{\partial \eta}} = 1. \tag{8}$$

The pressure and the density thus differ by a function of time t only.

4 Comoving coordinates

To obtain the comoving coordinates one has to solve the following partial differential equations

$$u^1\frac{\partial r}{\partial \xi} + u^4\frac{\partial r}{\partial \eta} = 0 \quad \text{and} \quad e^{2\mu}u^1\frac{\partial t}{\partial \eta} + e^{2\lambda}u^4\frac{\partial t}{\partial \xi} = 0. \tag{9}$$

We obtain the solution

$$r(\xi,\eta) = F_1\left[\frac{e^{2X}}{2X}\left(\frac{Z}{Y}\right)^{2L^2}\right] \quad \text{and} \quad t(\xi,\eta) = F_2\left[\frac{e^{2X}}{2X}\left(\frac{Y}{Z}\right)^{2L^2}\right], \tag{10}$$

where $Y = 2A\xi^{1/L}$, $Z = e^{\epsilon\eta/L}$. F_1 and F_2 are here arbitrary functions. To have a simple model we choose $F_1 = -\ln$ and $F_2 = \ln$.

With comoving coordinates we write the line element as follows:

$$ds^2 = C^2 dt^2 = B^2 dr^2 - R^2\left(d\theta^2 + \sin^2\theta d\phi^2\right). \tag{11}$$

We now obtain

$$8\pi\frac{\partial p}{\partial r} = 8\pi\frac{\partial \rho}{\partial r} = -\frac{1}{4R^2(1-2X)} = -4\pi\frac{\frac{\partial}{\partial r}(C^2)}{C^2}(\rho+p), \tag{12}$$

$$8\pi\frac{\partial p}{\partial t} = \frac{X^2}{(1-2X)R^2}, \qquad \frac{\partial R^2}{\partial t} = \frac{X}{2X-1}R^2, \tag{13}$$

$$8\pi\frac{\partial \rho}{\partial t} = \frac{X(2-3X)}{(1-2X)R^2} = -8\pi\left(\frac{1}{2}\frac{\frac{\partial}{\partial t}(B^2)}{B^2} + 2\frac{\frac{\partial}{\partial t}R}{R}\right)(\rho+p). \tag{14}$$

We thus have a contracting fluid with the pressure and the density increasing during contraction, and decreasing outwards.

5 Mass function

The mass function m is given by[4,5]

$$m = \frac{R}{2}\left[1 + \frac{\left(\frac{\partial R}{\partial t}\right)^2}{C^2} - \frac{\left(\frac{\partial R}{\partial r}\right)^2}{B^2}\right] = \frac{4}{3}\pi\rho R^3 + E = \frac{(1-2X)R}{4} \tag{15}$$

where E is interpreted as pure gravitational field energy[6].

We now find $E = \frac{1}{6}R$ and further

$$\frac{\partial m}{\partial r} = \frac{4}{3}\pi\rho\frac{\partial}{\partial r}(R^3) = \frac{R(1-6X)}{16(1-2X)}, \quad \frac{\partial m}{\partial t} = -\frac{4}{3}\pi p\frac{\partial}{\partial t}(R^3) = \frac{RX(1+2X)}{8(1-2X)}, \tag{16}$$

$$\frac{\partial E}{\partial r} = -\frac{4}{3}\pi R^3\frac{\partial \rho}{\partial r} = \frac{R}{24(1-2X)}, \quad \frac{\partial E}{\partial t} = -\frac{4}{3}\pi R^3(\rho+p)\left(\frac{\dot R}{R} - \frac{\dot B}{B}\right) = -\frac{RX}{12(1-2X)}. \tag{17}$$

From the last relation it is immediately clear that our model contains shear.

To examine if we can match this model with an exterior vacuum we put the pressure equal to zero and obtain $X_{\text{boundary}} = -\frac{1}{2}$ and we further find $m(r_{\text{boundary}}, t) =$ constant $\equiv M$. We thus have: $R(r_{\text{boundary}}, t) = 2M$ and also $\frac{\partial}{\partial t}R(r_{\text{boundary}}, t) = 0$. However, from Eq. 13 with $X = -\frac{1}{2}$ we obtain $\frac{\partial}{\partial t}R(r_{\text{boundary}}, t) = \frac{M}{4} \neq 0$. We conclude that a zero pressure surface does not exist.

For a trapped surface to exist we must have[4] $\frac{2m}{R} > 1$. However, for our model we have $\left(\frac{2m}{R}\right)_{\text{center}} = \frac{1}{2}$ and $\frac{\partial}{\partial r}\left(\frac{2m}{R}\right) = \frac{X}{2(2X-1)} < 0$.

The components of the Riemann–tensor $R^a{}_{bcd}$ which correspond to tidal forces are the components $R^i{}_{4j4}$. We find $R^1{}_{414} = 0, R^2{}_{424} = R^3{}_{434} = \frac{1}{16L^2}\left[(2X + 2L^2 - 1)^2 - 8X\right]$. Hence, there are no tidal forces in the radial direction.

6 Kinematic quantities of the velocity field

We find (semicolon denotes covariant differentiation)

$$\text{Acceleration} = u_{i;j}u^j = \left(\frac{2X - 2L^2 - 1}{4L\xi[1 - 2X]^2}, 0, 0, \frac{\varepsilon\,[2X + 2L^2 - 1]}{4L[1 - 2X]^2}\right) \tag{18}$$

$$\text{Expansion} = u^i{}_{;i} = \frac{2\sqrt{2}\varepsilon X(2 - 3X)}{R(1 - 2X)^{3/2}}. \tag{19}$$

Shear components $= \sigma_{ik}$:

$$\sigma_{11} = \frac{\sqrt{2}\varepsilon RX}{12L^2\xi^2(1 - 2X)^{5/2}}\left\{4L^2\left[2X - (L^2 + 1)\right] - (2X - 1)^2\right\}, \tag{20}$$

$$\sigma_{14} = \frac{\sqrt{2}RX}{12L^2\xi(1 - 2X)^{5/2}}\left[4L^2 - (2X - 1)^2\right], \tag{21}$$

$$\sigma_{44} = \frac{-\sqrt{2}\varepsilon RX}{12L^2(1 - 2X)^{5/2}}\left\{4L^2\left[2X + (L^2 - 1)\right] + (2X - 1)^2\right\}, \tag{22}$$

$$\sigma_{22} = \frac{1}{\sin^2\theta}\sigma_{33} = \frac{\sqrt{2}\varepsilon RX}{3(1 - 2X)^{3/2}} \tag{23}$$

and the shear invariant is given by

$$\sigma_{ik}\sigma^{ik} = \frac{X}{(2 - 3X)^2(1 - 2X)}. \tag{24}$$

Acknowledgements

The author would like to express his gratitude to K S Govinder for the skillful typing of the manuscript.

References

1. G. C. McVittie and R. J. Wiltshire, *Int. J. Theor. Phys.* **16** (1977) 121.

2. D. Kramer, H. Stephani, M. MacCallum and E. Herlt, *Exact Solutions of Einstein's Field Equations* (Cambridge University Press, Cambridge, 1980) p. 165–173.

3. S. W. Hawking and G. F. R. Ellis, *The Large Scale Structure of Space-time* (Cambridge University Press, Cambridge, 1973) p. 88–96.

4. W. C. Hernandez and C. W. Misner, *Astrophys. J.* **143** (1966) 452.

5. C. W. Misner and D. H. Sharp, *Phys. Rev.* **B136** (1964) 571.

6. E. N. Glass, *J. Math. Phys.* **20** (1979) 1508.

STATIC SPHERICALLY SYMMETRIC SOLUTIONS OF SCALAR GRAVITY WITH SU(2) GAUGE FIELD

GEORGE LAVRELASHVILI* and DIETER MAISON

Max-Planck-Institut für Physik
- Werner-Heisenberg-Institut -
P.O.Box 40 12 12, Munich (Fed. Rep. Germany)

ABSTRACT

A system of a $SU(2)$ Yang-Mills field coupled to a scalar graviton (dilaton) is studied. A discrete family of globally regular solutions of finite energy is found.

1. Introduction

We have performed a detailed numerical and analytical study of static spherically symmetric solutions of the Yang-Mills-Dilaton system for the gauge group $SU(2)$. The results show a striking similarity to those obtained for the Einstein-Yang-Mills (EYM) theory [1,2]. We find [3] a discrete family of globally regular solutions characterized by an increasing number N of zeros of the Yang-Mills (YM) potential. The mass M_N of the solutions is determined by the dilatonic coupling constant γ playing the rôle of Newton's constant. With increasing N the mass tends to a finite limit. All these solutions are unstable, supporting their interpretation as a kind of sphaleron.

As to be expected we found similar solutions for the combined Einstein-YM-Dilaton system [4]. Here we will restrict ourselves to solutions in flat space.

---

*On leave of absence from Tbilisi Mathematical Institute, 380093 Tbilisi, Georgia

2. Ansatz and Field Equations

The action for a dilaton coupled to the YM field is given by

$$S = \int \left[\frac{1}{2}(\partial\varphi)^2 - \frac{e^{\gamma\varphi}}{4g^2}F^2\right] d^4x \tag{1}$$

For the $SU(2)$ YM potential we make the usual (purely 'magnetic') spherically symmetric ansatz

$$W_0^a = 0, \quad W_i^a = \epsilon_{aik}\frac{x^k}{r^2}(W(r) - 1) \tag{2}$$

Substituting this ansatz into the action we obtain a reduced action (where a prime denotes $\frac{d}{dr}$)

$$S_{\rm red} = -\int \left[\frac{r^2}{2}\varphi'^2 + \frac{e^{\gamma\varphi}}{g^2}\Big(W'^2 + \frac{(W^2-1)^2}{2r^2}\Big)\right] dr \tag{3}$$

where γ resp. g denote the dilatonic resp. gauge coupling constant.

Rescaling $\varphi \to \varphi/\gamma$, $r \to \frac{\gamma}{g}r$ and $S \to g\gamma S$ removes the dependence on γ and g from S, hence we may put $\gamma = g = 1$ without restriction. The resulting field equations are

$$W'' = \frac{W(W^2-1)}{r^2} - \varphi' W' , \tag{4a}$$

$$(r^2\varphi')' = e^{\varphi}\Big(W'^2 + \frac{(W^2-1)^2}{2r^2}\Big) \tag{4b}$$

These equations are invariant under a shift $\varphi \to \varphi + \varphi_0$ accompanied by a simultaneous rescaling $r \to r e^{\varphi_0/2}$, hence globally regular solutions can be normalized to $\varphi(\infty) = 0$.

The total energy resp. mass M of the solutions is given by $M = -S_{\rm red}$. From the stationarity of the action under an infinitesimal scale transformation one obtains the virial theorem

$$\frac{M}{2} = \int \frac{r^2}{2}\varphi'^2\, dr = \int e^{\varphi}\Big(W'^2 + \frac{(W^2-1)^2}{2r^2}\Big)\, dr . \tag{5}$$

The points $r = 0$ and $r = \infty$ are singular points of Eqs 4. Hence only a subset of all solutions remains finite at these points. Employing the method described in [5] one can show that there is a 2-parameter family of regular solutions at $r = 0$ with the behaviour

$$W(r) = 1 - br^2 + O(r^4) , \tag{6a}$$

$$\varphi(r) = \varphi_0 + b^2 r^2 + O(r^4) . \tag{6b}$$

The generic solution with these initial conditions develops a singularity of φ at some value $r = r_s(b)$ with a finite value for $W(r_s)$ and $W'(r_s) = 0$. Only by fine-tuning of the parameters b and φ_0 can one hope to find globally regular solutions. The numerical results described in the following suggest that this yields an infinite discrete sequence of corresponding values of the parameters b and φ_0.

The asymptotic behaviour of solutions regular at $r = \infty$ is given by the 2-parameter family

$$W(r) = \pm(1 - \frac{c}{r} + O(r^{-2})) , \tag{7a}$$

$$\varphi(r) = \frac{d}{r} + O(r^{-4}) . \tag{7b}$$

Globally regular solutions have to interpolate smoothly between the asymptotic behaviour at $r = 0$ and $r = \infty$. It is easy to see that this is only possible, if the YM-potential W remains in the strip $-1 \leq W(r) \leq 1$.

3. Numerical Results

Using a suitably desingularized version of Eqs. 4 at $r = 0$ we have integrated these equations from $r = 0$ numerically, varying the parameters b and φ_0. For $b < b_1 \approx 31.9$ the function φ develops a singularity for some $r = r_s(b)$ with $W(r_s) < -1$. On the other hand for large enough values of parameter b the singularity of φ occurs before $W(r)$ has a zero. This behaviour is completely analogous to the one found for the EYM system. There it has been used to prove rigorously the existence of a globally regular solution with exactly one zero of W [6].

Fine-tuning the parameters b and φ_0 we found a discrete set $\{ b_N, \varphi_N^0 \}$ for which the YM-potential has N zeros and then approaches $(-1)^N$. Since with increasing N the asymptotic domain moves very rapidly to large values of r, we computed only solutions up to $N = 6$.

Due to the finite precision in the determination of $\{ b_N, \varphi_N^0 \}$ and numerical errors $W(r)$ eventually leaves the strip $|W| \leq 1$ and φ becomes singular. The numerically determined values of the parameters of the solutions up to $N = 6$ are collected in Tab. 1.

Tab. 1: Parameters for the solutions $N = 1 \ldots 6$

N	b	φ_0	M
1	$3.19839\ 10^1$	-3.422824	1.60751
2	$1.20505\ 10^3$	-6.747806	1.93112
3	$4.53391\ 10^4$	-10.31646	1.98863
4	$1.70639\ 10^6$	-13.93446	1.99814
5	$6.42034\ 10^7$	-17.56051	1.99970
6	$2.41551\ 10^9$	-21.18785	1.99995

4. Concluging Remarks

The solution we found was shown to be unstable in linearized perturbation theory against small time dependent spherically symmetric perturbations [3].

Note that the gauge configuration Eq. 2 with the asymptotic behaviour Eq. 7a has the Chern-Simons index equal to one half [7,8,9].

So a natural interpretation of these saddle point gauge configurations with half-integer Chern-Simons index is as a that of a (dilatonic) sphaleron.

5. Acknowledgements

We are indebted to P. Breitenlohner and P. Forgács for interesting discussions. One of us (G.L.) is grateful to M.C. Bento, O. Bertolami and J.M. Mourão for the invitation to the First Iberian Meeting on Gravity and for the warm hospitality extended to him during his stay in Évora.

References

1. R. Bartnik and J. McKinnon, *Phys. Rev. Lett.* **61** (1988) 141.
2. H.P. Künzle and A.K.M. Masood-ul-Alam, *J. Math. Phys.* **31** (1990) 928.
3. G. Lavrelashvili and D. Maison, *Phys. Lett.* **B 295** (1992) 67.
4. G. Lavrelashvili and D. Maison, *Regular and Black hole solutions of Einstein-Yang-Mills-Dilaton theory* , preprint MPI-Ph/92-115 (1992).
5. P. Breitenlohner, P. Forgács, and D. Maison, *Nucl.Phys.* **B383** (1992) 357.
6. J.A. Smoller, A.G. Wasserman, S.T. Yau, and J.B. McLeod, *Commun. Math. Phys.* **143** (1991) 115.
7. R. Dashen, B. Hasslacher and A. Neveu, *Phys. Rev.* **D 10** (1974) 4138.
8. N. Manton, *Phys. Rev.* **D 28** (1983) 2019,
   F. Klinkhamer and N. Manton, *Phys. Rev.* **D 30** (1984) 2212.
9. I. Moss, A. Wray, *Phys. Rev.* **D 46** (1992) R1215.

NAKED SINGULARITIES AND THE COSMIC CENSORSHIP

JOSÉ P. S. LEMOS

Departamento de Física, Instituto Superior Técnico,
Av. Rovisco Pais 1, 1096 Lisboa, Portugal, &
Departamento de Astrofísica, Observatório Nacional-CNPq
Rua General José Cristino 77, 20921 Rio de Janeiro, Brazil.

ABSTRACT

A summary of some important aspects related to the appearance of naked singularities in gravitational collapse is given. It is pointed out that the collapse of null radiation has the same properties of the collapse of dust matter. This has implications on the cosmic censorship conjecture.

Almost thirty years ago Wheeler[1] introduced into the theory of gravitational collapse the important concept of the issue of the final state. There is still no solution to this problem, we do not know what are the generic features of gravitational collapse. The singularity theorems of Penrose and Hawking[2,3] state, grosso modo, that if a black hole forms then a singularity necessarily forms. But what happens if a singularity forms first? Does a black hole appear afterwards? A striking hypothetical example was suggested by Penrose[4] in which the formation of a naked singularity is followed by the production of a shell of radiation of ∞ intensity which destroys the Universe.

That the issue of the final state is far from being settled, one can notice by the several conjectures related with gravitational collapse: (1) the hoop conjecture of Thorne[5] which states that a black hole forms when and only when mass M gets compacted into a region whose circumference in every direction is $C \lesssim 4\pi M$, (2) the trapped surface conjecture of Seifert[6] saying that massive singularities have to be trapped, (3) the event horizon conjecture of Israel[7] where an event horizon is formed when matter is sufficiently compressed, and (4) the compactness conjecture of Malec[8] stating that the formation of a black hole is possible if and only if matter is compressed in all three directions.

There is still (5) the cosmic censorship conjecture of Penrose[9] which states that spacetime singularities are enclosed by an event horizon and cannot be seen by external observers, i.e., there are no naked singularities. This is probably the most important conjecture, and the hardest to prove, because it splits the Theory of General Relativity. If cosmic censorship is valid General Relativity is a well-defined deterministic theory, otherwise the Universe may destroy itself by the production of an ∞ amount of uncontrollable radiation. An opposite way of thinking is that cosmic censorship might not be so significant and what we need is more quantitative statements related to observation. This was championed by Clarke [10] when he says:

"cosmic censorship is bound up with the question of wether or not we can observe quantum gravitational processes in the Universe".

Strictly speaking cosmic censorship does not work. There are now many counter-examples to this conjecture. Naked singularities form in the gravitational collapse of various different geometrical configurations and various different types of matter. Analytical studies show that they form in the spherical collapse of dust and of null radiation as we will discuss in detail in a moment, and in the collapse of highly elongated shells of null matter[11]. Naked singularities can also appear in the spherical collapse of matter with pressure and soft equation of state, as shown in numerical studies[12,13], as well as in analytical studies[14]. They also can appear in the prolate collapse of collisionless particles[15], although the nakedness of these singularities might be an artifact of the numerical codes used[16].

In this comunication I am going to focus attention on the unification of the collapses of dust and of null radiation in spherical symmetry[17]. I am going to analyze qualitatively the main features of the collapse. The reader may consult reference [17] for a quantitative analysis.

It is worth to have an historical perspective of the appearance of naked singularities within the theory of gravitational collapse. The work on spherical collapse was pioneered in 1934 by Tolman[18] who found the metric for spacetimes with dust, and was further extended by Bondi[19]. Another pioneering paper was the now famous Oppenheimer and Snyder[20] paper. They showed that in a homogeneous collapse a black hole forms and there are no naked singularities. Another development was given by Vaidya[21] in 1951 who found the metric for spherical symmetric spacetimes with null radiation.

Both Tolman-Bondi and Vaidya spacetimes admit naked singularities. In these spacetimes there are two types of naked singularities: shell crossing and shell focusing. Shell crossing singularities arise when shells of dust or radiation cross each other, while shell focusing appear when the shells collapse and stay at the centre. Historically, shell crossing were the first to appear[22–25]. The shell crossing naked singularities were not considered a serious treat to cosmic censorship since it was invoked they had Newtonian analogues. Cosmic censorship was also strengthen by the stability analysis on the Kerr-Newman family of black holes, which showed that small perturbations would destroy the interior naked singularities existent in that family. However, a curious model [26] where the Vaidya metric is used to drain energy from the collapsing cloud also produced a naked singularity as an endpoint. This model didnot have the shell-crossing problem.

A turning point was the work of Eardley and Smarr[27], where they showed for the first time that a central naked singularity could form in the marginally bound collapse of dust, with the blowing up of the curvature scalars. They coined these singularities, shell focusing singularities. Further work by Christodolou[28], Newman[29,30], Lake and collaborators[31–34], and others[35–39] enlarged the spectrum of naked singularities and their main features, through the study of bound, marginally bound and unbound dust collapses.

It was also found that shell focusing singularities would develop in the collapse of null radiation, notably by Hiscock, Williams and Earddley[40], Kuroda[41], Papapetrou[42], and developed by other authors[43–48].

In the marginally bound dust case, the formation of these singularities can be understood in terms of the inhomogeneity of the collapse[49]. If the collapse is homogeneous, (i.e., an Oppenheimer-Snyder collapse), or nearly homogeneous no naked singularity occurs. This is because all the matter collapses at the same comoving time, and thus, there is sufficient mass concentration for the appearance of a black hole, (i.e., for the appearance of trapped surfaces), prior to the formation of the singularity. On the other hand, if the collapse is sufficienltly inhomogeneous, such that the outer shells collapse much later than the central shells, then a naked singularity occurs. This means that light rays can emanate from the singularity and be seen by observers. The first light ray to appear is the Cauchy horizon. So, one can say that a naked singularity exists if the spacetime contains a Cauchy Horizon. Beyond the Cauchy horizon, to the future of the singulairity one cannot predict, and some form of determinism is lost even within classical General Relativity. Eardley and Smarr found that if the inhomogeneity parameter B satisfies the inequality, $B \geq \frac{16}{3} + 5\sqrt{3}$, then a Cauchy horizon develops and there is a naked singularity.

Now, in the marginally bound collapse the dust has zero energy per unit mass $\frac{E}{m}$ at ∞, or in other words, the matter starts to collapse at ∞ with zero velocity. In the unbound case the matter starts its collapse with a non-zero $\frac{E}{m}$. One finds[39] that to produce naked singularities, one needs now a higher inhomogeneity, (a higher B), in the collapse. This is expected. As the ratio $\frac{E}{m}$ increases one has that the inhomogeneity parameter necessary to produce a naked singularity is $B \gtrsim 16\sqrt{2E}$ asymptotically. When, eventually, $\frac{E}{m} = \infty$, the dust has zero rest mass, $m = 0$. In this limiting case the Tolman-Bondi metric transforms into the Vaidya metric[17]. A Cauchy horizon then develops when the inhomogeneity parameter for the Vaidya metric satisfies, $\mu \equiv \frac{B}{\sqrt{2E}} \geq 16$.

A criticism raised against the reality of these naked singularities is that the fluid approximation is not a fundamental description of nature, and when ∞ densities and curvatures are achieved one needs a more fundamental description in terms of fields. Now, we have just seen that the Vaidya collapse belongs to the Tolman-Bondi family of collapses. The collapse of null radiation is easier to treat than the collapse of dust matter. Thus generic features of the colapse of clouds of dust matter can be taken simply from the study of the collapse of null radiation. This has implications for the cosmic censorship conjecture. For instance, if one shows that the naked singularities arising in the Vaidya metric can be derived from more fundamental (massless) fields, then the naked singularities which form in Tolman-Bondi spacetimes can also be derived from more fundamental (massive) fields.

Acknowledgements – I thank JNICT for a grant to participate in the Évora meeting. I also thank the organizing committee for providing an excellent scientific atmosphere during the meeting.

References

1. J. A. Wheeler, (1964), in *Relativity, Groups and Topology*, eds. C. and B. DeWitt, (Gordon and Breach, New York).
2. R. Penrose, (1965),*Phys. Rev. Lett.*, **14**, 57.
3. S. W. Hawking, R. Penrose, (1970), *Proc. Roy. Soc. Lond.*, **314**, 529.
4. R. Penrose, (1978), in *Theoretical Principles in Astrophysics and Relativity*, eds. N. R. Lebovitz, W. H. Reid, P. O. Vandervoort, (Chicago University Press, Chicago).
5. K. S. Thorne, (1972), in *Magic without Magic*, ed. J. Klauder, (Freeman, S. Francisco).
6. H. J. Seifert, (1979), *Gen. Rel. Grav.*, **10**,1065.
7. W. Israel, (1986), *Phys. Rev. Lett.*, **56**, 789.
8. E. Malec, (1991), *Phys. Rev. Lett.*, **67**, 949.
9. R. Penrose, (1969), *Riv.Nuovo Cimento*, **1**, 252.
10. C. J. S. Clarke, (1987), in *Highlights in Gravitation and Cosmology*, eds. B. R. Iyer, A. Kembhavi, J. V. Narlikar, C. V. Vishveshwara, (Cambridge University Press, Cambridge).
11. C. Barrabés, W. Israel, P. S. Letelier, (1991), *Phys. Lett. A*, **160**, 41.
12. A. Ori, T. Piran, (1987), *Phys. Rev. Lett.*, **59**, 2137.
13. A. Ori, T. Piran, (1990), *Phys. Rev. D*, **42**, 1068.
14. P. S. Joshi, I. H. Dwiwedi, (1993) *Comm. Math. Phys.*,to appear.
15. S. L. Shapiro, S. A. Teukolsky, (1991), *Phys. Rev. Lett*, **66**, 994.
16. R. M. Wald, V. Iyer, (1991), *Phys. Rev. D*, **44**, 3719.
17. J. P. S. Lemos, (1992), *Phys. Rev. Lett.*, **68**, 1447.
18. R. C. Tolman, (1934), *Proc. Nat. Acad. Sci*, **20**, 169.
19. H. Bondi, (1947), *Mon. Not. R. astr. Soc.*, **107**, 410.
20. J. R. Oppenheimer, H. Snyder, (1939), *Phys. Rev. D*, **56**, 455.
21. P. C. Vaidya, (1951), *Proc. Indian Acad. Sci. A*, **33**, 264.

22. P. Yodzis, H. J. Seifert, H. Muller zum Hagen, (1973), *Comm. Math. Phys.*, **34**, 135.

23. P. Yodzis, H. J. Seifert, H. Muller zum Hagen, (1974), *Comm. Math. Phys.*, **37**, 29.

24. P. Szekeres, (1980), in *Gravitational Radiation, Collapsed Objects and Exact Solutions*, ed. C. Edwards, (Springer Verlag, Berlin).

25. P. Szekeres, (1992), in *Abstracts of Contributed Papers of the 13th International Conference On General Relativity and Gravitation*, eds. P. W. Lamberti, O. E. Ortiz, (Cordoba, Argentina).

26. B. Steinmuller, A. R. King, J. P. Lasota, (1975), *Phys. Lett. A*, **51**, 191.

27. D. M. Eardley, L.Smarr, (1979), *Phys. Rev. D*, **19**, 2239.

28. D. Christodoulou, (1984), *Comm. Math. Phys.*, **93**, 171.

29. R. P. A. C. Newman, (1986), *Class. Quantum Grav.*, **3**, 527.

30. R. P. A. C. Newman, (1986), in *Topological Properties and Global Structure of Space - Time*, eds. P. G. Bergmann, V. de Sabbata, (Plenum, New York).

31. B. Waugh, K. Lake, (1988), *Phys. Rev. D*, **38**, 1315.

32. B. Waugh, K. Lake, (1989), *Phys. Rev. D*, **40**, 2137.

33. K. Lake, T. Zannias, (1990), *Phys. Rev. D*, **41**, 3866.

34. K. Lake, *Phys. Rev. Lett.*, **68**, 3129.

35. V. Gorini, G. Grillo, M. Pelliza, (1989), *Phys. Lett. A*, **135**, 154.

36. G. Grillo, (1991), *Class. Quantum Grav.*, **8**, 739.

37. I. H. Dwiwedi, S. Dixtit, (1991), *Prog. Theor. Phys.*, **85**, 433.

38. R. N. Henriksen, K. Patel, (1991), *Gen. Rel. Grav.*, **23**, 527.

39. J. P. S. Lemos, (1991), *Phys. Lett. A*, **158**, 279.

40. A. Hiscock, L. C. Williams, D. M. Eardley, *Phys. Rev. D.*, **26**, 751.

41. Y. Kuroda, (1984), *Prog. Theor. Phys.*, **72**, 63.

42. A. Papapetrou, (1986), in *A Random Walk in Relativity and Gravitation*, eds. N. Dadich, J. K. Rao, J. V. Narlikar, C. V. Vishveswara, (Wiley, New York).

43. K. Lake, (1986), *Phys. Lett. A*, **116**, 17.

44. G. Hollier, (1986), *Class. Quantum Grav.*, **3**, L111.

45. K. Rajagopal, K. Lake, (1987), *Phys. Rev. D*, **35**, 1531.

46. K. Lake, (1992), *Phys. Rev. D*, **43**, 1416.

47. I. H. Dwivedi, P. S. Joshi, (1991), *J. Math. Phys.*, **32**, 2167.

48. I. H. Dwivedi, P. S. Joshi, (1991), *Phys. Rev. D.*, **45**, 2147.

49. D. M. Eardley, (1987), in *Gravitation in Astrophysics*, eds. B. Carter, J. B. Hartle, (Plenum, New York).

SYMMETRIC SOLUTIONS OF EINSTEIN-YANG-MILLS EQUATIONS

JOSÉ MOURÃO [1]

Departamento de Física
Instituto Superior Técnico
Av. Rovisco Pais
1096 Lisboa Codex, Portugal

ABSTRACT

Symmetric gauge fields and invariant metrics in homogeneous spaces are found. Their use for finding exact solutions of the Einstein-Yang-Mills (EYM) equations is discussed.

1. Introduction

The search for exact solutions in general relativity can usually be divided in two parts. The first part, which we call "kinematical", consists in the restriction to a subset of fields in the space of field configurations (i.e. in the making of appropriate ansätze).

This restriction is made in accordance with the physical problem in consideration and is achieved by considering only the configurations which are invariant (or symmetric) under the action of a symmetry group S. For example in the case of closed, flat and open Friedmann-Robertson-Walker cosmological models, S is respectively $SO(4)$, E^3, $SO(1,3)$ and plays the role of group of spatial homogeneity and isotropy.

In the second (or "dynamical") part of the search for exact solutions in General Relativity we study the equations implied by the original theory in the set of S-symmetric configurations. The number of independent variables in these equations is equal to the dimensionality of space-time minus the dimensionality of the orbits of the symmetry group.

Here we will describe, following [1,2], a method of finding invariant metrics and symmetric gauge fields used in the kinematical part of the search for solutions of the EYM equations (see also [3,4]).

2. Invariant Metrics and Symmetric Gauge Fields

Consider a space-time of the form

$$M = \tilde{M} \times S/R \quad , \tag{1}$$

[1] e-mail: JMOURAO@PTIFM.bitnet

where S and R are Lie groups, and assume that we are interested in S-symmetric solutions of the EYM equations. The gauge group K is assumed to be a compact Lie group.

Let us for further simplicity restrict ourselves to a homogeneous space-time

$$M = S/R \ . \tag{2}$$

The reader interested in the general case (1) is referred to the contribution by Kapetanakis and Zoupanos in the present volume and to the literature [1-5]. Notice that the S-invariant ansatz for the metric and the S-symmetric (i.e. invariant up to a gauge transformation) ansatz for the gauge field, discussed here, are always present in the general case.

We first chose a prefered moving frame in S/R and describe the S-invariant ansatz for the metric in S/R.

Consider the canonical ($\mathcal{S} \equiv Lie(S)$-valued) left-invariant one-form on the group manifold of S

$$\theta = \theta^\alpha T_\alpha = s^{-1} ds \ ,$$

where the T_α are the generators of S and θ^α are left-invariant one-forms on S. The form θ is very important for the study of the invariant geometry of the Lie group S and its pull back to S/R plays, as we shall see below, an analogous role in the study of the invariant geometry of the coset space S/R. The group S can be considered as a principal bundle over S/R with structure group R (we assume R to be a closed subgroup of S and S/R to be a reductive coset space). Let us chose a (local) section σ of this bundle in a neighborood of the origin $o \equiv [e] \equiv R$

$$\sigma : \mathcal{U} \subset S/R \longrightarrow S$$

and pull the form θ back to S/R with respect to σ:

$$\bar{\theta}_x = (\sigma^* \theta)_x = \sigma^{-1}(x) d\sigma(x) = \bar{\theta}^\alpha T_\alpha \ . \tag{3}$$

The $\mathcal{S}$-valued one-form $\bar{\theta}$ in (3) is called the Maurer-Cartan one-form in S/R. A local moving coframe can be obtained from the Maurer Cartan one form as follows. Let

$$\mathcal{S} = \mathcal{R} + \mathcal{M} \tag{4}$$

be a reductive decomposition of $\mathcal{S}$, where $\mathcal{R} \equiv Lie(R)$ and $\mathcal{M}$ is a reductive (i.e. $[\mathcal{R}, \mathcal{M}] \subset \mathcal{M}$) complementary subspace and

$$\{T_\alpha\} = \{T_a, T_{\dot{a}}\} \ \ , T_a \in \mathcal{M}, \ T_{\dot{a}} \in \mathcal{R} \tag{5}$$

be an adapted basis in $\mathcal{S}$. We have the natural decomposition of the Maurer-Cartan one-form associated with (5)

$$\bar{\theta} = \bar{\theta}_{\mathcal{M}} + \bar{\theta}_{\mathcal{R}} \equiv \bar{\theta}^a T_a + \bar{\theta}^{\dot{a}} T_{\dot{a}} \tag{6}$$

Unlike the group case S, the one-forms $\bar{\theta}^{\alpha}$ in S/R are, in general, not S-invariant. Nevertheless they have simple transformation laws under the action of S. For given $s \in S, x \in \mathcal{U}$ and σ, let $r_s(x;\sigma)$ be the element of the isotropy group R defined by

$$\sigma(sx) = s\sigma(x)r_s(x;\sigma) \quad .$$

Then the transformation law of $\bar{\theta}^{a}$ and $\bar{\theta}^{\dot{a}}$ under the action of $s \in S$ can be easily obtained from

$$s^*\bar{\theta}_{\mathcal{M}} = ad\ r_s^{-1}(x)\bar{\theta}_{\mathcal{M}} \tag{7a}$$

$$s^*\bar{\theta}_{\mathcal{R}} = ad\ r_s^{-1}(x)\bar{\theta}_{\mathcal{R}} + r_s^{-1}(x)dr_s(x) \quad , \tag{7b}$$

where ad denotes the adjoint representation of S. The reductiveness of the decomposition (4) implies that the restriction of ad to the subgroup R has $\mathcal{M}$ as an invariant subspace. The representation of R in $\mathcal{M}$ obtained in this way is called the isotropy representation and is denoted by ad_R. Let $\{X_a\}$ be the frame dual to $\{\bar{\theta}^a\}$. It is clear from (7a) that under s_* the vectors X_a in this frame transform in the same way the T_a do under the isotropy action of $r_s^{-1}(x)$ (i.e. $ad\ r_s^{-1}(x)$). Let $B(.,.)$, be an ad_R-invariant scalar product in $\mathcal{M}$ with components $B_{ab} = B(T_a, T_b)$. Then we conclude [6] that the metric in S/R with the same components B_{ab} in the frame $\{X_a\}$,

$$\gamma = B_{ab}\bar{\theta}^a \otimes \bar{\theta}^b \tag{8}$$

is S-invariant. Moreover any S-invariant metric in S/R can be constructed in this way. Therefore we have succeded in reducing the (differential geometry) problem of finding S-invariant metrics in the coset space S/R to the purely algebraic problem of finding ad_R-invariant scalar products in $\mathcal{M}$. The equation (8) with B_{ab} being an ad_R-invariant scalar product in $\mathcal{M}$ gives the ansatz for S-invariant metrics in S/R.

In an analogous way [1-5] the problem of finding S-symmetric gauge fields with gauge group K in S/R is reduced by Wang's theorem [6] to the algebraic problem of intertwining equivalent representations of the isotropy group. To see this recall that, according to Wang's theorem, the most general form of a S-symmetric gauge field in S/R is given by

$$A = \lambda(\bar{\theta}_{\mathcal{R}}) + \Lambda(\bar{\theta}_{\mathcal{M}}) \quad , \tag{9}$$

where

$$\lambda : \mathcal{R} \longrightarrow \mathcal{K} = Lie(K) \tag{10a}$$

is an homomorphism from the isotropy algebra $\mathcal{R}$ to the Lie algebra of the gauge group $\mathcal{K}$ and Λ is a mapping from $\mathcal{M}$ to $\mathcal{K}$

$$\Lambda : \mathcal{M} \longrightarrow \mathcal{K} \tag{10b}$$

satisfying the linear property

$$\Lambda(ad(r)u) = ad(\lambda(r))\Lambda(u) \quad , \qquad u \in \mathcal{M} \quad , \qquad r \in R \quad . \tag{10c}$$

This property means that the mapp Λ intertwines the representation of R in $\mathcal{M}$ (i.e. the isotropy representation) with the representation of $\lambda(R)$ in $\mathcal{K}$ (obtained by restricting the adjoint representation of K to $\lambda(R)$). The solutions of (10c) are therefore given by Schur's lemma.

By making the ansätze (8) and (9, 10) we have concluded the kinematical part of the search for S-symmetric solutions of the EYM equations.

Analagous symmetric ansätze for linear connections with torsion have been made in the framework of generalized Einstein-Cartan theories [7].

3. Acknowledgements

It is a pleasure to thank M.C. Bento, O. Bertolami, A.B. Henriques, Yu. A. Kubyshin, P.V. Moniz, R. Picken, P.M. Sá, J.N. Tavares and I.P. Volobuev for collaboration over the years and Yu. A. Kubyshin for usefull suggestions.

4. References

1. I.P. Volobuev and Yu.A. Kubyshin, *JETP Lett.* **45** (1987) 581; *Theor. Math. Phys.* **75** (1988) 509;

   Yu.A. Kubyshin, J.M. Mourão, G. Rudolph and I.P. Volobuev, *Lect. Notes in Physics* **349** (Springer Verlag, 1988);

   Yu.A. Kubyshin, J.M. Mourão and I.P. Volobuev, *Phys. Lett.* **B203** (1988) 349; *Nucl. Phys.* **B322** (1989) 531

2. O. Bertolami, J.M. Mourão, R. Picken and I.P. Volobuev, *Int. J. Mod. Phys.* **A6** (1991) 4149;

   P.V. Moniz and J.M. Mourão, *Class. Quant. Grav.* **8** (1991) 1815;

   O. Bertolami and J.M. Mourão, *Class. Quant. Grav.* **8** (1991) 1271;

   O. Bertolami, Yu.A. Kubyshin and J.M. Mourão, *Phys. Rev. D* **45** (1992) 3405;

   P.V. Moniz, J.M. Mourão and P.M. Sá, *Class. Quant. Grav.* in Press;

   M.C. Bento, O. Bertolami, P.V. Moniz, J.M. Mourão and P.M. Sá, *Class. Quant. Grav.* in Press

3. M. Henneaux, *J. Math. Phys.* **23** (1982) 830;

   Yu.A. Kubyshin, V.A. Rubakov and I.I. Tkachev, *Int. J. Mod. Phys.* **A4** (1989) 1409;

   R. Coquereaux and A. Jadczyk, *Lect. Notes in Physics* **16** (World Scientific, 1988);

   K. Yoshida, S. Hirenzaki and K. Shiraishi, *Phys. Rev. D* **42** (1990) 1973

   Yu. A. Kubyshin, J.I. Cadenas, *Preprint Nuclear Phys. Inst., Moscow State University*, Moscow (1993)

4. H.P. Kunzle, *Class. Quant. Grav.* **8** (1991) 2283

5. P. Forgacs and N.S. Manton, *Comm. Math. Phys.* **72** (1980) 15

I.P. Volobuev and Yu.A. Kubyshin, *Theor. Math. Phys.* **68** (1986) 788; **68** (1986) 885

F.A. Bais, K.J. Barnes, P. Forgacs and G. Zoupanos *Nucl. Phys.* **B291** (1986) 557;

G. Rudolph and I.P. Volobuev, *Nucl. Phys.* **B313** (1989) 95;

D. Kapetanakis and G. Zoupanos, *Phys. Lett.* **B249** (1990) 73; *Phys. Rep.* **C 219** (1992) 1;

S. Shabanov, *Phys. Lett.* **B272** (1992) 11

6. S. Kobayashi and K. Nomizu, "Foundations of Differential Geometry" (Interscience Publishers, 1969)

7. Yu. A. Kubyshin, O. Richter and G. Rudolph, *Rep. Math. Phys.* **30** (1991) 355; *Leipzig Univ. Preprint,* Leipzig (1992)

Yu. A. Kubyshin, V. Malyshenko and D. Marin Ricoy, *Preprint Nuclear Phys. Inst., Moscow State University,* Moscow (1993)

ON GENERALIZED KERR-SCHILD SPACETIMES AND NULL-GEODESIC ELECTROMAGNETIC POTENTIALS

ENER SALINAS
ITP, University of Stockholm,S-113 46 Stockholm,Sweden
and Facultad de Ciencias,UNI, Lima,Perú

Abstract

Using the optical scalars some properties of the field equations for Generalized Kerr-Schild spacetimes are deduced. Assuming a vacuum seed metric and a traceless energy-momentum tensor for the generated metric the field equations are simplified. The cases of vacuum, pure radiation and electromagnetic field are treated. In this last case, we consider electromagnetic 4-potentials which constitute a null-geodesic congruence, this implies the existence of hypersurfaces orthogonal to this congruence, produces physical constraints on the fields and generates a simple form of type N metric.

1 Introduction

Studying the possibility of propagation of information by Gravitational waves, in 1962 Trautman[1] considered metrics of the form

$$\boldsymbol{g} = \boldsymbol{\eta} + 2f\boldsymbol{l}\otimes\boldsymbol{l} \tag{1}$$

where $\boldsymbol{\eta}$ is the Minkowski metric, f is a scalar function and $\boldsymbol{l}$ is a null vector.

In 1967 Kerr and Schild[2] used this method to study algebraically special metrics. The Schwarzschild and Kerr metrics can be expressed in this form, in fact the Kerr metric was originally obtained in this way.

In 1981 Taub[3] generalized this idea: starting from a seed spacetime $(\mathcal{M},\boldsymbol{g})$, not necessarily restricted to being flat, one can generate another spacetime $(\hat{\mathcal{M}},\hat{\boldsymbol{g}})$ with the transformation:

$$\hat{\boldsymbol{g}} = \boldsymbol{g} + 2f\boldsymbol{l}\otimes\boldsymbol{l} \tag{2}$$

where f is a scalar field over $\mathcal{M}$ and $\boldsymbol{l}$ is a null and geodesic vector field with respect to both metrics. Spacetimes generated in this way are called generalized Kerr-Schild (GKS) spacetimes.

It is possible to obtain the connection coefficients, the Riemann, Ricci and Einstein tensors on $\hat{\mathcal{M}}$ in terms of the equivalent quantities on $\mathcal{M}$. The covariant derivative with respect to the seed metric will be denoted by a colon (:) and the one with respect to the generated metric by a semicolon (;).

We will study some properties of these tensors using the optical scalars of the null congruence $\boldsymbol{l}$: expansion Θ, twist (or rotation) ω and shear σ.

Considerable simplifications on the field equations are obtained when a vacuum metric on $\mathcal{M}$, and a traceless energy-momentum tensor on $\hat{\mathcal{M}}$ are assumed. Metrics representing vacuum, null fluids or electromagnetic fields can be generated in this way.

2 Some properties of the GKS spacetimes

Let us start by considering the expression for the Einstein tensor on the space $\hat{\mathcal{M}}$, originally obtained by Taub[3]

$$\begin{aligned}\hat{G}^u_{\ d} = G^u_{\ d} &- 2fl^w l^u(G_{wd} + \frac{1}{2}Rg_{wd}) + \delta^u_d f l^w l^v G_{wv} \\ &- [(fl^w l^u)_{:d} + (fl^w l_d)^{:u} - (fl^u l_d)^{:w} - (fl^w l^v)_{:v}\delta^u_d]_{:w}\,. \end{aligned} \qquad (3)$$

As we can observe, this expression is linear in f, which we can interpret as follows: *if we know the vector field* $\boldsymbol{l}$ *and the background metric* $\boldsymbol{g}$*, then by a proper choice of the energy-momentum tensor* $\hat{\boldsymbol{T}}$ *we can transform the Einstein equations to second order linear equations with respect to the scalar function f.*

It is also interesting to remark that to obtain (3) the shear-free condition (which Taub imposes on $\boldsymbol{l}$) has not been used (only the null and geodesic properties of $\boldsymbol{l}$). This freedom has important consequences when we deal with algebraically special spacetimes for vacuum to vacuum GKS metrics.

Using the null-geodesic condition on the vector field $\boldsymbol{l}$ we can prove that the optical scalars of this congruence are the same with respect to both metrics, i.e. $\hat{\Theta} = \Theta$, $\hat{\omega} = \omega$, $\hat{\sigma} = \sigma$. Using this result and the relation of the optical scalars to the Ricci tensor:

$$\Theta_{,a}l^a + \Theta^2 - \omega^2 + \sigma\bar{\sigma} = R_{ab}l^a l^b \qquad (4)$$

we obtain the following relations:

$$\hat{R}_{ab}l^a l^b = R_{ab}l^a l^b = \hat{G}_{ab}l^a l^b = G_{ab}l^a l^b. \qquad (5)$$

3 Vacuum seed metric

Here we study some interesting cases generated by a seed metric representing an empty (but in general curved) spacetime (i.e. $R_{ab} = 0$ on $\mathcal{M}$). This assumption implies that on $\hat{\mathcal{M}}$ the Ricci tensor and the scalar curvature have the simple forms:

$$\hat{R}^u_{\ d} = -[(fl^w l^u)_{:d} + (fl^w l_d)^{:u} - (fl^u l_d)^{:w}]_{:w} \qquad (6)$$

$$\hat{R} = -2[(fl^v)_{:v}l^w]_{:w}. \qquad (7)$$

The field equations on $\hat{\mathcal{M}}$ are:

$$\hat{G}^u_{\ d} = -[(fl^w l^u)_{:d} + (fl^w l_d)^{:u} - (fl^u l_d)^{:w} - (fl^w l^v)_{:v}\delta^u_d]_{:w} = 8\pi \hat{T}^u_{\ d}. \qquad (8)$$

In terms of the optical scalars of $\boldsymbol{l}$, we obtain the following relations:

$$\begin{aligned} \hat{R}^u_{\ d}l^d &= -(f_{,a,b}l^a l^b + 2f_{,a}l^a\Theta + 4f\omega^2)l^u = \lambda_1 l^u && (9)\\ -\frac{1}{2}\hat{R} &= f_{,a,b}l^a l^b + 4f_{,a}l^a\Theta + 2f\Theta_{,a}l^a + 4f\Theta^2 = \lambda_2 && (10)\\ \hat{T}^u_{\ d}l^d &= \frac{1}{4\pi}[(f\Theta)_{,a}l^a + 2f(\Theta^2 - \omega^2)]l^u = \lambda_3 l^u. && (11)\end{aligned}$$

Let us consider the most commonly used energy-momentum tensors in general relativity[4] and choose the ones which are suitable to use on $\hat{\mathcal{M}}$

$$\begin{array}{lll} \textit{dust}: & \hat{T}_{ab} = \mu u_a u_b \qquad u_a u^a = -1 & (12) \\ \textit{perfect fluid}: & \hat{T}_{ab} = (\mu + p) u_a u_b + p\hat{g}_{ab} & (13) \\ \textit{pure radiation}: & \hat{T}_{ab} = \rho k_a k_b \qquad k_a k^a = 0 & (14) \\ \textit{electromagnetic field}: & \hat{T}_{ab} = \hat{F}_{ac}\hat{F}_b{}^c - \frac{1}{4}\hat{g}_{ab}\hat{F}_{cd}\hat{F}^{cd}. & (15) \end{array}$$

Dust energy-momentum tensors can not be generated from vacuum by a GKS transformation because from expression (4) or (11) we have $\hat{T}_{ab}l^a l^b = 0$, this implies $u_a l^a = 0$ which is not possible since u_a is a timelike vector.

Using the same argument, from (11) and (13) we deduce that *perfect fluids* can not in general be obtained with a vacuum seed metric. The exception is $p = -\mu$, i.e. $\hat{T}_{ab} = -\mu\hat{g}_{ab}$ which represents vacuum energy or a de Sitter type of spacetime.

Pure radiation and *electromagnetic field* are in agreement with (9)-(11). Moreover, their energy-momentum tensors are traceless. From $-\hat{R} = 8\pi\hat{T}^u_{\ u}$ we can see that for these cases the scalar curvature $\hat{R}$ vanishes, as well as the last term of the Einstein tensor. Then the Einstein field equations are simply:

$$-\left[(fl^w l^u)_{:d} + (fl^w l_d)^{:u} - (fl^u l_d)^{:w}\right]_{:w} = 8\pi\hat{T}^u_{\ d}. \tag{16}$$

Assuming the tracelessness of $\hat{T}^u_{\ d}$ as an extra condition, we have the following useful cases:

$(\mathcal{M}, g)$	$\longrightarrow$	$(\hat{\mathcal{M}}, \hat{g}, \hat{T})$	$trace(\hat{T}) = 0$
vacuum	$\longrightarrow$	vacuum	
vacuum	$\longrightarrow$	*pure radiation*	
vacuum	$\longrightarrow$	*electromagnetic field.*	

4 Vacuum to vacuum

Here we have: $\lambda_1 = \lambda_2 = \lambda_3 = 0$. From (4) we have: $\Theta_{,a}l^a + \Theta^2 - \omega^2 + \sigma\bar{\sigma} = 0$ and $\lambda_3 = 0$ means: $(f\Theta)_{,a}l^a + 2f(\Theta^2 - \omega^2) = 0$, then we obtain the following conditions for the vanishing of the shear:

i) If $\Theta = 0$ then $\sigma\bar{\sigma} = 0$

ii) If $\Theta \neq 0$ and $[\ln(\frac{f}{\Theta})]_{,a}l^a = 0$ then $\sigma\bar{\sigma} = 0$

which means that if any of the above conditions is satisfied, then by the Goldberg-Sachs theorem[4] our metrics g_{vac} and $\hat{g}_{vac}$ are algebraically special.

5 Vacuum to pure radiation

Here the energy-momentum tensor is: $\hat{T} = \rho\, k \otimes k$, where k is null and $\rho > 0$.

We can prove using (11) that $\boldsymbol{k}$ is proportional to $\boldsymbol{l}$. If we absorb the proportionality factor in ρ, we have: $\hat{\boldsymbol{T}} = \rho\, \boldsymbol{l} \otimes \boldsymbol{l}$. Then another way to express our new metric is:

$$\hat{g} = g_{vac} + 2(\frac{f}{\rho})\hat{\boldsymbol{T}}. \tag{17}$$

6 Vacuum to electromagnetism, null-geodesic potentials

In the study of the Einstein-Maxwell equations for the generated space, it seems natural to look for quantities involving null vector fields in their structure.

Here we introduce electromagnetic 4-potentials with the properties of being null and geodesic: $A_a A^a = 0$ and $A_{a;b}A^b = 0$ $(F_{ab} = 2A_{[b;a]})$, this causes the vanishing of the two electromagnetic invariants $F_{ab}F^{ab}$ and $\tilde{F}_{ab}F^{ab}$.

Studying these potentials as a null congruence leads to the vanishing of its twist $\omega^2 \equiv \frac{1}{2}A_{[a;b]}A^{a;b} = 0$, which assures the existence of hypersurfaces perpendicular to this congruence.

Geometrically the null-geodesic conditions imposed on the 4-potentials imply that $A_a = \zeta\chi_{,a}$ (where $\chi_{,a}\chi^{,a} = 0$, $\zeta_{,a}\chi^{,a} = 0$ and $\zeta_{,a}\zeta^{,a} \equiv \alpha^2$), this leads to the decomposition of the Maxwell tensor in a wedge product of two exact forms $\boldsymbol{F} = \boldsymbol{d}\zeta \wedge \boldsymbol{d}\chi$.

The energy-momentum tensor is $\hat{T}_{ab} = \alpha^2\chi_{,a}\chi_{,b}$. The new metric (which is Petrov type N) has the simple form:

$$\hat{g} = g_{vac} + 2(\frac{f}{\alpha^2})\boldsymbol{d}\chi \otimes \boldsymbol{d}\chi. \tag{18}$$

7 Acknowledgments

I wish to thank Professor Bertel Laurent for his continuous encouragement and support and Dr. Nelson Zamorano for introducing to me the GKS method.

References

[1] Trautman, A.(1962), in : *Recent developments in general relativity*, Pergamon Press,p.459.

[2] Kerr.R.P., and Schild A. (1967), in: *Atti del convegno sulla relatività generale*, Firenze, p.222.

[3] Taub,A.H. (1981), Ann.Phys., NY. **134**, 326.

[4] D.Kramer,H.Stephani,M.Mac Callum,E.Herlt.(1980) *Exact solutions of Einstein field equations*. Cambridge.

CONSERVATION LAWS AND ENERGY MOMENTUM DEFINITION IN CURVED SPACE-TIMES

N. Umérez

Instituto de Astronomía y Física del Espacio
Casilla de Correo 67 - Sucursal 28, 1428 Buenos Aires - Argentina

Definitions of energy-momentum densities are obtained for any metric manifold, with or without torsion. We show that, with these definitions, it is possible to link supersymmetry and observable effects.

INFINITESIMAL GENERATORS

If we consider a family of parametric curves with tangent vector ϵ over the 4-dimensional manifold M, we can define a diffeomorphism over M according to the following transformation laws:

$$x^\mu \to x'^\mu = x^\mu + \epsilon^\mu \Delta\lambda$$

$$\phi(x) \to \phi'(x')$$

(where λ is the parameter and ϕ is any field) under which the action W transforms as

$$W = \int_{Z \subset M} \mathcal{L} d^4x \to W' = \int_{Z' \subset M} \mathcal{L} d^4x \tag{1}$$

Then, by direct computation:

$$\lim_{\lambda \to 0} \frac{W' - W}{\Delta\lambda} \equiv \mathcal{D}_\epsilon W = \mathcal{P}[\epsilon, \mathcal{L}, Z_+] - \mathcal{P}[\epsilon, \mathcal{L}, Z_-] \tag{2}$$

where the operator

$$\mathcal{P}[\epsilon, \mathcal{L}, Z_{+,-}] = \int_{Z_{+,-}} (\frac{\partial \mathcal{L}}{\partial \partial_\mu \phi} \pounds_\epsilon \phi - \mathcal{L} \epsilon^\mu) d\Sigma_\mu \tag{3}$$

with $\pounds$ is the Lie derivative, $Z_+ \cup Z_- = \partial Z$: boundary of Z and

$$[\phi; \mathcal{P}]_{PB} = \pounds_\epsilon \phi \qquad [\pi; \mathcal{P}]_{PB} = \pounds_\epsilon \pi \tag{4}$$

PB stands for Poisson brackets, π and ϕ are canonical conjugated variables. From 2 we see that $\mathcal{P}$ is the infinitesimal generator of the Lie derivative.

THE METRIC TENSOR

If we introduce the metric tensor $g_{\mu\nu}$ we can define the generators $\mathcal{P}[\mathcal{L}, \epsilon, S]$ and $\mathcal{P}[\mathcal{L}, \epsilon_T, S]$ over the space-like hypersurface S, where ϵ_T belongs to the tangent space of S and $\epsilon_T.\epsilon = 0$. These generators are nothing but the Hamiltonian and the momentum measured by observers with four-velocity ϵ who consider the hypersurface S as their fenomenological geometrical space (ϵ is timelike). Then, the vectorial density:

$$T^{\mu}[\epsilon\ (\epsilon_T)] = \frac{\partial \mathcal{L}}{\partial \partial_\mu \phi} \pounds_{\epsilon\ (\epsilon_T)} \phi - \mathcal{L}\epsilon^{\mu}\ (\epsilon_T) \tag{5}$$

is the energy (ϵ)-momentum (ϵ_T) density. If this vectorial density is divergentless, then $\mathcal{P}$ is a conserved magnitude.

The relation between the energy-momentum density of the matter fields and the energy-momentum tensor is given by:

$$T^{\mu}_{matter}(\epsilon) = T_{\mu\nu}\epsilon^{\nu}\sqrt{g} \tag{6}$$

where the Lagrangian is the matter Lagrangian. On the other hand, the energy density of the gravitational field is obtained considering $g_{\mu\nu}$ as a field ϕ and the free gravitational Lagrangian as the Lagrangian in eq.5.

Finally, taking into account that

$$\partial_\nu T^{\nu}(\epsilon) \equiv \lim_{V\to 0} \frac{1}{V} \int_{\partial V} \frac{T^{\mu}(\epsilon)}{\sqrt{g}} dS_\mu = \mathcal{F}(\epsilon), \tag{7}$$

and the physical meaning of T^{μ} we can interpret $\mathcal{F}(\epsilon = \hat{e}_0)$ as the energy flux in the reference system of a test particle with four-velocity $\hat{e}_0$, where $\hat{e}_0$ is the time-like versor of the tetrad (*vierbein*):

$$\hat{e}_A \hat{e}_B = \eta_{AB} \qquad A, B = 0, 1, 2, 3.$$

EQUIVALENCE PRINCIPLE AND ENERGY MOMENTUM FLUX

In a Riemannian geometry, the equivalence principle prescribes that it is always possible to choose an affine conection that is locally zero in a free falling frame. A geometrical consequence of this is the zero Lie derivative of the reference system versors ($\pounds_{\hat{e}_A} \hat{e}_B = 0$). As it is easy to see by direct computation, in the General Relativity Theory, this local holonomy of the free falling reference system, implies that the energy flux of the gravitational and matter fields will be zero. This is, of course, a nice property for free falling frames. Other is the case in non-Riemannian geometries. For example, if manifolds with torsion are considered, the local anhilation of the affine conection does not imply local holonomy. Then, in general, the gravitational and matter energy fluxes are different from zero. In consequence, we conclude that local holonomy is a more natural prescription for a free fallig reference system than the local affine conection anhilation [1]. These two prescrptions coincide only in zero torsion manifolds. We finally mention that if we demand anhilation of the conection for the equivalence principle, the trajectories will be selfparallels, but if we demand local holonomy the trajectories will be geodesics, even in non Riemannian cases. Then we can relate geodesics with zero total energy flux. On the other hand, particles with a non-geodesic trajectory must suffer changes in its energy density.

An example of this situation is studied in [2] where it is proved that in a supersymmetric background with torsion the spin 1/2 particles follow non-geodesic trajectories.

TOY MODEL

As an example we now consider the case of a spherically symmetric source of N=1 Supergravity, in which we suppose that we have non-zero torsion inicial conditions. We also suppose that the external solution for the metric tensor field is a quasi-flat solution ($g_{\mu\nu} = \eta_{\mu\nu}$). It is easy to prove that in the lowest order (Newtonian order) in $\bar{v}$ (typical velocity of the system), $\bar{M}$ (typical mass) and $\bar{r}$ (typical distance) [3], there exists the following non trivial solution of the dynamical equations for the gravitino field [4]:

$$\psi_i = \frac{\kappa}{16\pi}\epsilon_{ijk}\gamma_5 x^j \gamma^k r^{-3} \qquad \psi_0 = \frac{\kappa}{16\pi}\gamma_i x^i r^{-3} \tag{8}$$

x^μ are the coordinates and r is the Euclidian space distance. As in a supersymmetric background torsion is $S^\mu_{\nu\rho} = \kappa^2\bar{\psi}^\mu\gamma_\nu\psi_\rho$, this last equation tells us that we have torsion even at the Newtonian order.

If we now use as test matter the chirial Wess-Zumino multiplet (A, B, λ), the spin zero A, B particles follow geodesics but the spin 1/2 λ particle follows a non geodesic trajectory. Impossing adecuate inicial conditions we can consider that these kind of particles describe orbits with radious greater than the Schwartzchild radious. Then, by aplication of the third Kepler's law (see [5]), the formalism developed in [2] and the expression of the energy decay obteined in 7, we have:

$$\text{Energy decay} = -\frac{1}{2c}\frac{d\rho_\lambda}{dx_0} + \frac{1}{3}\frac{\rho_\lambda}{c}v^2 P^{-2}\frac{dP}{dx_0} = \frac{2\kappa}{\sqrt{g}}\frac{\partial \mathcal{L}_{WZ}}{\partial g_{\mu\nu}} f_{\mu\nu} \tag{9}$$

where ρ_λ is the energy density of the λ particle, P is the orbital period, $\mathcal{L}_{WZ}$ is the Wess-Zumino Lagrangian and $f_{\mu\nu}$ is an antisymmatric tensor function of the particle velocities, matter energy densities, gravitino field and the metric tensor.

Now we have the energy decay in terms of observable magnitudes. In order to obtain a numerical result we can use the P, dP/dx_0, v values of the binary pulsar PSR 1913+16. In this case we obtain that the temporal evolution of the energy density of the spinorial matter is:

$$\rho_\lambda = e^{-10^{-23}t/sec} \tag{10}$$

CONCLUSION

As we have seen it is possible to link supersymmetry and observable effects. But, as we know, supersymmetry is not observed now in nature. Neverthless, perhaps the universe shows supersymmetric vestiges. Then this property may be found in far astrophysical systems. All this enables us to suposse that more complete models will be useful in order to contrast the original supersymmetry of the universe hypotesis.

REFERENCES

[1] Von der Heyde, *Phys. Lett.* **58A** (1976) 141.

[2] M. Castagnino, G. Domenech, M. Levinas and N. Umérez, *Class. Quant. Grav.* **6** (1989) L1.

[3] S. Weinberg, *Gravitation and cosmology*, Wiley, New York (1972).

[4] M. Castagnino, G. Domenech, M. Levinas and N. Umérez, *Phys. Lett.* **B179** (1986) 83.

[5] C. Will, *Theory and experiment in gravitational physics*, Cambridge Univ. Press. (1981), pg. 298.

THE EXISTENCE OF EMPTY SPACE METRICS ADMITTING KILLING PAIRS

E.G.L.R. VAZ
Departamento de Matemática, Universidade do Minho,
Largo do Paço, 4700 Braga, PORTUGAL

ABSTRACT
We consider canonical forms of metrics obtained under the assumption that they admit a surface generating Killing pair, one member of which is hypersurface orthogonal and non null. Using the empty space field equations, we prove that for vacuum solutions the Killing pair is necessarily a trivial one.

1.Introduction

Let (M, g) be a space-time, i.e. a smooth, 4-dimensional paracompact manifold M, which we assume to be C^∞ for simplicity, together with a metric g of Lorentzian signature $(-,+,+,+)$. In a local chart, the coordinates assigned to each point are (x^1, x^2, x^3, x^4).

Consider a geodesic vector field so scaled that

$$\nabla_V V = 0,$$

where ∇ is the metric connection.

A pair of vector fields X and Y is a Killing pair[1] if the rational function

$$\Phi = \frac{X_i V^i}{Y_k V^k}$$

remains constant along each geodesic. According with Collinson[2], a necessary and sufficient condition for $\{X, Y\}$ to be a Killing pair is that

$$\nabla_{(i} X_{k)} = X_{(i} Z_{k)} \tag{1}$$

and

$$\nabla_{(i} Y_{k)} = Y_{(i} Z_{k)}, \tag{2}$$

where Z^i are the components of un unknown vector field and the round brackets denote symmeterization.

Here the degenerate case, when X is a multiple of Y, has been excluded. Notice that the first integral Φ remains invariant under a gauge transformation

$$X_i' = \alpha X_i, \quad Y_i' = \alpha Y_i, \quad Z_i' = Z_i + \alpha^{-1} \alpha_{,i},$$

α being an arbitrary scalar field and the comma being used for the partial derivative. If the gauge can be chosen to make $Z = 0$ then Eq.1 and Eq.2 simplify and the vector fields X and Y are simply Killing vectors, in which case Φ reduces to the ratio of two linear first integrals and the Killing pair is a *trivial* one.

Canonical forms of metrics admitting a Killing pair were obtained previously[2] under the assumptions that X is hypersurface orthogonal and that X and Y generate an integrable distribution. Two main cases were considered to perform the calculation, which arise when considering X non-null and X null separately. Assuming that X is non-null, the metrics obtained are tose considered in section 2, both arising for $g(X,Y) \neq 0$. The empty space field equations are then written (using the program "metric" developed by Pulham[4] and based on *Reduce*) for these metrics[5] in order to investigate the existence of empty spaces aditting such a Killing pair . However in both cases it is found that the Killing pair is necessarily a trivial one.

2.The empty space field equations

Coordinates can be chosen so that $X = \partial_{x^4}, Y = \partial_{x^3} + \beta\partial_{x^4}$, β being a scalar field. According to the results obtained previously[5], the metric of the spacetimes admitting a surface generating Killing pair $\{X,Y\}$ such that X is hypersurface orthogonal and non-null takes one of the following forms:

$$ds^2 = g_{\alpha\beta}dx^\alpha dx^\beta + g_{33}(dx^{3^2} + mdx^{4^2}) \tag{3}$$

$$ds^2 = g_{\alpha abeta a}dx^\alpha dx^\beta + g_{33}\left[dx^{3^2} + 2(x^{4^2} + fx^4 + h)^{-1}(1+\cosh x^3)dx^{4^2}\right], \tag{4}$$

where greek indices range from 1 to 2. Here m is a function of x^3 and x^4 and all the remaining functions depend only on x^1 and x^2. Additionally, it is found that $g(X,Y) \neq 0$ and that both canonical forms are invariant under the coordinate transformation $x^{\alpha'} = x^{\alpha'}(x^\beta)$.

For the metric given by Eq.3 and using a classical result which says that any two dimensional Riemannian space is conformally flat, one can rewrite it as

$$ds^2 = -2Pdudv - 2BQdzd\bar{z}, \tag{5}$$

the new coordinates beeing $(u, v, z, \bar{z})$ with $\bar{z}$ the conjugate complex of z. The functions B and P depend on u, v and Q is a function of z and $\bar{z}$ only. Equating to zero the components of the Ricci tensor we obtain the following empty space field equations:

$$2BPB_{,uu} - PB_{,u}{}^2 - 2BB_{,u}P_{,u} = 0 \tag{6}$$

$$2BP^2B_{,uv} - P^2B_{,u}B_{,v} + 2B^2PP_{,uv} - 2B^2P_{,u}P_{,v} = 0 \tag{7}$$

$$2BPB_{,vv} - PB_{,v}{}^2 - 2BB_{,v}P_{,v} = 0 \tag{8}$$

$$Q^3B_{,uv} + PQQ_{,z\bar{z}} - PQ_{,z}Q_{,\bar{z}} = 0. \tag{9}$$

Eq.(6) and Eq.(8) are integrated to yield

$$B_{,u}{}^2 = BP^2F^2 \tag{10}$$

and

$$B_{,v}{}^2 = BP^2G^2, \tag{11}$$

where F is a function of v and G depends only on u. It is found that either F vanishes or it is possible to set it equal to 1, using the freedom left in the metric, the same happening with G. Therefore the calculation follows considering three distinct cases, which arise when considering $F = 0 = G$, $F = 0$ and $G = 1$, $F = 1 = G$ respectively, the case $F = 1$ and $G = 0$ being transformed to the second one by interchanging u and v.

For the first two cases, Eq.9 simplifies and an integration gives that Q is a product of a function of z and a function of $\bar{z}$. Then the coordinate freedom of the metric can be used to set $Q = 1$, so that the metric components become independent of z and $\bar{z}$. This shows that two Killing vectors exist, the vectors belonging to the Killing pair being proportional to them, in which case the Killing pair is trivial.

In the third case, from Eq.10 and Eq.11 we obtain that B and P are functions of $(u+v)$ related by $\dot{B}^2 = BP^2$. This equation, together with Eq.7 and Eq.9, is used to prove that P must be zero and so the metric becomes singular. Therefore, this case cannot exist.

We now consider the metric of Eq.4 written in the form

$$ds^2 = -2Pdudv - B\left[dx^{3^2} + 2(x^{4^2} + fx^4 + h)^{-1}(1 + \cosh x^3)dx^{4^2}\right], \tag{12}$$

where P, B, f, h are functions of u and v. Separating the variable x^4 in the field equations $R_{13} = 0$ and $R_{23} = 0$ gives that f and h are constant. Defining

$$x^{4'} = \int (x^{4^2} + fx^4 + h)^{-1/2} dx^4$$

the metric is reduced to a form independent of $x^{4'}$ so that it admits a Killing vector. Under this transformation it can be shown that the vector field X of the Killing pair is proportional to the Killing vector and consequently the Killing pair is necessarily trivial.

4.References

1. N.M. Woodhouse, *Commun. Math. Phys.*, **44** (1975) 9.
2. C.D. Collinson, *Gen. Rel. Grav.*, **18** (1986) 207.
3. E.G.L.R. Vaz and C.D.Collinson, *Gen. Rel. Grav.*, **24** (1992) 405.
4. J. Pulham, private communication.
5. E.G.L.R. Vaz and C.D. Collinson, preprint, Universidade do Minho.

THE CONNECTION OF THE CLASSIFICATION PROBLEM OF THE PARTICLE MOTION IN THE KERR METRIC WITH THE CATASTROPHE THEORY

A.F.Zakharov

Institute of Theoretical and Experimental Physics,
B.Cheremushkinskaya, 25, Moscow, 117259, Russia

Abstract

Sets of constants of motion of a particle that correspond to different types of r-motion are considered. The topology of these sets is determined and a number of constants characterizing these sets are found. We examine possible finite motions of particles in the Kerr metric. We determine the topology of various sets of constants of motion of the particle and identify invariants that are independent from rotation parameter. We obtain structural stable and unstable results of the geodesics analysis and describe the connection some problems of the geodesics analysis with results of the catastrophe theory.

The equation of motion for radial variable in the Kerr metric is [1]:

$$\rho^4(dr/d\tau)^2 = R(r), \tag{1}$$

where for photons $R(r) = r^4+(a^2-\xi^2-\eta)r^2+2M[\eta+(\xi-a)^2]r-a^2\eta$, and for particles $R(r) = r^4+(a^2-\xi^2-\eta)r^2+2M[\eta+(\xi-a)^2]r-a^2\eta-r^2\Delta/E$, where $\rho^2 = r^2+a^2\cos^2\theta$, $\Delta = r^2-2Mr+a^2$, $a = S/M$. The constants S and M refer to the black hole, namely, S is the angular momentum and M is the mass of the black hole. The constants E, ξ and η refer to the particle, namely E is its energy at infinity, $\xi = L_z/E$ (L is the angular momentum of the particle about the axis of rotation of the black hole), and $\eta = Q/E$ (Q is given by $Q = p_\theta{}^2 + \cos^2\theta[a^2(\xi^2 - E^2) + L_z{}^2 \sin^{-2}\theta]$ and μ is the mass of the particle). It is readily verified that the radial motion of the particle depends on the following constants: $\hat{a} = a/M, \quad \hat{E} = E/\mu, \quad \hat{\xi} = \xi/M, \quad \hat{\eta} = \eta/M$. The radial motion of photons does not depend on the constant E. Instead of the coordinate r, we now introduce $\hat{r} = r/M$. (The $\wedge$-symbol will be omitted henceforth). Thus, the character of motion in the r-coordinate for given value of a is determined by the three constants E, ξ, η in the case of a moving particle, and by the two constants ξ and η in the case of photons.

It will be shown that a black hole in extreme rotation can have stable geodesics with any energy (in units of particle mass) in range $0 \leq E \leq 3^{-1/2}$ although it is well known that for a particle moving along a circular geodesics in the field of a black hole in a state of extreme rotation, the binding energy has been found capable of reaching the value $3^{-1/2}$. One readily finds [2] that if the particle is moving in the equatorial plane ($\theta = \pi/2$), $Q = 0$. The particle will then travel in a circular orbit if the next relations

$$R(r) = 0, \quad \partial R/\partial r = 0 \tag{2}$$

are satisfied , as well as the orbital stability criteria $\partial^2 R/\partial^2 r < 0$. Let us consider a black hole with an extreme value for the rotation parameter ($a = 1$). Assume that the particle orbits lie in the equatorial plane, with the following values for the constants of motion: $L = 2E, \quad E < (1/\sqrt{3}), \quad r = 1$. It is not hard to see that the criteria [3] will then hold true. Accordingly, as a particle moves through successive near-circular orbits, an energy of order 1 can be liberated. It also is noteworthy that for any values of the constant $Q (Q > 0)$ and the constants of particle motion that satisfy the conditions, the criteria Eqs.(2) will remain valid; that is stable orbits will exist that correspond to the motion of a particle along the surface r=const with a given energy and given angular momentum, but which are not circular. The maximum value of θ for such orbits will be determined by

$c_2\beta^2 - c_1\beta + c_0$. Thus, the set of constants to which the multiple root of polynomial $R(r)$, is semialgebraic submanifold of the algebraic manifold D_f, since there are the bounds: a) the multiple root must be have value $x \geq x_+ = r_+ + \beta$; b) $Q \geq 0$. It is clear that $c_2' > 0$ (when we consider the miltiple root of the polynomial $R(r)$ for $r \geq r_+$). Similarly, the algebraic curve on the manifold D_f connects with a set of the motion constants, to which a multiple root of polynomial corresponds. Since we introduce the notions

$$u = a^2 - \xi^2 - \eta, \quad v = 2[(a-\xi)^2 + \eta], \quad w = -a^2\eta.$$

Then $\xi = (2u+v)/(-4a), \eta = -w/a^2$. Hence, we can obtain the semialgebraic curve on the manifold D_f: $v = 2[(a + (2u+v)/(4a))^2 - w/a^2]$. If we add the bounds $w \leq 0, x \geq x_+$, then we obtain the semialgebraic curve, the is connected with the set of constants corresponding to third type of photon motion. Similarly, the semialgebraic curve on the manifold D_f connects with a set of the photon motion constants, to which the multiple root of polinomial $R(r)$ corresponds [3].

The connection of the catastrophe theory and gravitational lenses was described in the work [7].

I want to express my deep gratitude to Organizing Committee of First Iberian Meeting on Gravity for invitation and hospitality.

References

1. S. Chandrasekhar, *The Mathematical Theory of Black Holes.* (Oxford, Clarendon Press, 1983).
2. B.Carter, *Phys.Rev.* **D 174** (1968) 1559.
3. A.F. Zakharov, *Preprint ITEP*, **44** (Moscow) (1991).
4. D.C. Wilkins, *Phys.Rev.* **D 5** (1972) 814.
5. D. Christodoulou and R.Ruffini, *Phys.Rev.* **D 4** (1971) 3554.
6. T. Bröcker and L.Lander, *Differentiable Germs and Catastrophes* (London Math. Soc. Lect. Notes 17, Cambridge Univ. Press, 1975).
7. A.F. Zakharov, *Preprint ITEP*, **96** (Moscow, 1992).

Classical Cosmology

THE IDEAL FLUID DERIVED FROM QUANTUM FIELDS AS THE SOURCE OF INFLATION

G. Domenech[1,2], M. Levinas[1] and N. Umérez[1]

[1]*Instituto de Astronomía y Física del Espacio*
Casilla de Correo 67 - Sucursal 28, 1428 Buenos Aires - Argentina

[2]*Departamento de Física*
Facultad de Ciencias Exactas y Naturales, Universidad de Buenos Aires
Ciudad Universitaria - Pabellón I, 1428 Buenos Aires, Argentina

Supersymmetry may play an important role in giving a reasonable explanation of relevant cosmological features. We present a cosmological model in which matter is described as the classical limit of a supersymmetric group representation. Classical fluids are introduced as the limit of self-interacting matter supersymmetrically coupled to gravity. Using upper bounds of several red shift experimental data we give an estimate of the Hubble constant corresponding to the inflationary era.

The first non trivial linear representation of the superalgebra is the Wess-Zumino chiral multiplet (A, B, λ). To κ order ($\kappa^2 = 8\pi G,\ c = 1$) the matter Lagrangian density reads:

$$\mathcal{L} = \frac{\sqrt{-g}}{2}[(\partial_\mu A \partial_\nu A + \partial_\mu B \partial_\nu B)g^{\mu\nu} + \bar{\lambda} D[\omega(e^\mu_\nu)]\lambda - \kappa\bar{\psi}_\mu \partial(A + i\gamma_5 B)\gamma^\mu \lambda]$$

A is a scalar field, B is a pseudoscalar field, λ is a spin 1/2 and ψ_μ a spin 3/2. $D[w]$ is the covariant derivative built up with the spinorial connection without torsion. The WKB expansions of the fields read:

$$\phi = \sum_{n=0}^{\infty}(-i\hbar)^n \phi_n e^{iS_\phi(x)/\hbar} \qquad \phi = A + iB$$

$$\lambda_L = \sum_{n=0}^{\infty}(-i\hbar)^n \lambda_{L\ n} e^{iS_\lambda(x)/\hbar} \qquad \lambda = \begin{pmatrix}\lambda_L \\ \lambda_R\end{pmatrix} = \begin{pmatrix}\lambda_L \\ -\sigma_2 \lambda_L^*\end{pmatrix}$$

where λ is an Majorana spinor and S_i are real phases. We can think of S as the Hamilton principal function, then $P_{\mu(\phi,\lambda)} = \partial_\mu S_{(\phi,\lambda)}(x)$.

In [1] we show that for a free scalar field A in a curved background the WKB expansion gives:

$$T^{(\hbar^{-2})\mu\nu} = \frac{g^{\mu\nu}}{\hbar^2}[2A_o^2 P_{\rho A} P_A^\rho \sin^2(S_A/\hbar) + 2mA_o^2 \cos^2(S_A/\hbar)] - 4\frac{A_o^2}{\hbar^2} P_A^\mu P_A^\nu \sin^2(S_A/\hbar)$$

For the free case in which $P^\mu = p^\mu = mu^\mu$, i. e. it is the ordinary momentum,

$$T^{oo}_{rest} = \rho_A = -2\frac{A_o^2}{\hbar^2}m_A^2$$

We also show that for a free spin 1/2 massive field λ, the lowest order of the WKB expansion is:

$$T^{(\hbar^{-1})\mu\nu} = i\mathring{\bar{\lambda}}\gamma^\mu\gamma_5\mathring{\lambda}P^\nu_\lambda/\hbar \quad \text{where} \quad \mathring{\tilde{\lambda}} = \mathring{\lambda}_L e^{[iS_\lambda(x)/\hbar]}$$

γ are the Dirac matrices. Using the Dirac field equation, the rest energy reads

$$T^{(\hbar^{-1})oo}_{rest} = \rho_\lambda = -im_\lambda\mathring{\bar{\tilde{\lambda}}}\mathring{\tilde{\lambda}}/\hbar$$

Using the previous expressions, the spin 1/2 Lagrangian and the energy momentum tensor can be written as

$$\mathcal{L}^{(\hbar^{-1})}_\lambda = -\frac{1}{2}\sqrt{-g}[\rho_\lambda u^\mu u^\nu g_{\mu\nu} - \rho_\lambda] \qquad T^{(\hbar^{-1})\mu\nu} = \rho_\lambda u^\mu u^\nu$$

In order to obtain ideal fluids we need to introduce complex (not real) scalar fields. Precisely this is the case for the Wess-Zumino action if $\phi = A + iB$. The lowest order contributions to the energy-momentum tensor for A, B, λ in the Wess-Zumino multiplet are then

$$T^{\mu\nu} = \rho_{(\phi)}u^\mu_{(\phi)}u^\nu_{(\phi)} + \rho_{(\lambda)}u^\mu_{(\lambda)}u^\nu_{(\lambda)} + \frac{\kappa}{4}g^{\mu\nu}\bar{\psi}_\sigma\gamma_\theta u^\theta_{(\phi)}[\rho^{1/2}_{(A)} + i\gamma_5\rho^{1/2}_{(B)}]\gamma^\sigma\gamma_\delta\gamma_5\mathring{\lambda}u^\delta(\lambda)$$

Here u^μ is the timelike vector ortogonal to the hypersurface of constant phase S, $u^\mu u_\mu = -1 + O(\hbar)$ and the proper densities are

$$\rho(\phi) = \frac{\phi_o\phi_o^*}{\hbar^2}m^2(\phi) = (A_o^2 + B_o^2)\frac{m^2(\phi)}{\hbar^2} \equiv \rho(A) + \rho(B) \qquad \rho(\lambda) = -im(\lambda)\frac{\mathring{\bar{\tilde{\lambda}}}\mathring{\tilde{\lambda}}}{\hbar}$$

Field equations are obtained from the complete action $S_T = S_M + S_{HE} + S_{RS}$ (the sum of the matter action, the Hilbert-Einstein action and the spin 3/2 Rarita-Schwinger action). They read:

$$G_{\theta\nu} = \kappa^2\rho_{(\phi)}u_{(\phi)\theta}u_{(\phi)\nu} + \kappa^2\rho_{(\lambda)}u_{(\lambda)\theta}u_{(\lambda)\nu} + \frac{\kappa^2}{2\sqrt{-g}}g_{\mu\theta}\epsilon^{\lambda\mu\eta\rho}\bar{\psi}_\lambda\gamma_5\gamma_\nu D_\eta\psi\rho +$$
$$g_{\theta\mu}(\nabla_\rho S^{\mu\ \rho}_{\nu} + \nabla_\rho S^{\mu\rho}_\nu + \nabla_\rho S^{\rho\mu}_\nu - 2\nabla_\nu S^{\mu\rho}_\rho) - 2g_{\theta\nu}\nabla_\sigma S^{\sigma\rho}_\rho + \kappa^3\text{terms}$$

$$\epsilon^{\mu\nu\rho\sigma}\gamma_5\gamma_\nu D_\rho\psi_\sigma = \kappa\gamma_\nu u^\nu_{(\phi)}[\rho^{1/2}_{(B)} + i\gamma_5\rho^{1/2}_{(A)}]\gamma^\mu\gamma_5\gamma_\sigma\mathring{\lambda}u^\sigma_{(\lambda)}$$

$$u^\mu_{(\phi)}\nabla_\mu u_{(\phi)\theta} = 0$$

$$u^\mu_{(\lambda)}\nabla_\mu u_{(\phi)\theta} = -\kappa u^\nu_{(\lambda)}\{\partial_{[\theta}[\bar{\psi}^\pi(\rho^{1/2}_{(A)}u_{(A)} + i\gamma_5\rho^{1/2}_{(B)}u_{(B)})\gamma_\pi\gamma_{\nu]}\gamma_5\mathring{\lambda}/\rho_{(\lambda)}]\}$$

In [2] we show that the WKB expansion, which implies free behaviour for the spin 0 field, is equivalent to:

$$\frac{\rho_{A/B}}{\rho_\lambda} \gg \kappa^2 \hbar \bar{\psi}\psi \frac{m_\lambda}{m_{A/B}} \qquad (I), \qquad\qquad \frac{\rho_{A/B}}{\rho_\lambda} \sim \frac{m_\lambda}{\kappa^2 \hbar \bar{\psi}\psi} \qquad (II)$$

Combining (I) and (II) for the case of the complex field ϕ yields $\rho_\phi/\rho_\lambda \gg m_\lambda/m_\phi$. So, in the case of soft supersymmetry breaking (small non-degeneracy in mass) $\rho_\phi \gg \rho_\lambda$ which is consistent with the geodesic motion of ϕ and the non geodesic motion of λ. Neglecting the κ^3 terms in the "Einstein" field equations, and taking into account that from the 3/2 field equation $D\psi$ is of order κ

$$G^\mu_\alpha(\{\}) \sim e^\sigma_\alpha[\nabla_\rho S^\mu{}_\sigma{}^\rho + \nabla_\rho S^{\mu\rho}{}_\sigma + \nabla_\rho S_\nu{}^{\nu\mu} - 2\nabla_\sigma S^{\mu\nu}{}_\nu] - 2e^\mu_\alpha \nabla_\sigma S^{\sigma\rho}{}_\rho + \kappa^2 e^\nu_\alpha \rho_{(\phi)} u^\mu_{(\phi)} u_{(\phi)\nu}$$

with $g^{\mu\nu} = e^\mu_\alpha e^\nu_\beta \eta^{\alpha\beta}$.

We apply these results to the study of a Robertson-Walker k=0 metric in a U_4 space-time. All the components of torsion are space constants with a typical value S. In a comoving system ($u^o = -1$, $u^i = 0$) the Einstein equations are ($\dot{} \equiv \partial/\partial t$, $H = \dot{a}/a$ where $a(t)$ is the R-W scale factor):

$o - o:$ $\quad 3H^2 \sim \kappa^2 \rho_\phi - 6HS - 12\dot{S} \qquad o - i: \quad \dot{S} \sim -(7HS)/3a^3$

$i - j:$ $\quad -a^2[2\dot{H} + 3H^2] \sim HS(4a^2 - 2) + \dot{S}(1 - 8a^2)$

Combining the last equations

$$3H^2 \sim \kappa^2 \rho_\phi - S(6H + \frac{28H}{a^2}) \equiv \kappa^2 \rho_{eff}$$

$$-[2\dot{H} + 3H^2] \sim HS[4 + \frac{50}{3}a^{-2} - \frac{7}{3}a^{-4}] \equiv \kappa^2 P_{eff}$$

If $S \to 0$, we recover the standard Einstein equations with $\rho_{eff} = \rho_\phi$, $P_{eff} = 0$ and the Wess-Zumino represent nothing but free particles with non interacting term. So, $a(t) \sim t^{2/3}$, an era that is characterized by a negligible torsion and the gravitational field is only represented by one of the components of the supergauge gravitational field: $g_{\mu\nu}$.

If this is not the case:

$$P_{eff} \sim H\rho_{eff} \frac{4 - (50/3)a^{-2} - (7/3)a^{-4}}{\kappa^2 g - 6H - 28Ha^{-2}}$$

where $\rho_\phi \sim gS$. So we find inflation for $H = \kappa^2 g/2 \quad \to \rho_{eff} \sim 3H^2 = -P_{eff} \equiv -\Lambda_{eff}$ [3]. In this case, $H_{(infl)} = \kappa^2 g/2$.

In [4] we have obtained upper bounds for torsion using gravitational red-shift data from different systems which may be applied to find lower bounds for g in the case of classical supersymmetric matter. From the Rarita Schwinger eq.,

$$\sqrt{-g}G^\mu_\alpha = \frac{1}{2}\kappa^2 \bar{\psi}_\nu \gamma_5 \gamma_\alpha \epsilon^{\nu\mu\rho\sigma}(D_\rho\psi_\sigma - D_\sigma\psi_\rho)$$

$$\epsilon^{\mu\nu\rho\sigma}\gamma_5\gamma_\nu D_\rho\psi_\sigma = 0$$

and expanding the spin 3/2 field supposing a system in slow motion,

$$\psi^\mu = \psi^{(o)\mu} + \psi^{(1)\mu} + \psi^{(2)\mu} + \ldots \quad \text{where} \quad \psi^{(n)\mu} \sim \frac{\bar{M}^{1/2}}{\bar{r}}\bar{v}^n$$

(with $\bar{v}^2 \ll 1$ and the typical values of velocity, mass and distance of the system related through $\bar{v}^2 \sim G\bar{M}/\bar{r}$), the following non-pure gauge solution in the $x^i\psi_i = 0$ supergauge is obtained:

$$\psi_i^{(o)} = \frac{S}{16\pi}\epsilon_{ijk}\gamma_5 x^j \gamma^k r^{-3} \qquad \psi_o^{(o)} = \frac{S}{16\pi}\gamma_i x^i r^{-3}$$

where S is an integration constant of order $M^{1/2}\bar{r}$). In order to find upper bounds to torsion we must attribute the whole dispersion in the values of the g_{oo} data to the effects of torsion The Newtonian order of the Einstein equation, using this flat solution for ψ, is

$$R_{oo}^{(2)}(\{\}) = \frac{1}{2}\nabla^2 g_{oo}^{(2)} = -8\pi G[\frac{1}{(16\pi)^2}\bar{S}\gamma_i\gamma^o S\gamma_5\gamma^n\epsilon_{jmn}x^j(\frac{x^i}{r^6})_{,m}]$$
$$+ \text{ terms proportional to } [\frac{G\bar{S}S}{r^3}(\frac{x^i x^j}{r^3})_{,i}]$$

with $g_{oo}^{(2)} \sim \bar{v}^2$. So. in the surface of a star where R is the radious,

$$T(r = R) \sim \frac{G}{128\pi}\frac{\bar{S}S}{R^4}$$

which is of the same order as the Newtonian connection. Using the red shift data for different systems we find a lower bound for constant g. For the white draft (WD) 40 Eridani B the Newtonian potential is $g_{oo}^{(2)} \sim 10^{-5}$ with $\Delta g_{oo}^{(2)} \sim 10^{-5}$; so, the maximun permissible order of torsion is $T \sim 10^{-16}cm^{-1}$. For a Newtonian star (NS), the same quantities give $g_{oo}^{(2)} \sim 10^{-1}$, $\Delta g_{oo}^{(2)} \sim 10^{-1}$, and $T \sim 10^{-9}cm^{-1}$; so, values of S and density for white drafts are $S_{WD} \leq 10^{-16}cm^{-1}$ and $\rho_{WD} \sim 10^6 g/cm^3$. For the case of Newtonian stars, $S_{NS} \leq 10^{-9}cm^{-1}$ and $\rho_{NS} \sim 10^{14}g/cm^3$, which give $g \geq 10^{22} - 10^{23}g/cm^2$. Then the lower bound for the Hubble constant during inflation is $H > 10^{26}sec^{-1}$. If we want $H\tau \sim 65$, we conclude that the time for inflation is $\tau < 10^{-25}sec$.

REFERENCES

1. M. Castagnino, G. Domenech, M. Levinas and N. Umérez, *Class. Quant. Grav.* **6** (1989) L1.
2. G. Domenech, M. Levinas and N. Umérez, *Phys. Lett.* **A137** (1989) 17.
3. G. Domenech, M. Levinas and N. Umérez, *Gen. Rel. Grav.* **23** (1991) 683.
4. M. Castagnino, G. Domenech, M. Levinas and N. Umérez. *Phys. Lett.* **B179** (1986) 83.

Primordial Cosmology in Jordan-Brans-Dicke Theory: Nucleosynthesis

José Pedro Mimoso
Astronomy Centre, University of Sussex
Brighton BN1 9QH , United Kingdom
email: *jpom@star.susx.ac.uk*

Abstract

In this work we investigate nucleosynthesis bounds on the coupling parameter of Jordan-Brans-Dicke scalar-tensor theory when that parameter is allowed to be different in different epochs, and when the early domination of the scalar field is taken into account.

1. Introduction

The proposal of *extended inflation* by La and Steinhardt[1] in 1989 has generated a renewed interest in scalar-tensor theories of gravity. Their prescription, based on Brans-Dicke (BD) theory, revives Guth's influential idea of a stage during which the expansion of the universe is governed by a false vacuum configuration, as the outcome of a first-order phase transition[2].

Scalar-tensor theories are characterized by the basic action[3]

$$S = \frac{1}{16\pi}\int \sqrt{-g}d^4x\left[\phi R - \frac{\omega(\phi)}{\phi}\phi_{,\mu}\phi^{,\mu} + 2\phi\lambda(\phi)\right] + S_M \quad , \tag{1}$$

where the term S_M represents the action for the matter fields. Gravity is coupled to the scalar field ϕ and this feature is responsible for the effective variation of the gravitational constant G ($G \propto 1/\phi$). The strength of the coupling depends on the value of the adimensional parameter ω, and Einstein's General Relativity (GR) is recovered when ω approaches ∞. Additionally, the equivalence principle is preserved by independently imposing $T^{ab}{}_{;b} = 0$. The BD theory corresponds to the case where ω is a constant and λ is set to vanish.

It was soon realized that La and Steinhardt's scenario overlooked a difficulty. The value of the adimensional coupling parameter of BD theory, ω, although compelled by solar system experiments[5] to be $\omega \geq 500$, should be of the order of tens (certainly well below 100) during inflation[6]. And this mismatch is emphasized when the recent reassessments of the bounds placed on the coupling parameter by primordial nucleosynthesis[7] are taken into account. These indicate that constraints similar to those of post-newtonian tests hold well before the latter apply.

In the present work, we confirm these bounds from an angle which differs from previous work, and which allows for different values of the coupling parameter in different epochs. It also takes into consideration the fact that BD solutions exhibit in general an early stage dominated by the kinetic energy of the scalar field.

2. The JBD solution

We consider a 3-epochs isotropic cosmological solution, for the flat model, with the following profile for the scale factor $a(t)$

$$0 \leq t \leq t_c : \quad a(t) = A_1\, t^q \tag{2}$$

$$t_c \leq t \leq t_{eq} : \quad a(t) = A_2\,(t - t_{20})^{1/2} \tag{3}$$

$$t_{eq} \leq t \leq t_0 : \quad a(t) = A_3\,(t - t_{30})^{\frac{2\omega+2}{3\omega+4}} \tag{4}$$

where A_1, A_2, A_3, t_{20}, t_{30} are constants which have to satisfy some junction conditions guaranteeing the smoothness of the solution, and where q is given[8] by

$$q \equiv \frac{\omega}{3\left(\omega + 1 \mp \sqrt{\frac{2\omega+3}{3}}\right)} \quad . \tag{5}$$

In BD, the departure from the exact solution depends on the choice of the transition time t_c, and this encapsulates the arbitrariness of the initial condition[9] for $\dot{\phi}\, a^3(t)$ .The first two epochs, provide an approximation to the exact radiation solution of Gurevich et al[10] (1973), and distinguish the earlier BD scalar field dominated phase from the latter "matter" dominated behaviour[11]. The third epoch corresponds to the pressureless matter dominated phase.

Notice that the values of ω, though constant during each epoch, can be different in each of them.

3. Nucleosynthesis

Nucleosynthesis[12] is the result of the competition between the weak interactions and the expansion of the universe. Roughly, while the rate of expansion is slower than the rate of the weak interactions, these succeed in keeping the particles in thermal equilbrium. When the identity between the two rates occurs, equilibrium is destroyed, and the neutron to proton ratio is *frozen out*. Subsequently, the chain of reactions yielding helium-4 and the next light elements begins.

The generic temperature dependence of the weak interactions is given by $t_{weak} = C_W\, T^{-5}$ where $C_W \simeq 10^{50}\ \mathrm{GeV}^5 \times \mathrm{sec}$, and the time-temperature relations for the JBD solution are: (a) for $0 \leq t \leq t_c$, $t = t^{(GR)}\, (\theta_1/T)^{\frac{1-2q}{q}}$, and (b) for $t_c \leq t \leq t_{eq}$, $t - t_{20} = t^{(GR)}\, \theta_2$, with $t^{(GR)} = 0.301\, m_p\, g_*^{-1/2}\, T^{-2}$. In this latter expression m_p is the *Planck mass* and g_* represents the total number of spin states. The parameters θ_1 and θ_2 are

$$\theta_1 \equiv \left(0.301\, m_p\, g_*^{-1/2}\right)^{\frac{1}{2}}\; t_c^{-1/2}\; \left[\theta_2\, (2q)\right]^{\frac{1}{2(1-2q)}}$$

and

$$\theta_2 \equiv \left({\rho_0}^{\rm R}/{\rho_0}^{\rm M}\right)^{\frac{1}{2(\omega+1)}} \simeq \left(4\times 10^4 \Omega_0\, h_0^2\right)^{-\frac{1}{2(\omega+1)}}$$

(with Ω_0 the present ratio ρ/ρ_{crit} and h_0 the uncertainty in the present value of the Hubble parameter in units of 100 km Mpc^{-1} s^{-1}). Clearly, θ_1 is a temperature characterizing the departure from the GR behaviour, whereas θ_2 is just an adimensional parameter which scales the time-temperature relation due to the higher value of the gravitational constant in JBD theory. For the first epoch, we derive

$$T_f^{(JBD)} \equiv T_f^{(GR)} \left(\frac{\theta_1}{T_F^{(GR)}}\right)^{\frac{2q-1}{5q-1}} \quad . \tag{6}$$

In this expression, $T_f^{(JBD)}$ and $T_f^{(GR)}$ represent, respectively, the JBD and the GR freezing temperatures, the latter given by $T_f^{(GR)} \simeq 10^{10}K$. During the second epoch, the freezing temperature is

$$T_f^{(JBD)} = T_f^{(GR)}\; {\theta_2}^{-\frac{1}{3}} \quad . \tag{7}$$

It is possible to see that the JBD freezing temperature cannot be "tuned" to be the same as in GR, though at first sight that would seem to be feasible (given the two-parameter freedom – we can vary both t_c and ω – implicit in the selection of θ_1). During the second epoch that is clearly impossible, since the factor multiplying the GR freezing temperature is always bigger than one implying a higher freezing temperature. The underlying reason for this general impossibility is easily devised. In the JBD model, both the higher value of the effective gravitational constant and the faster rate of expansion during the two epochs under consideration contribute to yield a higher freezing temperature.

Now, it is possible to see, that during the first epoch, the lower value of T_f which occurs at $t = t_c$ is always higher than the freezing temperature during the second epoch. Thus, the least adverse situation for JBD clearly corresponds to the freezing to fall within the second epoch. In this event, there is a condition which t_c must fulfill for the freezing time to be well within the second epoch. That condition is $t_c < (2q\, \theta_2)^{\frac{5}{3}}\, t_f^{(GR)}$ which implies $\theta_1 \leq T_f^{(GR)}\, (\theta_2\, 2q)^{\frac{5q-1}{2(1-2q)}}$.

Having shown that in the most favourable case the freezing occurs when the matter component dominates the radiation solution, the bounds derived in previous works hold valid. Using the observational data on the abundances of ^{4}He, of D+^{3}He and ^{7}Li to define η_{10}, $N_\nu = 3$, and $\tau_n = 889.6 \pm 5.8$ s, the (2σ) bound for ω is $\omega > 320$ when $\Omega_0\, h_0^2 = 0.25$.

4. Discussion

We have shown that the constraints on the coupling parameter found in previous works[7] hold even when this parameter is allowed to vary in a simple way. For that matter, the model considered is justified on the grounds that it exhibits an early regime of scalar field domination, and on its simplicity. We conclude that if nucleosynthesis yields are not to be affected, ω has to increase swiftly before freezing. By the same token, if that is the case, any insight over its earlier variation has to be gathered from the investigation of the effects of that variation on other data. Presumably, from the spectra of density and gravitational perturbations, or from the anisotropy induced by bubble collisions[13] during inflation.

As for the BD results, our approach in this work contrasts previous works in two ways. It differs from Casas et al[7] in that it takes full account of the early domination of the JBD scalar field on the dynamics, and, from Serna et al[7], the difference is that a *forwards* approach is adopted, whereas theirs was based on running a nucleosynthesis code backwards from present time. The results of the latter, however, take numerically into account the case of non-vanishing curvature.

More complex scenarios based more general variations of ω are presently under analysis.

5. Acknowledgements

The author is grateful to John D. Barrow for many helpful discussions. He also acknowledges a studentship from INIC (Portugal).

6. References

1. D. La and P. J. Steinhardt, *Phys. Rev. Lett.* **62** (1989) 376
2. A. H. Guth, *Phys. Rev.* **D23** (1981) 347
3. C. M. Will, *Phys. Rep.* **113** (1984) 345
4. C. Brans and R. H. Dicke, *Phys. Rev.* **124** (1961) 925
5. R. D. Reasenberg et al, *Astrophys. J* **234** (1979) L219
6. E. Weinberg, *Phys. Rev.* **D40** (1989) 3950
7. J.A. Casas, J. García-Bellido, M. Quirós, *Phys. Lett.* **B278** (1992) 94
J.A. Casas, J. García-Bellido, M. Quirós, *Mod. Phys. Lett.* **A7** (1992) 447
A. Serna, R. Domíinguez-Tenreiro, G. Yepes, *Astrophys. J.* **391** (1992) 433
8 J. O'Hanlon, B. O. J. Tupper, *Il Nuovo Cimento* **7**, (1972) 305
9. S. Weinberg, *Gravitation and Cosmology* (Wiley, New York, 1972).
10. L. E. Gurevich, A. M. Finkelstein, V. A. Ruban, *Astrophys. Space Sci.* **22** (1973) 231
11. H. Nariai, *Progr. Theor. Phys.* **42** (1968) 544
12. E.W. Kolb and M.S. Turner, *The Early Universe*, (Addison-Wesley, New York, 1990)
13. A. R. Liddle and D. Wands, *Phys. Rev.* **D45** (1992) 2665

Qualitative Analysis of the Cosmic No-Hair Conjecture in Kantowski-Sachs Universes

P. Vargas Moniz *†
Departamento de Física, Universidade de Lisboa
Campo Grande, Ed. C1, piso 4, 1700-Lisboa, PORTUGAL

Abstract

In this talk we address the *generality* of an inflationary scenario in initially expanding Kantowski-Sachs spacetimes with a positive cosmological constant Λ. From a qualitative analysis of the Einstein equations, it is shown that the *majority* of the Kantowski-Sachs universes asymptotically approach a de Sitter stage, with the exception of a zero probability class of solutions which recollapse. Thus, we conclude that the Cosmic No-Hair Conjecture is widely valid in these Kantowski-Sachs cosmologies.

Cosmological inflation is a very popular model for the evolution of the early universe [1]. If the inflationary scenario constitute a natural phenomenon, it is important to show whether or not inflation is a generic cosmological feature, occuring from a wide range of initial conditions [2,3]. Related with this problem, there is the so called Cosmic No-Hair Conjecture (CNHC) [4]–[6]. It asserts that in the presence of a positive effective cosmological constant, Λ, the solutions of the Einstein equations evolve asymptotically towards the de Sitter solution. However, it has been argued that the CNHC is only true under certain conditions [6]. Thus, it is meaningful to know accurately which initial conditions are needed for a generic universe to approach a local de Sitter state.

The CNHC has been elegantly verified by R. Wald [5] for the case of spatially homogeneous universes which belong to the several Bianchi types [7]. All initially expanding Bianchi cosmologies of types I–VIII evolve towards a de Sitter solution. Concerning initially expanding Bianchi type-IX models, they approach a de Sitter stage *only if* at some time t they satisfy $\Lambda >^{(3)} \mathcal{R}/2$, where ${}^{(3)}\mathcal{R}$ is scalar curvature of the spatially homogeneous hypersurfaces. However, not all the spatially homogenous models are of the Bianchi type, bearing a three-dimensional isometry group acting simply transitively on the homogeneous hypersurfaces.

There is still the special case of the Kantowski-Sachs (KS) models [7]. The purposes of this talk are to re-examine the dynamics of vacuum KS models in the presence of a positive cosmological constant, Λ and to analyse explicitly the propriety of the cosmic no-hair conjecture with respect to universes of the KS type. The metric of the KS spacetimes is usually written in the form [7]

$$ds^2 = -dt^2 + A^2(t)\ dr^2 + B^2(t)\ [d\theta^2 + \sin^2(\theta) d\varphi^2], \tag{1}$$

*E-MAIL:FPVMONIZ@PTEARN.BITNET

†Work supported in part by the J.N.I.C.T. graduate scholarship BD/138/90-RM and by funds provided by the project STRIDE/FEDER JN.91.02 (D.F.F.C.U.L. - J.N.I.C.T./C.E.R.N.)

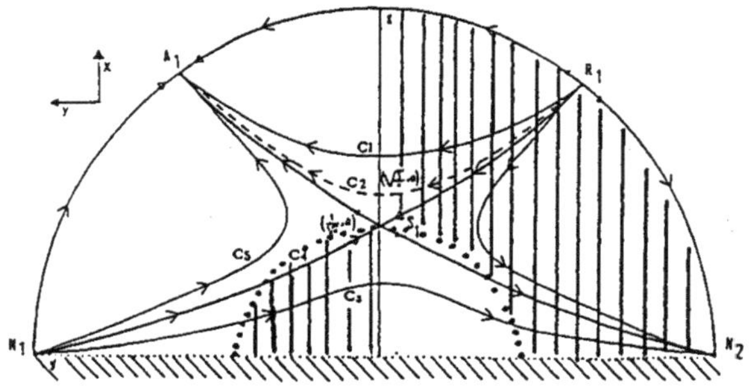

Figure 1: The phase space which corresponds to the system (2).

where $A(t)$ and $B(t)$ are the scale factors.

As a starting point of our discussion, let us write the Einstein equations for vacuum KS spacetimes with an effective positive cosmological constant. Such models correspond to a situation where the stress-energy tensor components which do not contribute to a large Λ-term can be neglected in face of its value. Hence, the Einstein equations may be written as $G_{ab} \approx -\Lambda g_{ab}$, which for definiteness we transcribe in the form of an equality as $R_{ab} = \Lambda\, g_{ab}$. From the metric (1) we write them as follows:

$$2\frac{\dot{A}}{A}\frac{\dot{B}}{B} + \frac{1+\dot{B}^2}{B^2} = 2\frac{\ddot{B}}{B} + \frac{1+\dot{B}^2}{B^2} = \frac{\ddot{A}}{A} + \frac{\ddot{B}}{B} + \frac{\dot{A}}{A}\frac{\dot{B}}{B} = \Lambda. \tag{2}$$

From the equations (2) we conclude that $A(t) = \alpha\dot{B}(t)$, where α is a constant. Therefore, information on the late time behaviour of the solutions of equations (2) is achieved through a qualitative analysis [9] of Einstein's equations. (For details, the dedicated reader can consult ref.[10]). The relevant information is compiled in the following table:

Points	Eigenvalues	Eigenvectors	Classification
S_1	$\sqrt{\Lambda}$	$\frac{\partial}{\partial x} + \sqrt{\Lambda}\frac{\partial}{\partial y}$	Saddle
$(x = \sqrt{\frac{1}{\Lambda}}, y = 0)$	$-\sqrt{\Lambda}$	$\frac{\partial}{\partial x} - \sqrt{\Lambda}\frac{\partial}{\partial y}$	
A_1	$-\sqrt{\frac{\Lambda}{3}}$	$\frac{\partial}{\partial \rho}$	Sink
$\left(\rho = 0, \theta_{A_1} = \arcsin\sqrt{\frac{\Lambda}{3+\Lambda}}\right)$	$-\sqrt{3\Lambda}$	$\frac{\partial}{\partial \theta}$	(stable proper node)
R_1	$\sqrt{\frac{\Lambda}{3}}$	$\frac{\partial}{\partial \rho}$	Source
$(\rho = 0, \theta_{R_1} = -\theta_{A_1})$	$\sqrt{3\Lambda}$	$\frac{\partial}{\partial \theta}$	(unstable proper node)

On the line $x = 0$ we obtain two equilibrium points at $y = \pm\infty$ and designated by N_1 and N_2, respectively. N_1 is an unstable node and N_2 is a stable one. Thus, using standard arguments in qualitative theory, the global phase portrait with inclusion of the infinite distant boundary is as shown in figure 1. The finite region of the phase portrait was previously depicted in ref.[11].

Additional information on the asymptotic behaviour of these Kantowski-Sachs models is obtained from the integration of equations (2). The steps of integration can be followed elsewhere [11,12] and provide the implicit solution $t = \int \frac{dB}{\sqrt{\frac{\Lambda}{3}B^2 - 1 + \frac{C}{B}}}$, $C = B_0\left(\dot{B}_0^2 + 1 - \frac{\Lambda}{3}B_0^2\right)$, where $B_0, \dot{B}_0$ are constants of integration corresponding to initial conditions on B and $\dot{B}$, respectively. In Fig.1 we have $C_1 < C_2 < C_3 < C_4 < C_5$ and $C_2 = 0$, $C_4 = 2/3\sqrt{\Lambda}$.

If we put $C = 0$, we straightforwardly obtain the solution

$$A(t) = \pm\sqrt{\frac{3}{\Lambda}}\,\sinh\left(\sqrt{\frac{\Lambda}{3}}\,t\right), \; B(t) = \sqrt{\frac{3}{\Lambda}}\,\cosh\left(\sqrt{\frac{\Lambda}{3}}\,t\right). \tag{3}$$

The phase trajectory which correspond to the solution (3) is depicted as a dashed line in figure 1. The "+" sign solution was presented in ref. [8]. Note that the solution (3) is *isotropic* [12] as it constitutes (part of) the de Sitter space! In fact, the solution (3) has a zero Weyl tensor and its entire isometry group has 10 independent Killing vectors (cf.ref.[12]). Thus, the solution (3) constitutes a non-zero shear foliation of the KS type of the De Sitter space. Note that the de Sitter solution (3) approaches asymptotically the states corresponding by the sink A_1 and the source R_1.

At this point, let us then address the propriety of the CNHC with respect to universes of the KS type. As one can see from fig. 1, only some of the phase trajectories evolve towards the de Sitter solution - stable node A_1. Using Einstein equations (2), it is easy to show that the dotted ellipsis in figure 1, with equation $3\dot{B}^2 + \Lambda B^2 = 1$, defines the region in phase space where the expansion factor value is $\Theta = 0$. In figure 1, the dashed areas correspond KS spacetimes that are recollapsing, i.e., $\Theta < 0$. Therefore, the CNHC should be analysed in the remaining undashed regions, where expansion is occuring. From the equation of the separatrices $N_1S_1A_1$ and $N_2S_1R_1$,

$$\dot{\underline{B}}^2 = \frac{\Lambda}{3}\underline{B}^2 - 1 + \frac{2}{3\sqrt{\Lambda}\underline{B}} \Longleftrightarrow C = \frac{2}{3\sqrt{\Lambda}}, \tag{4}$$

it is straightforward to conclude that if at some instant of time one the following inequalities

$$\underline{B} > \sqrt{\frac{1}{\Lambda}} \text{ or } \dot{\underline{B}}^2 - \frac{\Lambda}{3}\underline{B}^2 + 1 - \frac{2}{3\sqrt{\Lambda}\underline{B}} > 0, \tag{5}$$

(underline $\underline{B}$ and $\dot{\underline{B}}$ means we are taking values at the undashed regions) is satisfied, the KS solutions of eq.(2) approach a local de Sitter state. Inequalities (5) constitute a pair of requirements for the KS universes to escape a recollapse and are defined in terms of initial conditions. The first inequality can be rewritten as $\Lambda >^{(3)} \mathcal{R}/2$, since for KS spacetimes ${}^{(3)}\mathcal{R} = 2/B^2$ [7]. This means that if Λ is sufficiently large compared with the value of ${}^{(3)}\mathcal{R}$, we obtain inflation and escape a recolapse. Hence, this first condition is similar to Wald's criterion for Bianchi-IX cosmologies [5]. Moreover, the inequalities (5) constitute only sufficient conditions. If $\underline{B} < \frac{1}{\sqrt{\Lambda}}$ and $\dot{\underline{B}}^2 - \frac{\Lambda}{3}\underline{B}^2 + 1 - \frac{2}{3\sqrt{\Lambda}\underline{B}} < 0$, then the recollapse is inevitable.

Let us now evaluate the relative weight between the KS solutions which are expanding and progress towards a de Sitter stage and the ones that recollapse. We consider the simplest measure $dBd\dot{B}$ in the full non compactified phase space. If the area in the phase space covered by inflationary solutions is much more larger than the one which correspond to the recollapsing solutions, it seems reasonable to decide that inflation is a general phenomenon. Henceforth, we presume that the area covered by a particular class of solutions displays a qualitative measure of that class in the non compactified phase space.

Using the separatrix equation (4), we can evaluate the measure of the recollapsing solutions. This corresponds to the area under the curve N_1S_1, between $0 < B < (-1 + \sqrt{5})/2\sqrt{\Lambda}$, given by the integral $\int_0^{(-1+\sqrt{5})/2\sqrt{\Lambda}} \sqrt{\frac{\Lambda}{3}\underline{B}^2 - 1 + \frac{2}{3\sqrt{\Lambda}\underline{B}}}\, d\underline{B}$, which is finite. However, it is negligible as compared with the measure of the asymptotic De Sitter solutions, since this one is equal to infinity. Therefore we conclude that the vast majority of the solutions of eq. (2) evolve towards a De Sitter state, whereas only a small measure, depending on the value of Λ, recollapse.

Summarizing this work, the degree of propriety of the CNHC was established in vacuum KS cosmologies with $\Lambda > 0$ from a qualitative analysis of Einstein equations. Likewise, we determined a *pair* of suficient conditions for these models to inflate as well as another one for the recollapse. These inequalities constitute sufficient conditions in terms of initial values of the scale factors. Therefore, the *generality* of the initially expanding KS spacetimes evolve towards a de Sitter stage, with the exception of a zero probability class of solutions that recolapse.

Acnowledgments: The author is very grateful to O. Bertolami, P. Crawford, V. Frolov, J.S. Lemos, M. Madsen, V. Sahni and R. Wald for their useful comments and to J. Mourão for the many valuable discussions and suggestions. The author also whishes to thank M. Figueira, Ana Nunes and J. Vassalo Pereira for their helpful remarks on qualitative analysis of differential equations.

References

[1] A. Guth, Phys.Rev.**D23** (1981) 347; A.D. Linde, Phys. Lett.**B** 108 (1982) 389; A. Albrecht and P. Steinhardt, Phys. Rev. Lett. **48** (1982) 1220.

[2] V. Belinskii, L. Grishchuk, Ya. Zeldovich and I. Khalatnikov, Sov. Phys. JETP **62** (1985) 195.

[3] P. Moniz, J. Mourão and P.M. Sá, *"The dynamics of a flat FRW inflationary model in the presence of gauge fields"*, to be published in Class. Quantum Grav.; A. Barroso, J. Casasayas, P. Crawford, P. Moniz and A. Nunes, Phys. Lett. **B275** (1992) 264.

[4] G. Gibbons and S.W. Hawking, Phys. Rev. **D15** (1977) 2738.

[5] R. Wald, Phys. Rev **D28** (1983) 2118.

[6] J.D. Barrow, Nuc. Phys. **B296** (1988) 697, J. Yokoyama and K.I. Maeda, Phys. Rev. **D41** (1990) 1047.

[7] M.P. Ryan and L.C. Shepley, Homogenous Relativistic Cosmologies, Princepton University Press (Princepton, 1975); C.B. Collins, J. Math. Phys. **18** (1977) 2116.

[8] E. Weber, J. Math. Phys. **25**, 3279 (1984); Grøn, Ø., 1986, J. Math. Phys., **27**, 1490.

[9] O.I. Bogoyavlenskii, Methods in the qualitative theory of Dynamical Systems in Astrophysics and Gas Dynamics, (Springer-Verlag, Berlin, 1985).

[10] P. Vargas Moniz, *"Kantowski-Sachs Universes and the Cosmic No-Hair Conjecture"*, DFFCUL Preprint 01- 10/1992, IFM Preprint 17/1992, submited to Phys. Rev. **D**.

[11] V. Sahni and L.A. Kofman, Phys. Lett. **A117** (1986) 275.

[12] P. Vargas Moniz and P. Crawford, *"Anisotropic Foliations of the De Sitter Space"*, In: **MG6** Proceedings, Kyoto, Japan, 23-29 June 1991, ed. H. Sato, World Scientific, 1992; *"A Bianchi type III foliation of the De Sitter Space"*, DFFCUL Preprint 02-5/1992, to be published in International Journal of Theoretical Physics, 1993.

FRIEDMANN-ROBERTSON-WALKER COSMOLOGIES IN THE PRESENCE OF MASSIVE VECTOR FIELDS

M. C. BENTO [a)b)], O. BERTOLAMI [a)b)], P. V. MONIZ [b)c)1],
J. M. MOURÃO [a)d)] and P. M. SÁ [b)2]

a) Departamento de Física, Instituto Superior Técnico
Av. Rovisco Pais 1, 1096 Lisboa Codex, Portugal

b) Centro de Física da Matéria Condensada
Av. Prof. Gama Pinto 2, 1699 Lisboa Codex, Portugal

c) Departamento de Física, Faculdade de Ciências de Lisboa
Campo Grande C1-P4, 1699 Lisboa Codex, Portugal

d) Centro de Física Nuclear
Av. Prof. Gama Pinto 2, 1699 Lisboa Codex, Portugal

ABSTRACT

The dynamics of flat Friedmann-Robertson-Walker (FRW) cosmologies in the presence of a triplet of massive vector fields with SO(3) global symmetry is studied. An E^3-symmetric ansatz for the vector fields that is compatible with E^3-invariant FRW metric is found. We study qualitatively the behaviour of the model and find, in particular, that the effective equation of state of the system changes gradually from radiation-dominated to a matter-dominated form and that the scale of transition depends on the mass of the gauge fields.

1. In treating the classical dynamics of the very early Universe with the help of a variational principle applied to actions derived from particle physics theories, one hopes to obtain a better understanding of the effective equations of state valid in the classical regime. On the other hand, this approach may be useful in order to bridge the gap between the rather distinct descriptions of the classical and quantum regimes that have been used sofar. In this context, it is quite natural from the particle physics viewpoint to study non-abelian gauge theories in a cosmological setting. In realistic particle-physics-motivated cosmological models, however, the vector bosons normally acquire a mass thereby breaking the gauge symmetry. As this process takes place in the very early Universe, it is of interest to investigate cosmological solutions of the resulting theory.

Here we investigate homogeneous and isotropic solutions to the Einstein-Yang-Mills (EYM) system after symmetry breaking. We shall assume that the gauge symmetry one starts with is $SO_I(3)$, which subsequently breaks down to a global $SO_I(3)$ and that, in this process, the gauge fields acquire a mass. We derive an E^3-symmetric ansatz for the triplet of massive vector fields with $SO_I(3)$ global symmetry that is compatible with the E^3-invariant (flat) FRW ansatz for the metric. Using qualitative methods of the theory of dynamical systems we compare the

[1]Work supported in part by a JNICT graduate scholarship BD/138/90-RM and by STRIDE/FEDER project JN.91.02.

[2]Work supported by a GTAE (Grupo Teórico de Altas Energias) grant.

dynamical behaviour of the system before and after symmetry breaking and find, in particular, that, in the latter phase, the equation of state changes with time, gradually from a radiation-dominated to a matter-dominated form and that the scale of the transition depends on the mass of the vector field.

2. The spacetime in flat FRW universes has the form

$$M^4 = R^4 = R \times E^3/SO(3), \tag{1}$$

where the six-dimensional euclidean group E^3 is the isometry group of the spatial hypersurfaces and $SO(3)$ is the isotropy group.

Since the isotropy representation has no singlets, we conclude that there are no E^3-invariant vector fields in R^3. Being physically obvious for R^3 (as it just means that there are no homogeneous and isotropic vector fields), this result is important as it applies for any homogeneous space G/H for which the isotropy representation has no trivial singlets. This is also the case for the spatial hypersurfaces S^3 and H^3 of closed and open FRW models which, therefore, do not have vector fields invariant under the action of the corresponding isometry group either. Here, we shall study the case (1) of flat FRW models (the more general case of a spacetime of the form $M^d = R \times G/H$ is analysed in ref. [1]).

If, instead of a single vector field, we consider a theory with a multiplet of (covariant) vector fields $A^a_\mu, a = 1, ..., N$, where a is an internal index, we may hope that, although the EMT corresponding to each field A^a_μ is not E^3-invariant, it would be possible to relate the asymmetries of the different vector fields in such a way that the total EMT is E^3-invariant. Let us then consider the case of a triplet of massive vector fields with global $SO_I(3)$ symmetry. The generators of the isometry group $G = E^3$, T_i and Q_i $(i = 1,2,3)$ span, respectively, an algebra of translations and rotations. The corresponding Killing vector fields are, respectively, $X_i = \partial/\partial x^i$ and $Y_i = -\epsilon_{ijk}x^j\partial/\partial x^k$, where x^i are the cartesian coordinates in R^3 and $\partial/\partial x^i$ together with $\partial/\partial t$ form a (global) coordinate frame in $R^4 = R \times E^3/SO(3), \{X_\mu\} = \left\{\frac{\partial}{\partial t}, \frac{\partial}{\partial x^i}\right\}$, with the dual coframe being given by $\{\omega^\mu\} = \{dt, dx^i\}$.

The conditions of spatial homogeneity and isotropy read, infinitesimaly

$$\mathcal{L}_{X_i}A = 0, \tag{2}$$

$$\mathcal{L}_{Y_i}A = 0, \tag{3}$$

where $A = A^a_\mu(t, \vec{x})\omega^\mu L_a$, the L_a being the generators of the internal group $SO_I(3)$, $[L_a, L_b] = \epsilon_{abc}L_c$. The only solution of (2) and (3) is, as we anticipated, the trivial one, $A = A^a_0(t)dtL_a$. Condition (3) is, however, too restrictive and we now consider a weaker version of it by allowing the field A to be such that infinitesimal spatial rotations generate internal $SO_I(3)$ rotations

$$\mathcal{L}_{Y_i}A = -\left[L_i, A\right]. \tag{4}$$

It is clear that, if (2) and (4) take place, then, although A is not E^3-invariant, it generates an EMT that is invariant under the action of E^3. Indeed, the action of E^3 in the EMT is equivalent to an internal rotation with respect to which the EMT is invariant $\mathcal{L}_{Y_i}EMT(A) = -\delta_{L_i}EMT(A) = 0$. The conditions for spatial homogeneity, eqs. (2), imply that the components A^a_μ depend only on the time variable t. In order to solve (4), let us consider the linear space of vector fields with constant coefficients: $C = \{X = a^\mu X_\mu, a^\mu \in R\}$. $X \in C$ transforms as $\underline{1} \oplus \underline{3}$ under $SO(3)$ spatial rotations. On the other hand, the one-form A can be considered, for any t, as a linear mapping from C to the internal space (i.e. to the Lie algebra of $SO_I(3)$), $A(X) = A(a^\mu X_\mu) = A^a_\mu(t)a^\mu L_a$, in which the triplet representation $\underline{3}$ of $SO_I(3)$ is realized. Rewritting (4) in the form $A(\mathcal{L}_{Y_i}X) = [L_i, A(X)]$, we see that A intertwines the representations $\underline{1} \oplus \underline{3}$ and $\underline{3}$ of $SO(3)$ and $SO_I(3) \overset{isom}{\cong} SO(3)$, respectively. Schur's lemma then implies that for A to be a solution of (4) it must vanish between subspaces of non-equivalent representations

$$A(\partial/\partial t) = 0 \Longleftrightarrow A_0 = 0, \tag{5}$$

and be proportional to the identity operator between subspaces with equivalent representations:

$$A(\partial/\partial x_i) = \chi(t)L_i \Longleftrightarrow A^a_i(t) = \chi(t)\delta^a_i, \tag{6}$$

where $\chi(t)$ is an arbitrary function. Alternatively, the E^3-symmetric ansatz for A satisfying (2) and (4) can be written as $A = \chi(t)dx^i L_i$.

3. Consider now the action for the EYM system with $SO_I(3)$ gauge group

$$S' = \int_{M^4} d^4x\sqrt{-g}\left(\frac{1}{2\kappa^2}R + \frac{1}{8e^2}Tr\left[F_{\mu\nu}F^{\mu\nu}\right]\right), \tag{7}$$

where $\kappa^2 = 8\pi G$ and e is the gauge coupling. Our conventions and notation for the gravitational part of the action correspond to those of ref. [2]. The gauge field strenght is $F_{\mu\nu} = \partial_\mu A_\nu - \partial_\nu A_\mu + [A_\mu, A_\nu]$, where $A_\mu = A^a_\mu L_a$. If a mass term for the gauge fields is added[3], the action

$$S = S' + \int_{M^4} d^4x\sqrt{-g}\frac{1}{2}\,m^2 Tr[A_\mu A^\mu] \tag{8}$$

is no longer gauge invariant and only a global $SO_I(3)$ symmetry remains.

The FRW ansatz, for flat models, gives the most general form of an E^3-invariant metric

$$ds^2 = -N^2(t)dt^2 + a^2(t)\left((dx^1)^2 + (dx^2)^2 + (dx^3)^2\right), \tag{9}$$

where the lapse function $N(t)$ and the scale factor $a(t)$ are arbitrary non-vanishing functions.

We are interested in spatially homogeneous and isotropic cosmologies associated with the action (8). Hence, we restrict ourselves to the set of E^3-symmetric configurations for the fields $g_{\mu\nu}$ and A_μ. Substituting these ansätze into (8), we obtain

[3]Such a mass term would arise, e.g., via the Higgs mechanism, with a complex doublet of scalar particles acquiring a vacuum expectation value.

for the effective action

$$S_{eff} = \int_{t_1}^{t_2} dt \left[-\frac{3}{\kappa^2}\frac{\dot{a}^2 a}{N} + \frac{3a}{4e^2 N}\left(\frac{\dot{\chi}^2}{2} - \frac{N^2}{a^2}V_{gf}\right) - \frac{3}{4}Nm^2\chi^2 a \right]. \tag{10}$$

The equations of motion, derived upon variation with respect to $N(t)$, $a(t)$ and $\chi(t)$ are, respectively, in the "gauge" $N = 1$:

$$H^2 = \frac{\kappa^2}{3}\rho, \tag{11}$$

$$\dot{H} + H^2 = -\frac{\kappa^2}{6}(\rho + 3p), \tag{12}$$

$$\ddot{\chi} + \dot{\chi}H = -\frac{1}{2a^2}\chi^3 - 2m^2e^2\chi, \tag{13}$$

with ρ and p as given by

$$\rho = \frac{3}{4a^2e^2N}\left[\frac{\dot{\chi}^2}{2} + \frac{N^2\chi^4}{8a^2}\right] + \frac{3}{4}m^2\left(\frac{\chi}{a}\right)^2, \tag{14}$$

$$p = \frac{1}{4a^2e^2N}\left[\frac{\dot{\chi}^2}{2} + \frac{N^2\chi^4}{8a^2}\right] - \frac{1}{4}m^2\left(\frac{\chi}{a}\right)^2. \tag{15}$$

In new (dimensionless) variables, $x = \kappa\frac{\chi}{a}$, $y = \frac{\kappa^2}{2\sqrt{2}}\frac{\dot{\chi}}{a}$, $z = \kappa H$, $\tau = t/\kappa$, the system of eqs. (11)–(13) can be written as

$$x_\tau = 2\sqrt{2}y - xz, \tag{16}$$

$$y_\tau = -\frac{\sqrt{2}}{2}\mu^2 x - 2yz - \frac{\sqrt{2}}{8}x^3, \tag{17}$$

$$z_\tau = -2z^2 + \frac{1}{4}\mu^2 x^2, \tag{18}$$

$$z^2 = y^2 + \frac{1}{32}x^4 + \frac{1}{4}\mu^2 x^2, \tag{19}$$

where $\mu^2 = m^2\kappa^2$ and we have set $e = 1$; the index τ denotes the derivative with respect to τ. Substituting the constraint eq. (19) into eqs. (16) and (17), we are left with a two-dimensional dynamical system in the variables x, y

$$x_\tau = 2\sqrt{2}y - x\left(y^2 + \frac{1}{32}x^4 + \frac{1}{4}\mu^2 x^2\right)^{1/2}, \tag{20}$$

$$y_\tau = -\frac{\sqrt{2}}{2}\mu^2 x - 2y\left(y^2 + \frac{1}{32}x^4 + \frac{1}{4}\mu^2 x^2\right)^{1/2} - \frac{\sqrt{2}}{8}x^3, \tag{21}$$

where we have considered the positive root of the constraint (19) as we are interested in expanding models.

In the study of this system of differential equations we have used the methods of the qualitative theory of dynamical systems (see [3] and refs. therein). The phase space is shown in Fig. 1.

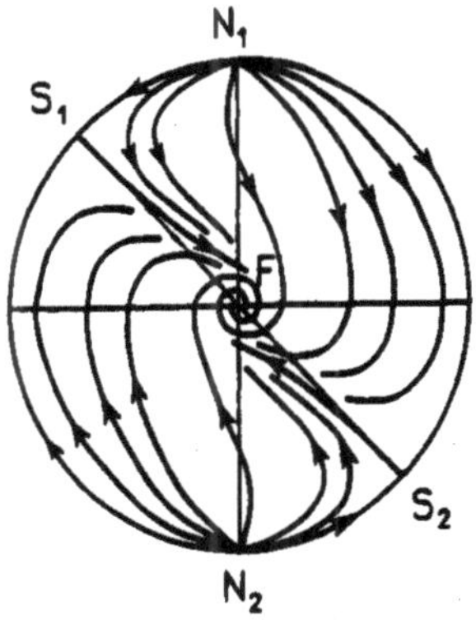

Figure 1. All trajectories of the phase diagramm of the model asymptotically approach the focus F; N_1, N_2, S_1 and S_2 are critical points at infinity.

We now turn our attention to the analysis of the effective equation of state for the system near the critical points. In the vicinity of S_1 and S_2, we get the following solutions

$$a = a_0 t^{1/2}, \qquad \chi = -2a_0(2\kappa^2)^{-1/4} \qquad (t \to 0) \tag{22}$$

Eqs. (14)–(15) then imply $\rho \simeq 3p$, indicating that the effective equation of state near S_1 and S_2 is approximately that of a radiation fluid. In fact, this is true everywhere except in the vicinity of F. This is to be expected since, were it not for the mass term, the equation of state would be $\rho = 3p$ throughout all regions of phase space and it is near the origin that the mass term most affects the trajectories.

Near F the solutions are:

$$a \sim t^{2/3}, \qquad \chi \sim t^{-1/3}\sin(t/\sqrt{\kappa}) \qquad (t \to +\infty), \tag{23}$$

implying that $p \simeq 0$, corresponding to the equation of state for dust matter.

In addition, we find that the time of transition between the regimes described by the different equations of state depends on the value of m^2 such that the transiton occurs earlier for more massive vector fields. Before the transition the energy density and pressure evolve as expected in a radiation-dominated Universe and, as the system approaches the focus F, the Universe becomes matter-dominated since the average value of the pressure approaches zero due to rapid oscillations.

4. References

1. M. C. Bento, O. Bertolami, P. V. Moniz, J. M. Mourão and P. M. Sá, *On the Cosmology of Massive Vector Fields with SO(3) Global Symmetry*, to appear in *Class. Quant. Grav.* (1993).

2. P. V. Moniz and J. M. Mourão, *Class. Quant. Grav.* 8 (1991) 1815.

3. P. V. Moniz, J. M. Mourão and P. M. Sá, *The Dynamics of a Flat Friedmann-Robertson-Walker Inflationary Model in the Presence of Gauge Fields*, to appear in *Class. Quant. Grav.* (1993).

Post-COBE predictions for Inflationary Gravity Wave and Density Perturbation spectra

VARUN SAHNI † and TARUN SOURADEEP *

Inter-University Centre for Astronomy and Astrophysics
Post Bag 4, Ganeshkhind, Pune 411 007
INDIA

ABSTRACT

We assess the relative contribution to the COBE - measured microwave anisotropy arising both from relic gravity waves as well as primordial density perturbations originating during inflation. We show that the gravity wave contribution to the CMBR anisotropy depends sensitively upon n – the primordial spectral index ($|\delta_k|^2 \propto k^n$), increasing as n deviates from a Harrison - Zeldovich spectrum (n = 1). As a result, for $n < 0.84$ the contribution from gravity waves towards $\delta T/T$ is greater than the corresponding contribution from density perturbations, whereas for $n > 0.84$ the reverse is true. ($n = 0.84$ corresponds to an expansion index $p = 13.5$ in models with power-law inflation $a \propto t^p$.) Our results show that for a scale-invariant Harrison-Zeldovich spectrum generated by chaotic inflation, gravity waves contribute approximately 24% to the CMBR anisotropy measured by COBE. Applying our results to the cold dark matter scenario for galaxy formation, we find that in general CDM models with tilted power spectra ($n < 1$), require the biasing parameter to be greater than unity, on scales of $16h_{50}^{-1}Mpc$. We also obtain an expression for the COBE - normalised amplitude and spectrum of the stochastic gravity wave background and compare it with the sensitivity of planned laser-interferometer gravity wave detectors.

email : † varun@iucaa.ernet.in
* tarun@iucaa.ernet.in

Both gravity waves and adiabatic density perturbations, characterised by a power spectrum $|\delta_k| \propto k^n$, are generically predicted by the Inflationary Universe scenario. On entering our horizon during the epoch of matter-radiation decoupling, density perturbations as well as gravity waves cause distortions in the cosmic microwave background radiation (CMBR) due to the well known Sachs-Wolfe effect. The anisotropy in the CMBR recently observed by COBE measures the amplitude of gravity waves as well as density perturbations providing thereby a direct probe of the inflationary epoch of our universe.

The close similarity between gravity waves and massless scalars, in a Friedman - Robertson - Walker (FRW) background originally pointed out by Grishchuk[1], makes it possible to express each polarisation state of the graviton in terms of solutions to the massless, minimally coupled Klein-Gordon equation: $h_{\times(+)} = \sqrt{8\pi G}\ \phi_k\, e^{-i\mathbf{k}.\mathbf{x}}$

$$\ddot{\chi}_k + [k^2 - \frac{\ddot{a}}{a}]\chi_k = 0 \tag{1a}$$

where $\chi_k = \phi_k \times a(\eta)$, $a(\eta)$ is the scale factor of the Universe: $ds^2 = a^2(\eta)(d\eta^2 - d\mathbf{x}^2)$, and k is the comoving wavenumber $k = 2\pi a/\lambda$. Eq.(1a) closely resembles the Schroedinger equation in quantum mechanics, with $\ddot{a}/a$, playing the role of the potential barrier 'V'. (The form of $V \equiv (\ddot{a}/a)$, is shown in Figure 1, for inflation followed by a matter dominated epoch.) Solutions of equation (1a) have some interesting properties, for instance, small wavelength solutions of (1a) are adiabatically damped:

$$\phi_k^+(\eta) \equiv \frac{\chi_k}{a(\eta)} \underset{k\eta \gg 2\pi}{=} \frac{1}{\sqrt{2k}a(\eta)} \exp(-ik\eta), \tag{1b}$$

whereas long wavelength modes asymptotically approach the form :

$$\phi_k(\eta) \equiv \frac{\chi_k}{a(\eta)} \underset{k\eta \ll 2\pi}{=} A(k) + B(k) \int \frac{d(\frac{\eta}{\eta_0})}{a^2}\ , \tag{1c}$$

where A and B are constants. The form of (1c) implies that the amplitude of a fluctuation (equivalently – gravity wave), freezes to a constant value when its wavelength (during inflation), becomes larger than the corresponding Hubble radius. Since modes are continuously being pushed outside the Hubble radius during inflation, a mode with a fixed wavenumber "k", which left the Hubble radius during inflation ($k = 2\pi\ \eta_1^{-1}$), will re-enter it later during the epoch of matter domination ($k = 2\pi\ \eta_0^{-1}$). The amplitude of the mode between these two epochs is super-adiabatically amplified, which in field-theoretic language corresponds to the quantum creation of gravitons[1] (see fig. 1).

The possibility that relic gravity waves may be generated during inflation, was first investigated by Starobinsky[2], who showed that the spectrum of gravitons is scale invariant in a fairly broad wavelength interval, if inflation proceeds exponentially and is followed by a radiation dominated epoch. The quantum creation of relic gravity waves was subsequently extended to other inflationary models by Abbott & Wise[4] and Sahni[6]. (The effect of a post inflationary matter dominated epoch on the graviton spectrum was studied by Allen[5], and also by Sahni[6].) In particular it was demonstrated that, for power law expansion $a = (t/t_0)^p \equiv (\frac{\eta}{\eta_0})^{\frac{1-2\nu}{2}}$ $(\nu - \frac{3}{2} = \frac{1}{p-1} = -3\frac{1+w}{1+3w}$, where $w = \frac{P}{\epsilon}$ is the equation of state of matter) Eq.(1a) can be solved exactly, with the following solution describing positive frequency waves in an FRW Universe in the adiabatic limit:

$$\phi_k^+(\eta) = \sqrt{\frac{\pi\eta_0}{4}}\left(\frac{\eta}{\eta_0}\right)^\nu H_\nu^{(2)}(k\eta) \qquad (1d)$$

($\nu = \frac{3}{2}$ corresponds to exponential inflation, $w = -1$; $\nu > \frac{3}{2}$ to power law inflation, $-1 < w < -\frac{1}{3}$; $\frac{1}{2} < \nu < \frac{3}{2}$ to super-exponential inflation (pole - driven inflation), $w < -1$; and $-\frac{3}{2} \leq \nu \leq 0$ to matter dominated expansion, $0 \leq w \leq 1$.) From (1c) and (1d) it can be shown that, the amplitude of gravity waves on scales greater than the Hubble radius, has the time independent form[6]

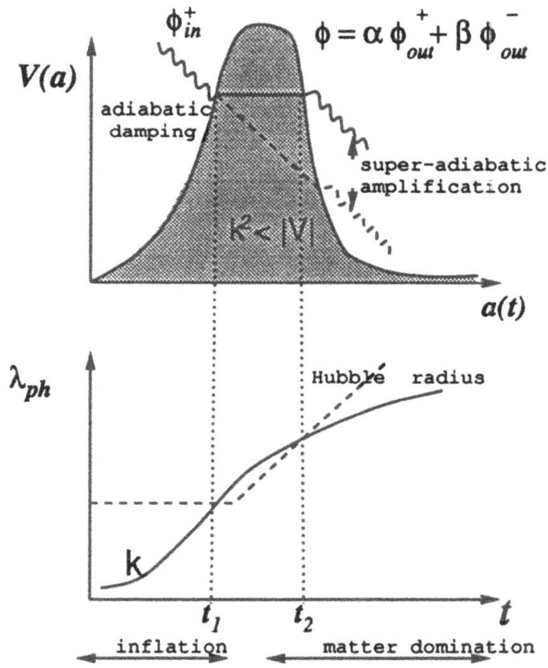

Fig. 1 The superadiabatic amplification of gravity waves is shown for a "potential barrier" $V[a(\eta)] = (\ddot{a}/a)$, for inflation $(a = (\eta_o/\eta))$ followed by radiation and matter domination [3] $(a = a_o\eta(\eta + \bar{\eta}_o))$. At early times $t < t_1$, the scalar field is in its vacuum state $\tilde{\phi}_k^{(+)}$. At late times $t \approx t_0$, the scalar field will in general be described by a linear superposition of positive and negative frequency solutions of eq. (1) : $\tilde{\phi}_k(\eta) = \alpha\tilde{\phi}_k^{(+)} + \beta\tilde{\phi}_k^{(-)}$. A mode with comoving wavenumber $k = (2\pi a/\lambda)$ is shown to leave the Hubble radius during inflation (t_1), and re-enter it during matter domination (t_0). (Figure not drawn to scale.)

$$h_{+,\times} = \sqrt{16\pi G}\frac{k^{\frac{3}{2}}}{\sqrt{2\pi}}|\phi_k|\Big|_{k|\eta_1|=2\pi} = \sqrt{16\pi}A(\nu)\left(\frac{H_1}{m_p}\right),$$
$$A(\nu) = \Gamma(\nu)\pi^{-\nu}(\nu-\frac{1}{2})^{-1}, \tag{2}$$

where we have summed over both polarisation states of the graviton. The value of ν is related to the spectral index n of density fluctuations as $\nu = (4-n)/2$ and H_1 is the value of the Hubble parameter at a time $k|\eta_1| = 2\pi$, when the given mode left the Hubble radius. For exponential inflation $\nu = \frac{3}{2}$ and $h_{+,\times} = (2/\sqrt{\pi})(H_1/m_p)$. ($m_p$ is the planck mass, $m_p = 1.2 \times 10^{19} GeV$.)

Both long wavelength gravity waves as well as adiabatic density perturbations generated during inflation, create potential fluctuations at the surface of last scattering, when their wavelength becomes smaller than the horizon size during the matter dominated epoch. This leads to distortions in the cosmic microwave background, described by the well known Sachs - Wolfe effect [4,7] $(\delta T/T) \sim \frac{1}{3}(\delta\varphi/c^2)$. Writing $(\delta T/T)$ in terms of a multipole expansion:

$$\frac{\delta T}{T} = \sum_{l,m} a_{lm} Y_l^m(\alpha, \delta), \tag{3}$$

it can be shown that, for gravity waves[4,8]

$$a_2^2 \equiv \langle |a_{2m}|^2 \rangle = 0.145\, h_{+,\times}^2\, g(n), \quad g(n) = 5770.3 \times \int_0^\infty db \left(\frac{b}{2}\right)^{n-2} [I_2(b)]^2, \tag{4a}$$

where[8]

$$I_l(b) = \int_{k\eta_{rec}}^{b} \frac{dy}{(b-y)} \frac{J_{l+\frac{1}{2}}(b-y)}{(b-y)^{\frac{1}{2}}} \frac{J_{\frac{5}{2}}(y)}{y^{\frac{3}{2}}}, \quad b = k\eta_0. \tag{4b}$$

We have numerically integrated (4b) to obtain $g(n)$ ($k\eta_{rec} \approx 0$), which is well approximated by the fitting formula $g(n) = \exp[0.6(n-1)]$ over the range $0.5 \le n \le 1$. The corresponding value of the *rms* quadrupole amplitude is

$$Q_T^2 = \frac{5}{4\pi}\, a_2^2\, F_2 = 2.9\, A^2(\nu) \left(\frac{H_1}{m_p}\right)^2 g(n)\, F_2\,. \tag{4c}$$

(For exponential inflation $\nu = \frac{3}{2}$, and $Q_T^2 = (2.9/4\pi^2)(H_1/m_p)^2\, F_2$.) (The subscript "T", denotes the fact that the quadrupole anisotropy is induced by *tensor* waves, gravity waves being quadrupolar, do not contribute to a dipole component a_1.) F_2 incorporates the finite beam width of the COBE-DMR instrument ($F_2 \approx 0.99$) .

Scalar density perturbations generated during inflation, are characterised by a spectrum $|\delta_k|^2 = \mathcal{A}\, k^n$, the scale invariant Harrison - Zeldovich spectrum $n = 1$, is predicted by inflationary models with exponential expansion. Models with power law inflation ($a(t) \propto t^p, p > 1$), which arise in a number of theories including extended inflation[9], predict a more general value for the spectral index[10] , $n = (p-3)/(p-1) < 1$. ($n > 1$, is predicted for super-exponential inflation[28].)

The amplitude of density perturbations at horizon crossing ($k = H_0/c$) is given by[11]

$$\left(\frac{\delta\rho}{\rho}\right)_H^2 \equiv \frac{k^3|\delta_k|^2}{2\pi^2} = \frac{\mathcal{A}}{2\pi^2}H_0^{n+3}. \tag{5}$$

(we assume $c = 1$, for simplicity.) The associated fluctuations in the microwave background are given by (for $l \geq 2$)

$$a_l^2 \equiv \langle |a_{lm}|^2\rangle = \frac{H_0^4}{2\pi}\int_0^\infty \frac{dk}{k^2}|\delta_k|^2 j_l^2(kx) \tag{6a}$$

where $x = (2/H_0)$, is the present day horizon size ($j_l^2(kx)$ are spherical Bessel functions). For power law spectra $|\delta_k|^2 = \mathcal{A}\ k^n$, we obtain[12]

$$a_2^2 = \frac{\mathcal{A}H_0^{n+3}}{16}\ f(n) = \frac{\pi^2}{8}f(n)\left(\frac{\delta\rho}{\rho}\right)_H^2, \quad f(n) = \frac{\Gamma(3-n)\Gamma(\frac{3+n}{2})}{\Gamma^2(\frac{4-n}{2})\Gamma(\frac{9-n}{2})}. \tag{6b}$$

Values of n in the range $0.5 \leq n \leq 1.7$, are consistent with the recent COBE results[13]. For $n = 1$, (6b) reduces to $a_2^2 = (\pi/12)\ (\delta\rho/\rho)_H^2$.

The related value of the *rms* quadrupole amplitude is given by

$$Q_S^2 = \frac{5}{4\pi}\ a_2^2\ F_2 = \frac{5\pi}{32}f(n)\left(\frac{\delta\rho}{\rho}\right)_H^2\ F_2 \tag{7}$$

(the subscript "S" refers to *scalar* perturbations.)

As shown by a number of authors, scalar field fluctuations which left the Hubble radius during inflation, upon re-entering the horizon during radiation (matter) domination, give rise to a density fluctuation, whose *rms* value is given by[14]

$$\left(\frac{\delta\rho}{\rho}\right)_H = b\ \frac{H_1}{\dot\phi}\ \delta\phi \tag{8}$$

where H_1 is the Hubble parameter at a time t_1, when a scale which reenters the horizon at t_0, left the Hubble radius during the inflationary epoch (see Figure 1). For modes entering the horizon during radiation (matter) domination, $b = 4(\frac{2}{5})$. (The main contribution to the observed quadrupole anisotropy however, comes from modes entering the horizon during matter domination, we shall therefore assume $b = \frac{2}{5}$ in the ensuing discussion.) $\delta\phi$ arises because of quantum fluctuations in the scalar field during inflation. For modes entering the horizon today $\delta\phi$ is given by $\delta\phi = A(\nu)H_1$ (see (2)), from which we recover the standard result $\delta\phi = \frac{H_1}{2\pi}$, for exponential inflation[14,15].

For power law inflation, the inflaton potential has the form $V(\phi) = V_0\exp(-\sqrt{16\pi/p}\ (\phi/m_p))$ which results in the following exact solution[16] to the field equations: $H = (p/t)$, $(H/\dot\phi) = (\sqrt{4\pi p}/m_p)$, as a result

$$\left(\frac{\delta\rho}{\rho}\right)_H = A(\nu)\ \frac{\sqrt{16\pi p}}{5}\ \frac{H_1}{m_p}. \tag{9}$$

Combining Eqs. (7) and (9), the *rms* quadrupole amplitude of the temperature anisotropy

$$Q_S^2 = \frac{\pi^2}{10}\, p\, A^2(\nu)\left(\frac{H_1}{m_p}\right)^2 f(n)\, F_2 \tag{10}$$

($\nu - \frac{3}{2} = (p-1)^{-1} = (1-n)/2)$), for power law inflation. (See also Fabbri, Lucchin and Mataresse; Lyth & Stewart.)[17] The above result for power law inflation can be extended to cover models of slow-roll (quasi-exponential) inflation . As shown in Souradeep and Sahni [18], Q_S^2 for a chaotic inflation model with a $m^2\phi^2$ potential corresponds to a power law model with $p = 2N$, where $N = H\Delta t$ is the number of e-foldings (Δt is the duration of the inflationary epoch and $H \approx$ *constant* is the Hubble parameter during inflation). In terms of the number of e-foldings in a model of slow-roll inflation, the following relation

$$2N = p = \frac{n-3}{n-1}, \tag{11}$$

fixes the spectral index of density perturbations n. As a result slow-roll inflation (chaotic $m^2\phi^2$ model) with minimal e-foldings $N = 67$ is equivalent to a power law model with $p = 134$ which corresponds to a spectral index $n = 0.985$.

Since the coefficients a_{lm} in (3) are Gaussian random variables, the variances Q_T^2 and Q_S^2 given by (4c) and (10) must be added, when considering the combined effect of both gravity waves and adiabatic density perturbations on the anisotropy of the CMBR. As a result, the net quadrupole anisotropy will be given by

$$Q^2_{\text{COBE-DMR}} = Q_S^2 + Q_T^2 \tag{12}$$

where $Q_{\text{COBE-DMR}}$ is the *rms*-quadrupole normalised amplitude $Q_{rms-PS} = 5.86 \times 10^{-6}$ detected by COBE [13]. From (4c) and (10), we find that the ratio $(Q_T^2/Q_S^2) = (2.94\,(n-1)g(n)/(n-3)f(n))$ depends sensitively on n - the spectral index of density perturbations. We find that for $n \geq 0.84$, the contribution from gravity waves to the CMBR anisotropy dominates the contribution from scalar density perturbations, whereas for $n < 0.84$ the reverse is true. ($n = 0.84$ corresponds to $p = 13.5$.) The ratio (Q_T^2/Q_S^2) has been plotted against n in Figure 2 . For slow-roll (chaotic) inflation we find that the microwave distortion due to gravity waves amounts to $\simeq 31\%$ of the distortion caused by density fluctuations, in agreement with earlier work by Starobinsky [8] and contrary to recent claims by Krauss and White[19].

Substituting the values of Q_T^2 and Q_S^2 in Eq.(12), we obtain

$$\left(\frac{H_1}{m_p}\right)^2 = \left[\frac{\pi^2}{10}\,\frac{n-3}{n-1}\, f(n) + 2.9\, g(n)\right]^{-1} [A(\nu)]^{-2}\, \frac{Q^2_{\text{COBE-DMR}}}{F_2} \tag{13}$$

for power law inflation. Setting $p = 2N$ in Eq. (13) we can recover the result for slow-roll inflation[18]. As a result Eq.(13) allows us to determine the Hubble parameter during inflation both for power law, as well as for quasi-exponential inflation.

The value of (H_1/m_p) allows us to determine the value of Q_S^2 - the fraction of the quadrupole anisotropy contributed by density perturbations, as well as $(\delta\rho/\rho)_H^2$ - the amplitude of density fluctuations at horizon crossing (see (5), (9)). The value of $(\delta\rho/\rho)_H$ fixes the normalisation $\mathcal{A}$ of the power spectrum of density fluctuations to be

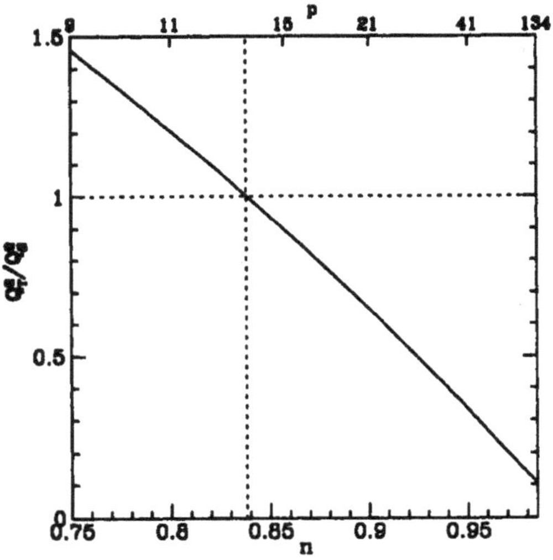

Fig. 2 The ratio of the Quadrupole anisotropy from tensor and scalar waves is plotted against both spectral index n and p (in $a \propto t^p$) for power law inflation. The maximum value $p = 134$ corresponds to $n = 0.985$ and is equivalent to ~ 67 e-foldings of slow-roll inflation.

$$\frac{A}{2\pi^2} = \left[\frac{72.5}{16\pi}\,\frac{n-1}{n-3}\;g(n) + \frac{5\pi}{32}\;f(n)\right]^{-1}\left(\frac{c}{H_o}\right)^{n+3}\frac{Q^2_{\rm COBE-DMR}}{F_2}\,, \tag{14}$$

(we have reintroduced c in the above expression for completeness and as noted earlier (Eq. 4c), the finite beam width factor $F_2 \approx 0.99$ for COBE-DMR). We now use the COBE-DMR results to normalise the spectrum and predict the value of the *rms* mass fluctuation on a given scale

$$\left(\frac{\Delta M}{M}\right)^2(R) = \frac{1}{2\pi^2}\int k^2 dk|\delta_k|^2 W^2(kR), \tag{15}$$

where $W(kR)$ is the *top hat* window function [20].

Our results for cold dark matter models with power law primordial spectra : $|\delta_k|^2 = A\,k^n\,T_{k,\rm cdm}$, (where $T_{k,\rm cdm}$ is the CDM transfer function given in Appendix G of Bardeen et al.[21]) are shown in

Figure 3, for different values of the parameters n and h_{50} ($\Omega_{cdm} \simeq 0.9, \Omega_{baryonic} \simeq 0.1$, is assumed). (Values of n in the range $0.5 \leq n \leq 1.7$, are consistent with the COBE - DMR results [13].) For a scale-invariant spectrum $n = 1$, our results agree with those of Bond and Efstathiou[22]. Our results also show that $\frac{\Delta M}{M}(16h_{50}^{-1}Mpc) \leq 1$ for $n \leq 0.87$. Since mass need not trace light, in models with nonbaryonic dark matter, our results can be taken to mean that a biasing factor ($b_{16} = (\frac{\Delta M}{M})_{16}^{-1}$) mostly greater than unity, is required in order to reconcile theoretical models based on power law inflation, with observations. From Figure 3 it follows that the value of b_{16} is sensitive to both n as well as h_{50}, decreasing with increasing n and h_{50}. It may be noted that since gravity waves contribute predominantly to the CMBR anisotropy for $n \leq 0.84$, the biasing factor $b_{16h_{50}^{-1}}$ is much larger than it would have been had only the contribution from scalar density perturbations to the CMBR been considered. This makes CDM models with power law inflation less compatible with the excess galaxy clustering observed in the APM survey, than had previously been assumed [23].

Having normalised the amplitude of scalar density perturbations generated during inflation, we now proceed to evaluate the normalised spectrum of relic gravity waves. During inflation (*ie* for $\eta < \eta_0 < 0$), gravitons are described by Eq.(1d) which represents the state corresponding to the adiabatic vacuum . During the matter dominated epoch however, gravitons are described by a state which is a linear superposition of positive and negative frequency solutions of Eq.(1a):

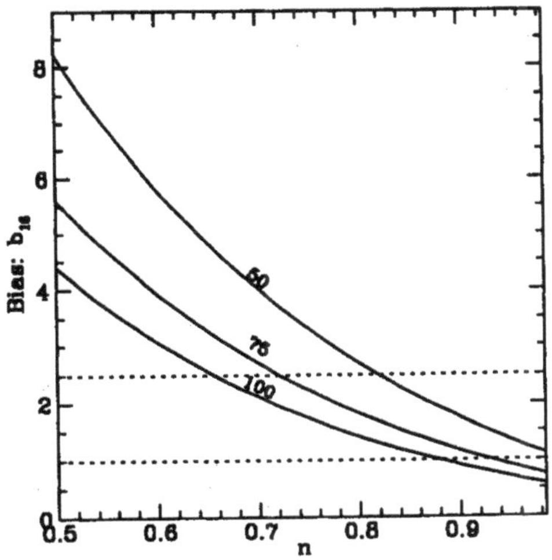

Fig. 3 The biasing factor $b_{16} = (\frac{\Delta M}{M})_{16}^{-1}$, is shown plotted against the spectral index n, for three values of the Hubble parameter: $H = 50$, 75 & 100 $kmsec^{-1}Mpc^{-1}$. The dotted lines correspond to $b_{16} = 1$ and $b_{16} = 2.5$.

$$\tilde{\phi}_k(\eta) = \alpha\tilde{\phi}_k^{(+)} + \beta\tilde{\phi}_k^{(-)} \tag{16a}$$

where

$$\tilde{\phi}_k^{(+,-)}(\eta) = \sqrt{\frac{\pi\eta_0}{4}}\left(\frac{\eta}{\eta_0}\right)^{\mu} H_{|\mu|}^{(2,1)}(k\eta) \tag{16b}$$

$\eta > |\eta_0|$. μ can be related to the equation of state of matter $\mu = \frac{3}{2}\frac{w-1}{1+3w}$ where $w = P/\epsilon$, ($-\frac{3}{2} \leq \mu \leq 0$ for reasonable equations of state for matter: $0 \leq w \leq 1$).

The Bogoliubov coefficients α and β can be obtained by matching (1d) and (16) at wavelengths larger than the horizon size via (1c). As a result we obtain for $k\eta_0 < 2\pi$,

$$\alpha \pm \beta = \pm i\gamma^{\pm 1}\left(\frac{k\eta_0}{2}\right)^{\mp(\nu+|\mu|)} \tag{17}$$

where $\gamma = \pi^{-1}\Gamma(\nu)\Gamma(1+|\mu|)$, and $|\alpha|^2 - |\beta|^2 = 1$, (see Sahni[6] (1990) for details). (For $k\eta_0 > 2\pi$ the adiabatic theorem gives $\alpha \simeq 1, \beta \simeq 0$.)

The energy density of created gravitons can now be determined exactly by substituting the expression for $\tilde{\phi}_k(\eta)$ in (16), with the values of the Bogoliubov coefficients given in Eq.(17) in the expression[29]

$$\epsilon_g = \langle T_0^0 \rangle = \frac{1}{2\pi^2 a^2}\int dk\, k^2\left(\left|\dot{\tilde{\phi}}_k\right|^2 + k^2|\tilde{\phi}_k|^2\right), \tag{18}$$

(both polarisation states of the graviton have been included in Eq.(18).)

Of greater relevence to us is the spectral energy density of gravity waves, $\epsilon(\omega) = \omega\frac{d\epsilon_g}{d\omega}$ which can be derived from Eq.(18) and has the simple form[6]

$$\epsilon(\tilde{\omega}) = b_2^2\tilde{\omega}^{1-2\tilde{\nu}}\left(\frac{H_1}{m_p}\right)^2\epsilon_m \quad \text{for } 1 < \tilde{\omega} < \frac{3}{4\pi}\Omega_r^{-\frac{1}{2}},$$

$$\epsilon(\tilde{\omega}) = b_1^2\tilde{\omega}^{3-2\tilde{\nu}}\left(\frac{H_1}{m_p}\right)^2\epsilon_r \quad \text{for } \frac{3}{4\pi}\Omega_r^{-\frac{1}{2}} < \tilde{\omega} \tag{19}$$

where $\tilde{\omega} = \frac{k\eta}{2\pi} = \frac{\lambda_h}{\lambda}$, is the dimensionless wavenumber expressed in units of the horizon scale (λ being the physical wavelength and λ_h being the present scale of the horizon, $\lambda_h \simeq 2.10^{28}h_{50}^{-1}$ *cm.*). $1 < \tilde{\omega} < \frac{3}{4\pi}\Omega_r^{-\frac{1}{2}}$ corresponds to wavelengths larger than the horizon size at matter radiation equality, ϵ_m and ϵ_r are the energy densities of matter and radiation respectively; $\epsilon_m \approx \epsilon_{cr} = 4.2\,.10^{-9}h_{50}^2$ ergs/cm^3, $\Omega_r = \frac{\epsilon_r}{\epsilon_m} \simeq 10^{-4}h_{50}^{-2}$, h_{50} is the present value of the Hubble parameter in units of 50 km/sec/Mpc.

From Eq.(19) we see that the spectrum of the gravity wave background is uniquely specified once $(\frac{H_1}{m_p})$ – the dimensionless value of the Hubble parameter during inflation is established from Eq.(13). The ratio of the spectral energy density of gravity waves to the critical energy density $\Omega_g(\lambda) = \epsilon(\lambda)/\epsilon_{cr}$ and the dimensionless amplitude of gravity waves, $h_{\times+}(\lambda)$ have been plotted in Figure 4, for gravity waves created during exponential inflation, slow-roll (quasi-exponential) inflation and power law inflation. (For the corresponding analytical expressions describing $\Omega_g(\lambda)$, see Sahni[6] and Grishchuk & Solokhin [24].) The quantities $h_{\times+}(\lambda)$ and $\Omega_g(\lambda)$ are directly related by $h_{\times+}^2(\lambda) = 24(\lambda/\lambda_h)\Omega_g(\lambda)$. We find that the amplitude of gravity waves is significantly smaller than the sensitivity of the current generation of terrestrial bar and beam detectors [25]. The best

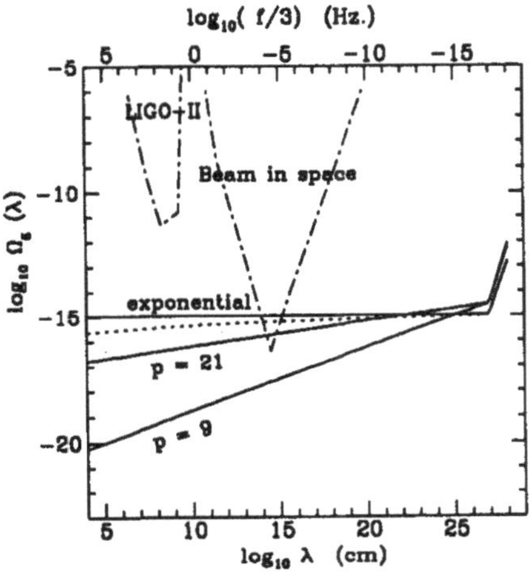

a

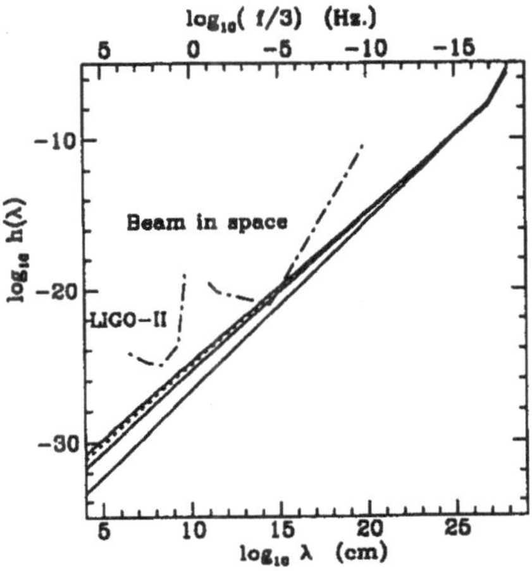

b

Fig. 4 The COBE-normalised spectral energy density (in units of the critical density) and the amplitude, $h_{\times +}$, of gravity waves is shown (in figures a and b respectively) as a function of the wavelength (and frequency f), for exponential inflation, quasi-exponential inflation (dotted line), and power law inflation $a \propto t^p$ with $p = 21$ and $p = 9$. For comparison, the expected sensitivity of the Laser Interferometer Gravitywave Observatory (LIGO), (Christensen 1992) and of the projected "Beam in space" (space - interferometer) has also been plotted (Thorne 1988).

hope for the detection of the stochastic gravity wave background appears to lie with the *space - interferometer*, whose development is still in the conceptual stage [26].

After completing this work, we became aware of papers[27] by Davis et al.; Salopek; Liddle & Lyth; Lucchin, Mataresse & Mollerach ; and Lidsey and Coles, in which results overlapping with those of this paper and ref [18] were obtained.

References

1. Grishchuk, L.P., *Zh. Eksp. Teor. Fiz.* **67** (1974) 825 [*Sov. Phys. JETP* **40**,(1975) 409]

2. Starobinsky, A.A., *JETP Lett.* **30** (1979) 719

3. Sahni, V., *Class. Quantum Grav.* **5** (1988) L113

4. Abbott, L.F. and Wise, M.B., *Nucl. Phys.* **135B** (1984) 279

5. Allen, B., *Phys. Rev. D* **37** (1988) 2078

6. Sahni, V., *Phys. Rev. D* **42** (1990) 453

7. Sachs, R.K. and Wolfe, A.M., *Astrophys. J.* **147** (1967) 73

   Rubakov, V.A., Sazhin, M.V. and Veryaskin, A.V., *Phys. Lett.* **115B** (1982) 189

   Fabbri, R. and Pollock, M.D., *Phys. Lett.* **125B** (1983) 445

8. Starobinsky, A.A., *Pis'ma Astron. Zh.* **11** (1985) 323 [*Sov. Astron. Lett.* **11** (1985) 133]

9. La, D. and Steinhardt, P.J., *Phys. Rev. Lett.* **62** (1989) 376

10. Lucchin, L.F. and Mataresse, S., *Phys. Lett.* **164B** (1985) 282

    Lucchin, L.F. and Mataresse, S., *Phys. Rev. D* **32** (1985) 1316

11. Kolb, E.W. and Turner, M.S., *The Early Universe*, (Addison - Wesley Publishing Company 1990)

12. Gradshteyn, I.S. and Ryzhik, M., *Table of Integrals, Series and Products*, (Acad. Press Inc., N.Y., 1980) p. 692

13. Smoot, C. F. et al., *Astrophys. J. Lett.* **396** (1992) L1

14. Starobinsky, A.A., *Phys. Lett.* **117B** (1982) 175

    Guth, A. and Pi, S.Y., *Phys. Rev. Lett.* **49** (1982) 1110

    Hawking, S.W., *Phys. Lett.* **115B** (1982) 295

    Bardeen, J.M., Steinhardt, P., and Turner, M., *Phys. Rev. D* **28** (1983) 679

15. Bunch, T. S. and Davies, P. C. W., *Proc. R. Soc.* **A360** (1978) 117

    Linde, A.D., *Phys. Lett.* **116B** (1982) 335

Vilenkin, A. and Ford, L. H., *Phys. Rev.* **D25** (1982) 1231

16. Burd, A.B. and Barrow, J.D., *Nucl. Phys.* **B308** (1988) 929

17. Fabbri, R., Lucchin, L.F. and Mataresse, S., *Phys. Lett.* **166B** (1986) 49

Lyth, D.H. and Stewart, E.D., *Phys. Lett.* **274B** (1992) 168

18. Souradeep, T. and Sahni, V., IUCAA preprint/July-07-92 (1992),

Souradeep, T. and Sahni, V., *Mod. Phys. Lett.* **A7** (1992) 3541

19. Krauss, L. M. and White, M., *Phys. Rev. Lett.* **69** (1992) 869

20. Peebles, P.J.E., *The Large Scale Structure of the Universe* (Princeton University, Princeton 1980)

21. Bardeen, J.M., Bond, J.R., Kaiser, N, and Szalay, A.S.,*Astrophys. J.* **30** (1986) 15

22. Bond, J.R. and Efstathiou, G., *Astrophys. J.* **285** L45, (1984) L45

Bond, J.R. and Efstathiou, G., *MNRAS* **226** (1987) 655

23. Liddle, A. R., Lyth, D. H., and Sutherland W. J. , *Phys. Lett.* **B274** (1992) 244

24. Grishchuk, L.P. and Solokhin, M., *Phys. Rev. D* **43** (1991) 2566

25. Christensen, N., *Phys. Rev. D* **46** (1992) 5250

26. Thorne, K.S., in *300 Years of Gravitation*, eds. S.W. Hawking and W. Israel, (Cambridge Univ. Press, Cambridge, 1988) p.330

27. Davis R.L., Hodges, H.M., Smoot, G.F., Steinhardt, P.J. and Turner, M.S., *Phys. Rev. Lett.* **69** (1992) 1856

Salopek, D. S., *Phys. Rev. Lett.* **69** (1992) 3602

Liddle, A. R. and Lyth, D. H., *Phys. Lett.* **B291** (1992) 391

Lucchin, F., Mataresse, S. and Mollerach, S., *Astrophys. J. Lett.* **401** (1992) L49

Lidsey, J.E. and Coles, P., *MNRAS* **258** (1992) 57p

28. Gasperini, M. and Giovannini, M., *Phys. Lett.* **B282** (1992) 36

Gasperini, M. and Veneziano, G., CERN preprint: CERN-TH.6572/92

29. Ford, L.H. and Parker, L., *Phys. Rev. D* **16** (1977) 1601

Cosmology, Oscillating Physics and Oscillating Biology

Pablo D. Sisterna
Departamento de Física, F.C.E.y N, Univ. Nac. de Mar del Plata
Funes 3350 - (7600) Mar del Plata, ARGENTINA

Hector Vucetich
Departamento de Física, Facultad de Ciencias Exactas, U.N.L.P.
C.C.67 1900 - La Plata, ARGENTINA

Abstract

According to recent reports there is an excess correlation and an apparent regularity in the galaxy one-dimensional polar distribution with a characteristic scale of 128 h^{-1} Mpc. This aparent spatial periodicity can be naturally explained by a time oscillation of the gravitational constant G. On the other hand, periodic growth features of bivalve and coral fossiles appear to show a periodic component in the time dependence of the number of days per year. In this letter we show that a time oscillating gravitational constant with similar period and amplitude can explain such a feature.

Our theoretical description of nature presents a set of parameters, the so called fundamental constants, that have to be determined from experience. It is generally thougth that this is so because we lack a unified theory of all interactions. The time variation of any fundamental constant might be one of the few low energy phenomena which could manifest this "new physics". Actually, superstring theories [1] and Kaluza-Klein theories [2] have cosmological solutions in which the low-energy fundamental constants vary with time [3, 4, 5, 6, 7].

Broadhurst, Ellis, Koo and Szalay [8] combined data from four distinct deep pencil-beam surveys at the north and south Galactic poles to produce a well sampled distribution of galaxies by redshift on a linear scale extending to 2000 h^{-1} Mpc. They reported a periodicity in the galaxy distribution of 128 h^{-1} Mpc. Soon afterwards Morikawa [9] noted that this aparent spatial periodicity could be naturally explained by a time oscillation of the Hubble parameter. Hill, Steinhardt and Turner [10] proposed a scenario in which the oscillation of the gravitational constant G was the responsible for the Hubble oscillation.

Even if the periodicity reported by Broadhurst *et.al.* [8] were an artifact of small angle sampling of galaxy distribution, the problem of an oscillating gravitational constant is interesting enough to deserve a different observational analysis. There are indeed superstring models that predict oscillating fundamental constants. These are functions of massive dilaton fields associated with the size of extra dimensions [3, 4]. To account a period $c\tau \simeq 128h^{-1}$ Mpc the dilaton mass m must be $O(10^{-31}eV)$. In Ref. [10] $G = G_0 + G_1 \cos[2m(t-t_0)+\psi]$ and the equation for the cosmological scale factor is

$$H^2 = -\frac{\dot{G}}{G}H + \frac{8\pi G\rho}{3}, \tag{1}$$

where H is the Hubble parameter. The leading correction to the Hubble parameter is $H \simeq \bar{H}[1 - \frac{1}{2}(\dot{G}/G\bar{H})]$. Following Ref. [10], if the Universe is spatialy homogeneous, with a uniform galaxy density per comoving volume n_0, the number of galaxies dN in a solid angle $d\Omega$ with red shift between z and $z+dz$ is modulated compared to the distribution in the absence of oscillations in the following way:

$$\frac{dN}{z^2 dz d\Omega} = \left(\frac{dN}{z_0^2 dz_0 d\Omega}\right)\frac{dz_0}{dz} \tag{2}$$

where $\mathcal{A} = -mG_1/\bar{H}G_0$ and to lowest order $z^2 \simeq z_0^2$. We have then an apparent variation in the density of galaxies which is isotropic and has peaks lying on concentric spherical shells at periodically spaced radii. An amplitude $\mathcal{A} \simeq 0.5$ (or equivalently about 1% for G_1/G_0), a phase $\psi \simeq 0$ and a period $c\tau \simeq L = 128h^{-1}$ Mpc approximately reproduces the result of Ref. [8]. Such a period is similar to the time scales involved in the paleontological record.

It is well known that several taxons record growth rhythms in their skeletons; i.e. periodic markings locked to the astronomical cycles of day, month and year [11]. From these growth rhythms, the number of days per year, days per month and months per year have been obtained as functions of geological time [11, 12, 13]. This parameters can be simply expressed in terms of the Earth's rotation Ω, the mean motion of the Sun $n_\odot$ and the mean motion of the moon $n_)$. The effect of a varying G on the Earth-Moon system can be studied under the adiabatic hypothesis, as stated in references [15, 16], so the Keplerian equations of motion mantain their form, and G is replaced by the appropiate time function. Morover, in a reference system where G depends only on time but not on space, angular momentum is still conserved [15]. According to Lambeck[13] the palaeontological data suggest that there may be significant systematic errors in the counts, and it is convenient to introduce parameters $\Delta n_\odot$,$\Delta n_)$ and$\Delta\Omega$ accounting for such errors in the estimation of the planetary angular velocities. Then for a time oscillating G the final expressions for the observables are

$$\frac{N_{d/y}}{N^0_{d/y}} - 1 = \beta\frac{G_1}{G_0}[\sin(\omega\tilde{t} - \phi) - \sin(\phi)] - \left.\frac{\dot{\Omega}}{\Omega}\right|_t \tilde{t} - \frac{\Delta\Omega}{\Omega} + \frac{\Delta n_\odot}{n_\odot}, \tag{3}$$

$$\frac{N_{m/y}}{N^0_{m/y}} - 1 = \beta\frac{G_1}{G_0}[\sin(\omega\tilde{t} - \phi) - \sin(\phi)] + \left(\left.\frac{\dot{n}_)}{n_)}\right|_t - \left.\frac{\dot{\Omega}}{\Omega}\right|_t\right)\tilde{t} - \frac{\Delta\Omega}{\Omega} + \frac{\Delta n_)}{n_)}, \tag{4}$$

$$\frac{N_{d/m}}{N^0_{d/m}} - 1 = -\left.\frac{\dot{n}_{\rangle}}{n_{\rangle}}\right|_t \tilde{t} + \frac{\Delta n_{\rangle}}{n_{\rangle}} + \frac{\Delta n_{\odot}}{n_{\odot}}, \tag{5}$$

where $\beta \simeq 1.83$ depends on the mass and pressure distribution in the Earth's interior [15], $\dot{n}_{\rangle}|_t$ is the tidal acceleration of the lunar longitude, $-(2-\beta)\frac{\dot{G}}{G}$ is the change of the moment of inertia due to the change of G [15], $\tilde{t}$ is minus the geological (not ephemeris) time [15] and $\phi = \frac{\pi}{2} - \psi$.

We used the data carefully filtered by Lambeck[13] according to biological reliability criteria [12]. Originally Lambeck adjusted the data assuming that only (constant) dissipation mechanisms are the responsible for $\dot{\Omega}$ and $\dot{n}_{\rangle}$. The adjusted values of $\dot{\Omega}$ and $\dot{n}_{\rangle}$ were in good agreement with the modern astronomical values. The residuals showed no obvious systematic trends, indicating that the growth rings present a high degree of confidence. In our case we adjusted simultaneously the three curves with and without the oscillatory parameters (i.e. with and without an oscillatory G), using the Levenberg-Marquardt least-squares method. The χ^2 significances of the adjustments are shown in Table 1. There is still a debate on the correct values for standard deviations of the palaeontological number counts, so the absolute value of the χ^2 significance is not important. What it is really important is the *change* of the significance when we include the oscillatory parameters. We see that there is a conspicuous increase of significance when we include the oscillatory parameters. For this last adjustment the best fit values are shown in Table 2, together with independent 95% confidence limits. The upper confidence limit of G_1/G_0 is about 0.009, which is the desired amplitude accounting for the red-shift survey, i.e.,we have marginal consistency with the oscillating G hypothesis. The best fit value of the period of oscillation remarkably coincides with the galaxy distribution period, although the confidence limits weaken its relevance. The values for (tidal) $\dot{n}_{\rangle}/n_{\rangle}$,$\dot{\Omega}/\Omega$ and $\dot{n}_{\odot}/n_{\odot}$ agree with other estimates [13]. The phase ψ is consistent with the zero value proposed in [10], which implies a zero value for the present rate of change of G. Our fit is then consistent with the current upper bounds on the time variation of G based upon the Viking radar-echo experiments [14] [note that only upper bounds based on present observations are valid if G oscillates, so the much more stringent upper bounds of [15, 16] do not apply because they are also based in long-term (several oscillation periods) phenomena].. In order to test the sensibility of the solution, we made the same adjustments using original data from several sources, including bivalves, corals, cephalopods, brachiopods and estromatolites and totalizing 61 points. As shown in Table 1 in this case both significances are small, showing that the filtering of data as made by Lambeck introduces a bias towards the oscillatory hypothesis.

We conclude that our results do not exclude an oscillating gravitational constant inducing a periodic galaxy distribution. Indeed the significance of the Lambeck data adjustment suggests thet there is an oscillatory component in the time evolution of planetary orbit ratios. However, there are several uncertainties in our model that forbid a definite conclusion. In the first place, paleontological growth rhythms are

Table 1: Significance of the χ^2 test for all adjustments.

Data source	number of adjusted parameters	χ^2 significance
Original	5	below 10%
Original	8	below 10%
Lambeck	5	17%
Lambeck	8	85%

Table 2: Adjusted parameters of the curves fitting Lambeck data, together with 95% confidence limits.

Parameter	best fit value	confidence limits lower	upper
$\Delta n_{\odot}$	-1.3×10^{-2}	-2.5×10^{-2}	1.6×10^{-3}
$\Delta n_{)}$	9.3×10^{-3}	-4.1×10^{-2}	5.9×10^{-2}
$\Delta\Omega$	1.9×10^{-3}	-4.1×10^{-3}	7.9×10^{-3}
$\frac{\dot{n}}{\eta}[10^{-11}\mathrm{yr}^{-1}$	16.4	12	20
$\frac{\dot{\Omega}}{\Omega}[10^{-11}\mathrm{yr}^{-1}$	32.0	29	35
$\frac{G_1}{G_0}$	6.5 ×10^{-3}	3.8×10^{-3}	9.2×10^{-3}
period [10^6yr]	486	370	680
ψ	0.27	-0.53	0.83

subject to large variations, and should be handled with care [17, 18, 19]. Secondly, it should be noted that the changes in the resonance structure of the oceans due to continental drift provoke considerable variations of the Earth-Moon tidal torque within 100 million years time scales [20]. This fact could account as well for the time oscillation of the number of days per year. As a result we can only state that an upper bound for the oscillation amplitude of the gravitational constant is $G_1/G_0 < 0.01$ (taken from the upper confidence limit of Table 2). The Hubble oscillating hypothesis is strongly testable beacause it predicts the same periodic patterns in all directions. Also, as stated by Morikawa [9] both the density contrast and the distance to the nearest peak is not clear from the survey. Then it is worth making further deep-pencil surveys in the same and other directions in the sky. Summarizing, the coincidences which we observe in this work (coincidences may become consequences [21]) are significant enough so as to become a subject of further numerical and observational research.

References

[1] J. H. Schwartz, Phys. Rep. **89**, 223 (1982); M. B. Green J. H. Schwartz, Phys. Lett.**199B**, 117 (1984).

[2] T. Kaluza, Sitz. Preuss. Akad. Wiss. Phys. Math. **K1**, 966 (1921); O. Z. Klein, Phys. **37**, 895 (1926).

[3] Y. Wu and Z. Wang, Phys. Rev. Lett. **57**, 1978 (1986).

[4] K. Maeda, Mod. Phys. Lett. **A3**, 243 (1988);

[5] S. M. Barr and P. K. Mohapatra, Phys. Rev. **D38**,3011 (1988).

[6] A. Chodos and S. Detweiler, Phys. Rev. **D21**, 2167 (1980). M. Gleiser and J. G. Taylor, Phys. Rev. **D31**, 1904 (1985).

[7] W. J. Marciano, Phys. Rev. Lett. **52**, 489 (1984).

[8] T. J. Broadhurst, R. S. Ellis, D. C. Koo and A. S. Szalay, Nature (London) **343**, 726 (1990).

[9] M.Morikawa, Astrophys. J. **369**, 20 (1991).

[10] C. T. Hill, P. J. Steinhardt and M. S. Turner, Fermi National Accelerator Laboratory prepint 90/129-T (1990).

[11] C. T. Scrutton, in *Tidal Friction and the Earth's Rotation* (eds Brosche, P. Sundermann, J.) 154-196 (Springer, New York, 1978).

[12] K. Lambeck, *The Earth's Variable Rotation* (Cambridge University Press, New York, 1980).

[13] K. Lambeck, in *Tidal Friction and the Earth's Rotation*, ed. by P. Brosche and J. Sundermann, (Springer-Verlag, Berlin 1978).

[14] R. H. Hellings, P. J. Adams, J. D. Anderson, M. S. Keesey, E. L. Lau, E. M. Standish, V. M. Canuto and I. Goldman, Phys. Rev. Lett. **51**, 1609 (1983).

[15] P. D. Sisterna and H. Vucetich, Phys. Rev. **D41**, 1034 (1990).

[16] P. D. Sisterna and H. Vucetich, Phys. Rev. **D44**, 3096 (1991).

[17] R. A. Lutz and D. C. Rhoads, Science **198**, 1222 (1977).

[18] W. W. Hughes, Geophysical Surveys **7**, 169 (1985).

[19] G. D. Rosenberg, Geophysical Surveys **7**, 185 (1985).

[20] P. Brosche, in *Sun and Planetary System*, ed. by W. Fricke and G. Teleki (Reidel, D. Publishing Company).

[21] J.D.Barrow and F.J.Tipler, *The Anthropic Cosmological Principle* (Oxford University Press, Oxford, 1988).

Quantum Gravity and Quantum Cosmology

Coleman's Mechanism in Jordan–Brans–Dicke Gravity

Juan García–Bellido
Instituto de Estructura de la Materia, CSIC
Serrano, 123 E-28006 Madrid, Spain

and

Luis J. Garay
Instituto de Optica, CSIC
Serrano, 121 E-28006 Madrid, Spain

ABSTRACT

We consider the quantum gravity and cosmology of a Jordan–Brans–Dicke theory. Its constraint algebra is that of general relativity, as a consequence of the general covariance of scalar–tensor theories. We propose that boundary conditions must be imposed in the Jordan frame, in which particles satisfy the strong equivalence principle. We discuss both Hartle–Hawking and wormhole boundary conditions in the context of quantum cosmology. Wormholes may affect the constants of nature and, in particular, the Brans–Dicke parameter. Following Coleman's mechanism, we find a probability distribution for wormhole configurations which is strongly peaked at zero cosmological constant and infinite Brans–Dicke parameter. That is, we recover general relativity as the effective low energy theory of gravity.

Jordan–Brans–Dicke theory [1] is based on the idea of Mach that inertia arises from accelerations with respect to the general distribution of matter in the universe. Therefore, the inertial masses of elementary particles are not fundamental constants but represent the interaction of particles with a cosmic scalar field whose dynamics depend on the rest of the matter in the universe. Supposing that all matter particles have the same coupling to the scalar field, $m(\phi) = e^{\beta\phi}m$, the action in the so called Einstein frame can be written as [2]

$$\bar{S} = \frac{1}{16\pi G}\int d^4x\sqrt{-\bar{g}}\left(\bar{R} - \frac{1}{2}\bar{g}^{\alpha\beta}\bar{\nabla}_\alpha\phi\bar{\nabla}_\beta\phi\right) + \int e^{\beta\phi}m\, d\bar{s}. \tag{1}$$

Physics must be invariant under conformal redefinitions of the metric since they correspond to an arbitrary choice of measuring units. We are thus free to choose the conformal frame in which we want to describe physics [1]. The most natural choice is the so called physical frame, in which observable particles have constant masses, since in this frame particles follow geodesics of the metric, thus satisfying the strong equivalence principle [3]. From the cosmological point of view, the comoving frame, in which a fundamental observer sees the universe as homogeneous and isotropic, is the physical frame since those observers will follow geodesics of the metric. In our case, the conformal redefinition of the metric that allows us to describe the theory of gravitation (1) in the physical frame is

$$\bar{g}_{\alpha\beta} = e^{-2\beta\phi} g_{\alpha\beta} \equiv \Phi g_{\alpha\beta}. \tag{2}$$

This is the so called Jordan frame, in which the action takes the JBD form

$$S = \frac{1}{16\pi \mathrm{G}} \int d^4x \sqrt{-g} \left(\Phi R - \frac{\omega}{\Phi} g^{\alpha\beta} \nabla_\alpha \Phi \nabla_\beta \Phi \right) + S_m, \tag{3}$$

where Φ is a dimensionless scalar field and ω a constant given by $2\omega + 3 = (4\beta^2)^{-1}$. General relativity is recovered in the limit $\omega = \infty$, $(\beta = 0)$. These theories are not ruled out by post–Newtonian experiments [4] nor by primordial nucleosynthesis bounds [5], and in fact have recently recovered great interest since they have been proposed as the arena for extended inflation [6], a new inflationary scenario which could solve some of the traditional problems of previous schemes [7]. Furthermore, as we have mentioned, JBD theory may well be the Planck scale theory of gravity, as predicted by string effective actions [8]. We analyze the Jordan–Brans–Dicke quantum cosmology [9] with special emphasis made on the effect of quantum wormholes on the low energy coupling constants [10].

The Hamiltonian constraints can be explicitly written in both the Jordan and Einstein frames. Both expressions are related through the canonical transformation (2)

$$\mathcal{H} = e^{-\beta\phi} \bar{\mathcal{H}}, \qquad \mathcal{H}_i = \bar{\mathcal{H}}_i. \tag{4}$$

The constraint algebra is the same in both frames and corresponds to that of general relativity, expressing the fact that JBD theory is generally covariant.

Boundary conditions must be imposed in the physical frame, where particles satisfy the strong equivalence principle. The boundary conditions in any other frame can be obtained by transforming those in the physical frame through the corresponding conformal redefinition. In JBD quantum cosmology we shall be concerned with boundary conditions both for the universe and for wormholes. The Hartle–Hawking wave function for the universe [11] is given by the Euclidean path integral over all compact four–metrics in the Jordan frame, over all possible BD field configurations and all matter fields. The saddle point configuration, which gives the dominant contribution to the path integral, will have a curvature scalar

$$R = \frac{2\omega}{2\omega + 3} \frac{4\Lambda}{\Phi} + \omega \left(\frac{\nabla \Phi}{\Phi} \right)^2, \tag{5}$$

and therefore will be compact when both the cosmological constant and the BD parameter are positive.

Non–perturbative quantum gravity effects due to non–trivial topologies, *e.g.* wormholes [12], will change the effective value of the physical parameters of the theory, in

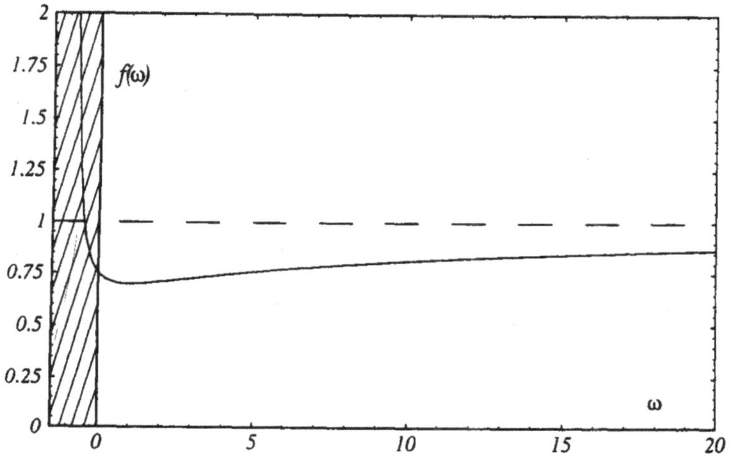

Figure 1. Plot of the function $f(\omega)$. The $\omega < 0$ region has been excluded since it corresponds to non-compact saddle points. This function acquires its maximum at $\omega = \infty$, where $f(\infty) = 1$, which corresponds to the general relativity result.

particular the Brans–Dicke parameter [10]. Wormholes may be interpreted, semiclassically, as throats joining two otherwise disconnected large regions of spacetime. In this picture one assumes the dilute wormhole approximation. JBD wormhole wave functions can be written as Euclidean path integrals over asymptotically flat spacetimes. One must also sum over all JBD field configurations whose Hamiltonian vanishes in the asymptotic region and similarly for the matter fields [13]. These boundary conditions have been imposed in the Jordan frame, although in the Einstein frame they take the same form, since the JBD field is constant at infinity.

Although Planck scale wormholes are not directly observable, they will produce effective interactions in the low energy physics that turn the coupling constants of nature into dynamical variables [14]. In particular, JBD wormholes will affect the kinetic term of the JBD field and therefore will modify the value of the low energy BD parameter.

Wormholes not only affect the constants of nature but also may provide a probability distribution for them. In general relativity this distribution is strongly peaked at zero cosmological constant. This is the so called Coleman's mechanism [15] for the vanishing of the cosmological constant. In JBD theory the probability distribution for the wormhole configurations α and thus for the coupling constants has the form [10]

$$\mathcal{Z}(\alpha) = \exp\left(\exp \frac{3\pi}{G_\alpha \Lambda_\alpha} f(\omega_\alpha)\right), \tag{6}$$

where $f(\omega_\alpha)$ is the function shown in fig.1, whose maximum value is obtained for

$\omega_\alpha = \infty$, where $f(\infty) = 1$, for which we recover the general relativity result. $\mathcal{Z}(\alpha)$ is strongly peaked at $G_\alpha \Lambda_\alpha = 0$. Since this probability distribution is not normalizable, a suitable choice of the cutoff is needed in order to determine its maximum. In the context of general relativity, different cutoffs have been considered [15, 16] that give different results. However, $G\Lambda$ seems to be the most natural choice for the cutoff since it is the adimensional vacuum energy, *i.e.* the cosmological constant in Planck units, which is observed to be less than 10^{-120}. With this choice for the cutoff, the probability distribution (6) acquires its maximum at $\omega_\alpha = \infty$. Thus we recover general relativity as the low energy effective theory of gravity, even though the fundamental high energy description may well be a scalar–tensor theory. This means that there can be no effective dynamical scalar field coupled to energy or matter. Note that this result can be generalized to any scalar–tensor theory with an arbitrary coupling $\omega(\Phi)$.

The usual interpretation [15] of this prediction is that Coleman's mechanism only ensures that the bottom line cosmological constant is zero. This does not exclude an inflationary universe with a non-zero false vacuum energy due to a certain phase transition. It just says that, whatever the effective potential for inflation is, the true minimum is at zero cosmological constant. However, it does preclude extended inflation since wormholes drive the BD parameter to infinity at all times and thus "freeze out" the evolution of the scalar field.

References

[1] P. Jordan, *Nature* (London) **164** (1949) 637; *Z. Phys.* **157** (1959) 112; C. Brans and R. H. Dicke, *Phys. Rev.* **124** (1961) 925; R. H. Dicke, *Phys. Rev.* **125** (1962) 2163; C. Brans, *Phys. Rev.* **125** (1962) 2194; S. W. Hawking, *Comm. Math. Phys.* **25** (1972) 167.

[2] See for instance, S. Weinberg, *Gravitation and Cosmology* (Wiley, New York, 1972).

[3] J. A. Casas, J. García–Bellido and M. Quirós, *Class. Quant. Grav.* **9** (1992) 1371.

[4] For a general review see C. M. Will, *Theory and Experiment in Gravitational Physics*, Cambridge University Press (1981); *Phys. Rep.* **113** (1984) 345.

[5] J. A. Casas, J.García–Bellido and M. Quirós, *Mod. Phys. Lett.* **A 7** (1992) 447; *Phys. Lett.* **B 278** (1992) 94.

[6] D. La and P. J. Steinhardt, *Phys. Rev. Lett.* **62** (1989) 376; *Phys. Lett.* **B 220** (1989) 375; E. J. Weinberg, *Phys. Rev.* **D 40** (1989) 3950; D. La, P. J. Steinhardt

and E. W. Bertschinger, *Phys. Lett.* **B 231** (1989) 231; J. García-Bellido and M. Quirós, *Phys. Lett.* **B 243** (1990) 45; *Nucl. Phys.* **B 368** (1992) 463.

[7] A. H. Guth, *Phys. Rev.* **D 23** (1981) 347; A. D. Linde, *Phys. Lett.* **B 108** (1982) 389; A. Albrecht and P. J. Steinhardt, *Phys. Rev. Lett.* **48** (1982) 1220; A. H. Guth and E. J. Weinberg, *Nucl. Phys.* **B 212** (1983) 321.

[8] J. A. Casas, J. García-Bellido and M. Quirós, *Nucl. Phys.* **B 361** (1991) 713.

[9] S. del Campo and A. Vilenkin, *Phys. Rev.* **D 40** (1989) 688; A. K. Sanyal and B. Modak, *Int. J. Mod. Phys.* **A 7** (1992) 4039.

[10] L. J. Garay and J. García-Bellido, "Jordan-Brans-Dicke Quantum Wormholes and Coleman's Mechanism", preprint IEM-FT-59/92, submitted to Nuclear Physics B.

[11] J. B. Hartle and S. W. Hawking, *Phys. Rev.* **D 28** (1983) 2960.

[12] S. W. Hawking, *Mod. Phys. Lett.* **A 5** (1990) 145; *Mod. Phys. Lett.* **A 5** (1990) 453.

[13] S. W. Hawking and D. N. Page, *Phys. Rev.* **D 42** (1990) 2655; L. J. Garay, *Phys. Rev.* **D 44** (1991) 1059.

[14] S. W. Hawking, *Phys. Rev.* **D 37** (1988) 904; S. W. Hawking and R. Laflamme, *Phys. Lett.* **B 209** (1988) 39; S. Coleman, *Nucl. Phys.* **B 307** (1988) 864.

[15] S. Coleman, *Nucl. Phys.* **B 310** (1988) 643.

[16] J. Preskill, *Nucl. Phys.* **B 323** (1989) 141; B. Grinstein and M. B. Wise, *Phys. Lett.* **B 212** (1988) 407; B. Grinstein, *Nucl. Phys.* **B 321** (1989) 439; S. W. Hawking, *Nucl. Phys.* **B 335** (1990) 155.

EXPLAINING THE LARGE SCALE STRUCTURE OF THE UNIVERSE USING QUANTUM GRAVITY

O. BERTOLAMI[†,‡], J. M. MOURÃO[§,‡] and J. PÉREZ-MERCADER[¶]

[†] *Centro de Física da Matéria Condensada*
[§] *Centro de Física Nuclear*
Av. Prof. Gama Pinto, 2
1699 Lisboa Codex

[‡] *Departamento de Física*
Instituto Superior Técnico
Av. Rovisco Pais
1096 Lisboa Codex

[¶] *Laboratorio de Astrofísica Espacial y Física Fundamental*
Apartado 50727
28080 Madrid

ABSTRACT

We show that generic asymptotic freedom (AF) in the gravitational coupling implies in more power in the two-point correlation function at large distances in comparison with $\Omega_0 = 1$ Cold Dark Matter (CDM) models.

1. A recent survey of distant galaxies performed by IRAS[1] in the infrared band reveals that, at scales larger than about 20 Mpc, the Universe is lumpier than predicted by the CDM Model (see Ref. [2] for a review and references therein). We show that[3] this difficulty can be solved by a modification in the usual theory of gravity without relying on the hypothesis the Universe is filled by a large amount of Dark Matter (DM). This modification arises when taking into account quantum effects in regimes where the theory exhibits AF in Newton's constant, G, and is effectively renormalizable[4] from which implies that metric fluctuations act over macroscopic distances[5].

In general quantum fluctuations require a scale–dependent redefinition of the physical parameters, and to achieve renormalizability at one-loop level the gravity theory must be of the so–called "higher derivative gravity" type[4]. Theories of this kind may be viewed as an effective version of a more fundamental theory such as the theory of strings.

Removal of the infinities generated by the quantum fluctuations in the physical parameters still leaves a dependence of the latter on some finite quantities whose particular values are arbitrary. Assignment of these values is done by specifying the value of the physical parameters at some momentum transfer or, equivalently, scale of length; once this is done, changes of scale are accounted by appropriate changes in the values of the physical parameters in a way ruled by the renormalization group equations (RGEs). Thus in the quantum field theory of gravity, the equations of motion for the theory are similar to the classical equations, but with the parameters substituted by their "improved" values, i.e. the solutions to their RGEs. However, since gravity couples coherently to matter and there is no screening mechanism for gravitation, quantum fluctuations of the gravitational degrees of freedom contribute on all scales; for this reason one must include the effects of these unavoidable quantum corrections into G, promoting it into a scale–dependent quantity at every length scale. Current models quantum gravity support that the coupling $G(\mu^2/\mu_*^2 \sim r_*^2/r^2)$ is AF, where μ_* is a reference momentum, meaning that G grows with distance[4]. A typical solution for $G(r_*^2/r^2)$ is of the form[5],

$$G(r_*^2/r^2) = G_{lab}(r_*^2/r_{lab}^2)\,\delta(r, r_{lab}), \tag{1}$$

where $G_{lab}(r_*^2/r_{lab}^2)$ is the value of G measured in the laboratory at a length scale r_{lab}, and $\delta(r, r_{lab})$ is a growing function of r equal to one at $r = r_{lab}$. A convenient choice for r_* is $r_* = r_{lab}$. In order for the AF of $G(\mu^2/\mu_*^2)$ to have impact, for instance on the dynamics of galaxies and their rotation curves, the function $\delta(r, r_{lab})$ should be close to one for $r < 1$ Kpc, growing significantly only for $r \geq 1$ Kpc. Naturally, this dependence of G with distance has also implications of cosmological nature.

Let us for simplicity assume that the Friedmann equation valid in this case is the improved one for a flat Universe:

$$H^2(\ell) = \frac{8\pi}{3} G(\frac{a_0^2 \ell_*^2}{a^2 \ell^2}) \rho_m \;,, \tag{2}$$

where $a = a(t)$ is the scale factor, the index "0" denotes present values, $H = \dot{a}/a$ is the Hubble parameter, ρ_m is the density of baryonic matter (i.e. no DM), ℓ is the comoving distance and ℓ_* is some convenient length scale.

General arguments ensure that ρ_m is independent of the scale; hence, Eq. (2) implies that the Hubble parameter is also scale-dependent (it grows with scale), a fact which may account for the observational difficulties in determining its value.

Although Eq. (2) implies by construction a physical density parameter at present, Ω_0^{phys} equal to one, the quantity which is usually referred to as density parameter is actually:

$$\Omega_0 = \frac{8\pi}{3} \frac{G \rho_{m_0}}{H_{0_*}^2} \tag{3}$$

where H_{0_*} is the present Hubble parameter for a given large scale distance, $r = r_*$, and the value of the product $G\rho_{m_0}$ is inferred for different scales with, for instance, the help of the Virial Theorem. This results in Ω_0 being a growing function of scale which is in reasonable agreement with (1) and with a constant ρ_{m_0}. Thus we see that macroscopic quantum effects are effectively the same as the presence of DM.

2. Let us now analyse the impact of the macroscopic quantum fluctuations on the problem of cosmological structure formation. It is reasonable to expect that an effective gravitational coupling that grows with distance will give more power at large distances than a constant G. This can be verified in three different ways; first, by studying the effect on the power spectrum, then by investigating the effects on the Jean's length and finally by analyzing the effect on the growth of density fluctuations in an expanding Universe. We here show the way a running gravitational coupling implies in a two-point correlation function with more power on large scales. In Ref. [3] for the other two checks are carried out.

As mentioned above recent survey by IRAS[1] reveals that the power spectrum, $P(k)$, of CDM models lack enough power *at large distances*. Let us consider here the effect of a scale dependent gravitational coupling on the two-point correlation function, or equivalently, on its Fourier-transform associated quantity $< |\delta_k|^2 > \equiv P(k)$. One obtains that $P(k)$ at large scales must behave as $P(k) \sim k$ due to the assumption that the power spectrum at these scales is related to scale invariant primordial energy density fluctuations and must change to $P(k) \sim k^{-3}$ for small scales. The best interpolation being given by[2]

$$P(k) = \frac{k}{[1 + (\alpha k + (\beta k)^{3/2} + (\gamma k)^2)^\nu]^{2/\nu}} \tag{4}$$

with $\alpha = 6.4(\Omega_0 h^2)^{-1}$ Mpc, $\beta = 3.0(\Omega_0 h^2)^{-1}$ Mpc, $\gamma = 1.7(\Omega_0 h^2)$ Mpc, $\nu = 1.13$ and h is the Hubble parameter in units of 100 km s^{-1} Mpc^{-1}. For this spectrum, the transition between the small scale and large scale regimes takes place at $\lambda_{eq} = ct_{eq} \simeq 10(\Omega_0 h^2)^{-1} Mpc$, where t_{eq} is the time when matter and radiation had the same energy density. Equation (4) shows explicitly that $P(k)$ depends only on the product $\Omega_0 h^2$. This feature was made use of in Ref. [6] to save CDM-scenarios: the present density parameter is chosen to be $\Omega_0 = 0.2$ and the remaining density up to one is provided by a cosmological constant $\Lambda \equiv 0.8 \times 3H_0^2$. It follows from Eq. (4) that the smaller the value of Ω_0 the more correlation there is at large scales. To check this we show that the value k_{max} of the point where the power spectrum has a maximum is a linear function of $\Omega_0 h^2$. Indeed, Eq. (4) can be rewritten as

$$P(k) = \Omega_0 h^2 \ f(k(\Omega_0 h^2)^{-1}) \ , \tag{5}$$

where $f(z)$ is a given function of its argument, $z = k(\Omega_0 h^2)^{-1}$. The maximum of $P(k)$ occurs at $z_{max} = z_0 = 0.098$, or

$$k_{max} = z_0 \Omega_0 h^2 \ . \tag{6}$$

Since the wavenumber is inversely proportional to the distance and z_0 is a constant, this equation shows that the smaller the value of Ω_0, the more extra structure one expects at larger scales. We shall see that this situation does not change when the "effective" density parameter (3) is smaller than one but not constant. For the sake of clarity, and being interested only in qualitative arguments, we assume that the form of the power spectrum is in this case also given by Eq. (5) but with Ω_0 substituted by $\Omega_0(k)$. Then

$$P(k) = \Omega_0(k)h^2 f(k(\Omega_0(k)h^2)^{-1}) \,, \tag{7}$$

and from the behavior of $G(k^2/k_*^2)$ (see Eq. (1)) it is clear that $P(k)$ has in this case still a single maximum. The maximum however occurs at a value of k such that $k_{max}^{\Omega_0(k)} < k_{max}^{\Omega_0=1}$. Indeed from the extremum condition, $\frac{dP(k)}{dk} = 0$, we obtain

$$\frac{dP(k)}{dk} = -A^2(k(z))f(z) + B^2(k(z))\frac{df(z)}{dz} = 0 \,, \tag{8}$$

where $k(z)$ is obtained by solving $k - z\Omega_0(k)h^2 = 0$ with respect to k. In view of the AF of G, together with (3), the functions $A(k)$ and $B(k)$ satisfy

$$A^2(k(z)) = -\frac{d\Omega_0}{dk}h^2 > 0 \,, \tag{9}$$

$$B^2(k(z)) = 1 - \frac{k}{\Omega_0(k(z))h^2}\frac{d\Omega_0}{dk}h^2 > 0 \,. \tag{10}$$

The zero, $z_{max}^{\Omega_0(k)}$, of (8) is smaller than $z_{max}^{\Omega_0=1} = z_0$, the maximum of $f(z)$. To see this, notice that the combination $B^2(k(z))\frac{df(z)}{dz}$ has the following behaviour:

$$B^2(k(z))\frac{df(z)}{dz} > 0 \qquad z < z_0 \,, \tag{11}$$

$$B^2(k(z))\frac{df(z)}{dz} < 0 \qquad z > z_0 \,, \tag{12}$$

such that the only effect of the term $A^2(k(z))f(z)$ in (8) is to shift down the graph of $B^2(k(z))\frac{df(z)}{dz}$, from which we inmediately see that

$$z_{max}^{\Omega_0(k)} < z_{max}^{\Omega_0=1} = z_0 \,. \tag{13}$$

Collecting the above results, we get

$$k_{max}^{\Omega_0(k)} = z_{max}^{\Omega_0(k)}\Omega_0(k)h^2 < z_0\Omega_0(k)h^2 < z_0 h^2 = k_{max}^{\Omega_0=1} \,. \tag{14}$$

Therefore we conclude that, as expected intuitively, an asymptotically free gravitational coupling leads to a two–point correlation function with more power at larger distances than the one for CDM without a cosmological constant ($\Omega_0 = 1$).

3. In summary, we have explored some of the implications that an AF gravitational coupling has for the power spectrum of density contrast. We have considered the kind of generic growth predicted by current theories of quantum gravity and shown that models with a gravitational coupling growing with distance and with a corresponding density parameter smaller than the value favoured by CDM models ($\Omega_0 = 1$), can account for the observed excess power in the two–point correlation function for distances greater than about 20 Mpc[1]. This conclusion is corroborated by the analysis, performed in Ref. [3], of the evolution of density perturbations in an almost homogeneous and isotropic Universe which indicates that the quantum gravity corrected Jeans' length is greater than in the classical ($G = constant$) case with DM, and therefore, without introducing more density than what is required by nucleosynthesis, one predicts gravitational collapse and structure at scales larger than for the CDM standard scenario. Furthermore the growth of G with distance leads to a faster evolution of density perturbations than in the classical theory without DM.

4. References

1. G. Efstathiou et al., *Mon. Not. R. Astron. Soc.* **247** (1990) 10;
   W. Saunders et al, *Nature* **349** (1990) 32.
2. M. Davis, G. Efstathiou, C.S. Frenk and S.D.M. White, *Nature* **356** (1992) 489.
3. O. Bertolami, J.M. Mourão and J. Pérez–Mercader, "Quantum Gravity and the Large Scale Structure of the Universe" Lisbon Preprint IFM 4/92.
4. J. Julve and M. Tonin, *Nuovo Cimento* **46B** (1978) 137;
   E.S. Fradkin and A.A. Tseytlin, *Nucl. Phys.* **B201** (1982) 469;
   E.G. Avramidy and A.O. Barvinsky, *Phys. Lett.* **B159** (1985) 269.
5. T. Goldman, J. Pérez–Mercader, F. Cooper and M. Martin Nieto, *Phys. Lett* **B281** (1992) 219.
6. G. Efstathiou, W.J. Sutherland and S.J. Maddox, *Nature* **348** (1990) 705.

MULTIDIMENSIONAL COSMOLOGY

ULRICH BLEYER
WIP Gravitation Project
Potsdam University
Rosa-Luxemburg-Str. 17a
D-O-1590 Potsdam, Germany

Kaluza-Klein type cosmological models (KKCM) are considered to describe early stages of the universe. The possibility of a *multidimensional stage* of the universe is actually discussed in connection with unification schemes in supersymmetry or string theory [1]. The main problem consists in the dynamical process leading from a stage with all dimensions developing on the same scale to the actual stage of the universe where we have only four external dimensions and all internal spaces have to be compactified and contracted to unobservable scales.

In general Kaluza-Klein theories start with a N-dimensional space-time manifold. After *compactification* the structure of this manifold is reduced to a product manifold of the four dimensional external space-time and one or some compact internal spaces. To make the internal dimensions unobservable the internal spaces have to be reduced to scales near the Planck length. This can be achieved by contraction of the internal spaces during cosmological evolution or by reduction to Planck scales during compactification (see e.g. [2, 3]).

Windows to such higher dimensional stages arise from the defects of topological compactification, e.g. massive monopoles, strings and so on, from the excitations in the internal space which occur in the external space as massive particles (pyrgons [4]), or from the amplification of perturbation modes during the contraction of internal spaces [5].

In any case the final stage has to be described by *static internal spaces* of Planckian size in order to make internal dimensions hidden at actual cosmic time. The role of static internal spaces consists in determining physical constants and fundamental interactions in the four-dimensional space-time. Physical constants appearing in four-dimensional space-time are induced by the fundamental constants during compactification (see e.g. [6]). If the internal space has a time dependent scale the values of the physical constants also vary in time. But experiments actually show that physical constants are independent on time. Therefore, *all higher dimensional comological models have to contain a mechanism to make the internal spaces static ones.*

We give a short collection of recent results concerning 1. *Classical solutions* to KKCM, 2. *Quantum effects on the classical background* of KKCM, and 3. *Canonical quantization* of KKCM.

1. CLASSICAL KKCM

Let us start with an N-dimensional space-time manifold $M = R \times M_0 \times \ldots \times M_n$ containing n compact internal spaces M_i of d_i dimensions. M_0 denotes our three dimensional space, but we do not restrict ourself to this special value and use $d_0 > 1$ as dimension. In general the line element is given by

$$ds^2 = g_{AB}dx^A dx^B \qquad A, B = 0, \ldots, N-1 \tag{1}$$

In our case of homogeneous spaces the line element is given by

$$ds^2 = -dt^2 exp(2\gamma(t)) + \sum_{i=0}^{n} a_i^2(t) ds_i^2 \tag{2}$$

where the $a_i = e^{\beta^i}$ denote the scale factors of the different factor spaces. The spacial metrics of these spaces are given by

$$ds_i^2 = h_{m_i n_i} dx^{m_i} dx^{n_i} \qquad m_i, n_i = 1, \ldots, d_i \tag{3}$$

or with the demand of homogeneity

$$ds_i^2 = \frac{\sum (dx^{n^i})^2}{(1 + \frac{1}{4} k_i \sum (x^{n^i})^2)^2} \tag{4}$$

where $k_i = 0, \pm 1$. Matter sources compatible with this symmetry can be described by an energy momentum tensor of a generalized ideal fluid: $T = diag\{\rho; p_1, \ldots, p_1; \ldots; p_n, \ldots, p_n\}$.

The gravitational action S for the metric (2) can be written in harmonic time gauge ($\gamma = 0$) in the form [7]

$$S = \int \mathcal{L} dt \qquad \text{where} \qquad \mathcal{L} = \frac{\mu}{2} G_{ij} \dot{\beta}^i \dot{\beta}^j - \frac{\mu}{2} V \tag{5}$$

The metric of minisuperspace reads

$$G_{ij} = d_i \delta_{ij} - d_i d_j \tag{6}$$

and the potential is given by

$$V = \rho(\beta^i) \exp\left(2 \sum_{i=0}^{n} d_i \beta^i\right) \tag{7}$$

Here ρ denotes the matter density which has for the metric (2) and linear equations of state the form [8]

$$\rho = M_{m_0 m_1 \ldots m_n} \exp\left(-\sum_{i=0}^{n} m_i \beta^i\right) \tag{8}$$

We have to mention that in ρ there is also included a term produced by the curvature of the product spaces $\Theta_i = k_i d_i (d_i - 1)$. μ is proportional to the product of the volumes of spaces with constant curvature. The Einstein field equations read

$$G_{ij} \ddot{\beta}^j = -\frac{1}{2} \frac{\partial}{\partial \beta^i} V \tag{9}$$

There properties where discussed e.g. in [9] where it is shown that we need special compensation effects between at least three different matter components in order to ensure that a state with static internal spaces becomes possible.

Of special importance is the case where curvature is the unique source and the potential takes the form

$$V = -\sum_{i=0}^{n} \Theta_i e^{-2\beta^i} \exp\left(2 \sum_{i=0}^{n} d_i \beta^i\right) \tag{10}$$

As it is shown in [10] using the analogy of these models with Toda-like systems these KKCM are integrable only in the case with one non Ricci-flat space. For this special type of KKCM new classical solutions are found for the Lorentzian region as well as classical wormhole solutions for the Euclidean one [11].

2. SCALAR FIELD ON KALUZA-KLEIN BACKGROUND

In [12] it was shown that the internal spaces of a model with external space of constant curvature can be only static if all internal spaces are Ricci flat. On this background a scalar field was considered with the action

$$S_m = -\frac{1}{2}\int \prod_{A=0}^{D} dx^A (-g)^{\frac{1}{2}} \left[g^{AB} \frac{\partial \phi}{\partial x^A} \frac{\partial \phi}{\partial x^B} - 2V(\beta_j, \phi) \right] \tag{11}$$

where

$$V(\beta_j, \phi) = V(\phi) + \frac{1}{2}\xi R \phi^2 \tag{12}$$

and $V(\phi)$ is the usual potential of minimal coupling with self interaction

$$V(\phi) = \frac{1}{2}m^2\phi^2 + \frac{\lambda}{4}\phi^4 \tag{13}$$

If we use the conformal time, a as the scale factor of the external space, and the conformally transformed scalar field

$$\phi = \tilde{\phi} a^{\frac{1-d_0}{2}} \tag{14}$$

we find the scalar field equation

$$\begin{aligned} &\ddot{\tilde{\phi}} - \Delta_2^{(d_0)} \tilde{\phi} - a^2 \sum_{j=1}^{n} \frac{1}{a_j^2} \Delta_2^{(d_j)} \tilde{\phi} \\ &\quad + \left[m^2 a^2 + \xi d_0 (d_0 - 1) k_0 + d_0 (\xi - \xi_{conf}) \left(2\tfrac{\ddot{a}}{a} - (d_0 - 3) \tfrac{\dot{a}^2}{a^2} \right) \right] \tilde{\phi} \\ &\qquad + \tfrac{\lambda}{4} a^{3-d_0} \tilde{\phi}^2 \tilde{\phi} = 0 \end{aligned} \tag{15}$$

Here $\Delta_2^{(d_j)}$ denotes the Lapace-Beltrami operator in a d_j dimensional space and $\xi_{conf} = (d_0 - 1)/4d_0$ is the value of ξ for conformal coupling.

Now, we find from the discussion of the excited states of the scalar field the induction of effective masses due to the compactness of the internal spaces. In the ground state of the scalar field there are found the *conditions for symmetry breaking* in dependence on the value of the coupling parameter ξ. They are *much more general* than expected up to now (for details see [12]).

3. CANONICAL QUANTIZATION

One possible scheme to take gravitational quantum effects into account is to consider the multidimensional generalization of the Wheeler-DeWitt equation (WDW), which is the quantum version of the Hamiltonian constraint of the canonical Hilbert action [7].

The WDW gives us the possibility to reduce the quantization procedure to a system with a finite number of degrees of freedom. This leads in special cases to integrable models.

If we make in the Lagrangian (5) the usual substitution (putting $\hbar = 1$)

$$\dot{\beta}^i \Rightarrow \frac{1}{i} G^{ij} \frac{\partial}{\partial \beta^j} \tag{16}$$

we get a simple form of the Wheeler-DeWitt equation

$$\left(\frac{1}{2} G^{ij} \frac{\partial}{\partial \beta^i} \frac{\partial}{\partial \beta^j} - V \right) \Psi = 0 \tag{17}$$

where Ψ denotes the wave function of the universe.

We can diagonalize the minisuperspace metric in special coordinates $(z^0, z^1, \ldots, z^\alpha)$ and find the hyperbolic WDW equation

$$\left(-\frac{\partial^2}{\partial z^{0\,2}} + \Delta - e^{\lambda z^0} V(z^0, z^1, \ldots, z^\alpha)\right) \Psi = 0$$

This equation can be discussed for several special potentials [13]. For the special case of a potential containing only curvature the Weeler-De Witt equation can be solved explicitly by seperation of variables and the WDW operator can be splitted into two parts which combine to the Schrödinger equation form with energy $E \geq 0$. This gives the possibility to introduce time into the quantum KKCM and to define the states $E = 0$ as appropriate ground states. Further results are e.g. quantum wormhole solutions and solutions describing quantum tunneling [14, 15]

REFERENCES

1. H. C. Lee (ed.), *An Introduction to Kaluza-Klein Theories* (World Scientific, Singapore 1984)
2. U. Bleyer, in *Relativistic Astrophysics and Cosmology*, eds. S. Gottlöber, J.P. Mücket, V. Müller (World Scientific, Singapore 1992)
3. P.G.O. Freund, M.A. Rubin, *Phys. Lett.* **97B** (1980)233
4. E.W.Kolb, R. Slansky, *Phys. Lett.* **135B** (1984) 378
5. U. Bleyer, D.-E. Liebscher, A. G. Polnarev, in *Proceedings of the FIFTH SEMINAR ON QUANTUM GRAVITY, Moscow 1990*, eds. M. A. Markov, V. A. Berezin, V. P. Frolov (World Scientific, Singapore 1991) p. 613
6. M. J. Duff, B. E. Nielson, and C. N. Pope, *Phys. Rev.* **130** (1986) 1
7. V. D. Ivashchuk, V. N. Mel'nikov, and A. I. Zhuk, *Nuov. Cim.* **B104** (1989) 575
8. U. Bleyer, D.-E. Liebscher, in *Proceedings of the THIRD SEMINAR ON QUANTUM GRAVITY, Moscow 1984*, eds. M. A. Markov, V. A. Berezin, V. P. Frolov (World Scientific, Singapore 1985) p. 662
9. U. Bleyer, D.-E. Liebscher, A. G. Polnarev, *Class. Quant. Grav.* **8** (1991)
10. V. D. Ivashchuk, *Phys. Lett.* **A170** (1992) 16
11. A. Zhuk, *Sov. Journ. Nucl. Phys.* , in press.
12. U. Bleyer, A. Zhuk, On Scalar Field in Multidimensional Cosmology, preprint FUB-HEP/92-23, subm. to GRG
13. U. Bleyer, D.-E. Liebscher, H.-J. Schmidt, A. I. Zhuk, *Wiss. Zeitschr. Univ. Jena* **1** (1990) 20
14. A. Zhuk, *Phys. Rev.* **D45** (1992) 1192, *Sov. Journ. Nucl. Phys.* **55** (1992) 149
15. A. Zhuk, *Non Singular Boundary Conditions in Multidimensional Cosmology*, preprint FUB-HEP/92-20

WORMHOLE SOLUTIONS IN COUPLED EINSTEIN-YANG-MILLS-AXION SYSTEM

E.E. DONETS and D.V.GAL'TSOV
Departament of Theoretical Physics, Physics Faculty,
Moscow State University, 119899 Moscow, Russia

ABSTRACT

Non-abelian wormhole solutions within the scope of the 4-dimensional heterotic string effective action are considered. The possibility of a non-trivial Kalb-Ramond field coupled to an homogeneous SU(2) gauge field is qualitatively discussed.

Non-abelian Einstein-Yang-Mills wormholes were discussed recently [1-3]in connection with the idea of a Big Fix. To have a finite action one is forced to introduce a cosmological constant that leads to the quantization of the topological charge. However for a specific value of the cosmological constant a continuous family of wormholes[4] exists that interpolate between the instanton and meron solutions. Here we generalize this result to include a Kalb-Ramond field and discuss some new features due to the presence of an axion.

We start with the following bosonic part of the 4-dimensional heterotic string euclidean effective action with the cosmological constant Λ in the Einstein frame

$$S_E = \frac{1}{16\pi} \int \sqrt{g}\,[\,M_{Pl}^2\,(-R - \frac{1}{2}\partial_\mu\Phi\partial^\mu\Phi + 2\Lambda) + F_{a\mu\nu}F_a^{\mu\nu} + \Phi\, F_{a\mu\nu}\tilde{F}_a^{\mu\nu}\,]d^4x\ , \quad (1)$$

where Φ is Peccei-Quinn axion field, $F_{a\mu\nu}$ is Yang-Mills curvature corresponding to $SU(2)$ gauge group (generalization to higher groups is straightforward[5], and the (constant) dilaton field is excluded.

We are interested in the homogeneous and isotropic configurations with the following line element in the euclidean space-time

$$ds^2 = N^2\,d\tau^2 + a^2(d\chi^2 + \sin^2\chi\,(d\theta^2 + \sin^2\theta d\varphi^2)). \quad (2)$$

The corresponding $SU(2)$ Yang-Mills connection parametrized by a single real-valued function σ in the temporary gauge reads

$$\begin{aligned} A_0 &= 0,\ gA_j = \frac{i}{2}\,(\sigma(\tau) + 1)\,U\partial_j U^{-1} \\ U &= \exp\,[i\chi\,(\sin\theta(\sigma_x\cos\varphi + \sigma_y\sin\varphi) + \sigma_z\cos\theta)], \end{aligned} \quad (3)$$

where g is the gauge coupling constant. After dimensional reduction and elimination of a total divergence the action Eq.1 will read

$$\begin{aligned} S_E &= 3\pi/4 \int d\tau [M_{Pl}^2(-aa'^2/N - aN + \Lambda a^3 N/3 - \Phi'^2 a^3/(12N)) \\ &+ g^{-2}(\sigma'^2 a/N + (\sigma^2-1)^2 N/a + 4\Phi\sigma'(\sigma^2-1))], \end{aligned} \tag{4}$$

where primes denote the derivatives with respect to τ. Variation of Eq.4 with respect to Φ leads to the following equation

$$\Phi' = (4R_g^2 N/a^3)[P - 2(\sigma^3 - 3\sigma - 2)], \tag{5}$$

where $R_g = 1/(M_{Pl}g)$ is the characteristic length, and P is the integration constant, which could be interpreted as an axion charge in absence of the axion–Yang–Mills coupling. Variation of S_E in Eq.4 with respect to the lapse metric function N gives the Hamiltonian constraint which after using Eq.5 and putting $N = a$ (conformal gauge) reads

$$\begin{aligned} a'^2 &- R_g^2\sigma'^2 + U(a,\sigma) = 0, \\ U(a,\sigma) = -a^2 + (\Lambda/3)a^4 &+ R_g^2(\sigma^2-1)^2 + (4/3a^2)R_g^4[P - 2(\sigma^3 - 3\sigma - 2)]^2. \end{aligned} \tag{6}$$

The system of equations for metric $a(\tau)$ and Yang-Mills $\sigma(\tau)$ functions obtaines as follows

$$\begin{aligned} 2aa'' - a'^2 - a^2 + \Lambda a^4 - (4R_g^4/a^2)[P - 2(\sigma^3 - 3\sigma - 2)]^2 + R_g^2(\sigma'^2 - (\sigma^2 - 1)) = 0, \\ \sigma'' - 2\sigma(\sigma^2 - 1) + (8R_g^2/a^2)(\sigma^2 - 1)[P - 2(\sigma^3 - 3\sigma - 2)] = 0, \end{aligned} \tag{7}$$

with appropriate boundary conditions

$$a'(0) = a'(T) = \sigma'(0) = \sigma'(T) = 0, \tag{8}$$

where T is a period of the solution (in euclidean time) . We are looking for the solutions of system Eq.6, Eq.7, Eq.8 with finite euclidean action, e.g. static and periodic ones.

For some particular cases this system was studied rather extensively during the recent years. Putting $\Phi = 0$ in Eq.6, Eq.7 one gets a coupled Einstein–Yang–Mills system, which was analized in Ref.[1-3]. In this case the system of EYM equations reduces to a two–dimensional separable non–linear oscillator system. For some discrete values of separation constant $C \in (0,1)$ the periods of two oscillators are in a rational relation and the boundary conditions Eq.8 hold. Hence, a discrete set of wormhole solutions with the quantized (through the separation constant) values of YM topological charge exists between the self–dual instanton on a S^4 background ($C = 0$) and the static Hosoya-Ogura ($C = 1$) meron solutions. For a critical value of the cosmological constant $\Lambda = (3/4)M_{Pl}^2 g^2$ both oscillators have equal periods for all values of separation constant $C \in (0,1)$ and a continuous family of EYM wormholes[4] exists interpolating between the instanton and the meron solutions. These wormholes can possess any value of YM topological charge $Q \in (0,1)$ and the corresponding euclidean action is exactly zero, the positive YM contribution being canceled by the negative contribution of the gravitational field. They corresponds to the (a, σ) trajectories connecting two turning points under potential barrierrs and hence can be interpreted as describing the quamtum tunneling from the hot Friedman–Robertson–Walker Universe to the De Sitter one[5].

As was pointed out in Ref.[6], the entire EYMAxion system reproduces the Hosoya-Ogura solution of the EYM system $\sigma = 0$, $a = \sqrt{(3/2\Lambda)}$ for a special value $P = -4$ of an integration constant, because the Kalb–Ramond axion field strength is exactly zero in this case ($\Phi' = 0$ in Eq.5).

Let us discuss qualitatively the possibility of solutions with a non–trivial axion field. Due to the Hamiltonian constraint Eq.6 and the boundary conditions Eq.8 all such solutions are field configurations connecting two points in the configuration space (a_i, σ_i) and (a_j, σ_j) with both zero momenta and such that $U(a_i, \sigma_i) = U(a_j, \sigma_j) = 0$. Different configurations can be classified as follows.

First, for some values of the parameters other static configurations $(a_i, \sigma_i) = (a_j, \sigma_j)$ can arise similar to those found in Ref.[6]. For an arbitrary value of P (not only for the special value of $P = -4$) the function $U(a, \sigma)$ can possess up to five points of local extrema. If the Hamiltonian constraint $U(a_i, \sigma_i) = 0$ is imposed on some (a_i, σ_i), this will provide strong relations between the parameters of theory R_g and the cosmological constant Λ.

The terms a'^2 and σ'^2 have different signs in the constraint equation Eq.6, hence all saddle points of the function $U(a, \sigma)$ are the points of local minima of the euclidean action functional S_E Eq.4, and vice versa, all points of local minima of $U(a, \sigma)$ are saddle points of the euclidean action S_E.

Dynamic configurations connecting two different points of local minima of $U(a, \sigma)$ can be solutions if the Hamiltonian constraint holds for both of them. Note that only for the value of $P = -4$ as in Ref.[6] different YM vacua $\sigma = -1$ and $\sigma = 1$ give the same contribution in to the energy–momentum tensor and can both lie on the same (zero) level of the potential $U(a, \sigma)$. One then can have instanton–like configuration with the integer topological charge which is not self–dual and can be interpreted as a "closed handle" wormhole.

Next possibility is provided by configurations which start at some point of local minima of the potential and terminate at the turning point of a zero level of $U(a, \sigma)$. One can interpret such configurations as wormholes describing the splitting off the baby Universe (contracting or expanding) like in Ref.[7,8] in other context. For an arbitrary value of P the YM vacua $\sigma = -1$ and $\sigma = 1$ always live on different levels of the potential $U(a, \sigma)$ (only $\sigma = -1$ is true YM vacuum with zero energy-momentum tensor). Hence, no instanton–like configuration with the integer topological charge are allowed in the case of arbitrary value of P.

Finally, the solutions of Eq.6, Eq.7, Eq.8 can connect two different turning points in the vicinity of the same or different points of local minima of $U(a, \sigma)$. They should describe the tunneling under the two-dimensional barrier from an initial hot Universe into De Sitter one similarly to the case of the pure Einstein–Yang–Mills system.

A numerical work is now in progress in order to check which of the above options really lead to physically interesting solutions.

References

1. A.Hosoya and W.Ogura, *Phys. Lett.* **B225** (1989), 117.
2. Y.Verbin, A.Davidson, *Phys. Lett.* **B229** (1989), 364; *Nucl. Phys.* **B339** (1990), 545.
3. S.J.Rey, *Nucl. Phys.* **B336** (1990), 146.
4. E.E.Donets, D.V.Gal'tsov, *Phys. Lett.* **B296** (1992), 311.
5. O.Bertolami, J.M.Mourao, R.F.Picken, I.P.Volobujev, *Int.J. Mod. Phys.* **A6** (1991), 4149.
6. A.Das, J.Maharana, *Phys. Rev.* **D41** (1990), 699.
7. S.Giddings, A.Strominger, *preprint HUTP-88/A006* (1988).
8. V.A.Rubakov, P.G.Tinyakov, *Phys. Lett.* **B214** (1988), 334.

The Hilbert space of wormholes*

Luis J. Garay[†]
Instituto de Optica
Consejo Superior de Investigaciones Científicas
Serrano 121, E-28006 Madrid, Spain

ABSTRACT

Wormhole boundary conditions imply that the wormhole wave function must be damped for large three-geometries and regular for small ones; it must also be damped for large matter fields. However, these conditions are incompatible with the minisuperspace structure. This leads us to consider the maximal analytic extension of minisuperspace as the configuration space in which the quantization process can be carried out properly. Then, the wormhole wave function must be square integrable in this maximal analytic extension of minisuperspace. This boundary condition suitably contains the previous ones and also defines the Hilbert space of wormholes. Although this Hilbert space is gauge dependent, physical results such as effective interactions do not depend on the gauge choice.

Wormholes may affect the constants of nature through low energy effective interactions [1, 2]. In particular they may drive the cosmological constant to zero [2] and select general relativity as the low energy theory of gravity among Jordan–Brans–Dicke theories [3]. In order to determine which are these interactions, it is necessary to have a well defined Hilbert space structure for wormholes [1]. The quantum wormhole wave function is given by the path integral over all possible asymptotically Euclidean spacetimes and over all matter fields defined on them whose energy momentum tensor vanishes at infinity [4, 5]. Formally, this wave function satisfies the Wheeler–DeWitt equation and the diffeomorphism constraints. Therefore, in order to find wormhole wave functions, one can equivalently solve these equations with appropriate boundary conditions which can be derived from those for the path integral formulation [4, 6]. They will determine the wormhole Hilbert space. In this work, we address the problem of precisely defining the wormhole boundary conditions and consequently the wormhole Hilbert space. Since spacetime is asymptotically Euclidean, the wave function will be exponentially damped for large three–geometries. On the other hand, for small three–geometries no singularities are expected and therefore, the wave function should be regular at these configurations in some sense, so that the absence of singularities be properly reflected [4]. Likewise, the wormhole wave function must be regular at any field configuration and any three–geometry. Finally, as it happens in ordinary quantum mechanics, the wave function must vanish at infinite field values, since these configurations cannot dominate the wave function. All these conditions

*Work partially supported by DGCYT under contract PB91-0052
[†]Supported by a Basque Country Grant. E-mail address: **Garay@cc.csic.es**

suggest that the wave function must be square integrable in superspace.

Due to the well known difficulties in studying superspace, we shall concentrate in some minisuperspace models. Although the generalization of the results thus obtained to full superspace is not straightforward at all, it may be expected that some qualitative results will still hold. In minisuperspace, the Euclidean action takes the form (see, for instance, [7])

$$I = \int_0^\infty d\tau N \left(\frac{1}{N^2} f_{\mu\nu} \dot{q}^\mu \dot{q}^\nu + V(q) \right),$$

where q^μ, $\mu = 1, 2, \ldots, n$ are the degrees of freedom that represent the three-geometry and the matter fields; N is the lapse function which ensures the invariance under time reparametrizacións; $f_{\mu\nu}$ is the metric in minisuperspace, so that the line element in minisuperspace is $d\mathcal{S}^2 = (f_{\mu\nu}/N) dq^\mu dq^\nu$; and $V(q)$ is the Wheeler–DeWitt potential. As particular examples, we shall consider a homogeneous massless scalar field either minimally or conformally coupled to a Friedmann–Robertson–Walker spacetime $ds^2 = (2G/3\pi) a^2(\tau) (N^2 d\tau^2 + d\Omega_3^2)$.

The square integrability condition in minisuperspace ensures that the wormhole wave function is damped for large values of the scale factor and the scalar field. It also implies regularity of the wave function in the whole minisuperspace except on its boundary. In particular, it does not imply that the wave function be regular when the three–volume is vanishingly small. This fact leads us to study the boundary of minisuperspace. According to Vilenkin [8], we shall define it as the set of all configurations which are singular in a general sense. On the one hand, it may contain configurations whose singularities are due to the slicing procedure but are non truly singular. This part of the boundary will be called regular boundary. In particular, the configurations associated with vanishing three–volume belong to the regular boundary. On the other hand, the so called singular boundary will consist of real singularities in minisuperspace such as those corresponding to arbitrarily large three–geometries or infinite scalar field values.

The square integrability boundary condition concerns to the singular boundary, but it says nothing about the regular boundary. In this respect, two questions arise concerning the regular boundary: (*i*) Can we properly impose wormhole boundary conditions on it? and (*ii*) Is it a real boundary of minisuperspace? The answer to both questions is negative. The Wheeler–DeWitt operator is hyperbolic and, for this kind of operators, the boundary value problem is well posed only if boundary conditions are imposed on null surfaces of the configuration space [9]. It is very easy to see that the regular boundary is not a null surface in general. For instance, in the minisuperspace model with a conformal scalar field the regular boundary is the surface $a = 0$ (it is a line in this case, since the minisuperspace is two–dimensional) which is not null. This answers the first question. With respect to the second, we have seen that the regular boundary is made of 'coordinate singularities' rather than true singularities.

It plays a similar role to the Schwarzschild horizon in black hole physics. It may be preventing us from gaining access to other regions of the configuration space.

For the reasons explained above, one is naturally led to consider the maximal analytic extension[10] of minisuperspace as the configuration space in which the quantization procedure can be properly accomplished. We can now state the wormhole boundary conditions as saying that the wormhole wave functions must be square integrable in the maximal analytic extension of minisuperspace. In other words, the Hilbert space of wormholes consists of all normalizable solutions of the Wheeler–DeWitt equation. The inner product is given by the natural measure in the extended minisuperspace $d\bar{q}\sqrt{\bar{f}}$ where the overbar denotes quantities in the extended minisuperspace and $\bar{f}$ is the determinant of the extended metric. This boundary condition gives full meaning to the wormhole boundary conditions formulated previously. Indeed, the boundary of the maximal extension of minisuperspace is truly singular and therefore the wormhole wave function will vanish there by virtue of square integrability. Besides, the regular boundary of minisuperspace (and in particular the configurations that represent vanishing three-geometries) is in the interior of the extended minisuperspace and thus no additional boundary conditions are required there. Regularity of the wave function in these configurations is automatically guaranteed.

We have been able to build up a wormhole Hilbert space. It can be seen that an orthonormal basis which is parametrized by the wormhole throat radius can be found, at least in the minisuperspace models we have considered here [4, 11]. Any other wormhole state can be written as a linear combination of these basis elements with fixed radius, the coefficients being square integrable as functions of the throat radius [11]. This means, as expected, that a general quantum wormhole state will not have a well defined throat radius and therefore will not have a classical counterpart. Even more, since the coefficients are square integrable, they must vanish for large values of the throat radius. This implies that large wormholes are suppressed (exponentially) and therefore will not overcrowd spacetime [11]. This conclusion agrees with the semiclassical result found by Giddings and Strominger [12].

There is however an issue that we have not mentioned yet. This whole formulation (the analytic extension, the inner product and, consequently, the Hilbert space itself) is gauge dependent since different choices for the lapse function will lead to different structures. This is not new. Hilbert spaces are not physical. They are just mathematical constructions that are extremely useful in the calculation of physical quantities which are of course gauge invariant. The wormhole Hilbert space will be different for different choices of the lapse function. Even more, for some 'inadmissible' choices, there will not even exit a Hilbert space since every wormhole wave function will be non normalizable in the induced inner product [11]. When such an inadmissible choice is made, one cannot use the canonical formalism in the calculation of the effective interactions. Rather, one must use the path integral expression for the propagator through the wormhole. Otherwise, this interaction can also be found by inserting the identity operator in the middle of the propagator between the two asymptotically Euclidean regions joined by the wormhole [1]

$$\langle 0|\,\phi(x_1)\phi(x_2)\cdots\phi(y_1)\phi(y_2)\cdots|0\rangle = \sum_{wh}\langle 0|\,\phi(x_1)\phi(x_2)\cdots|wh\rangle\,\langle wh|\,\phi(y_1)\phi(y_2)\cdots|0\rangle$$

where the sum ranges over all elements of an orthonormal basis of the wormhole Hilbert space. Although the Hilbert space structure depends on the gauge choice, the propagator does not. This is in the end a consequence of the fact that these interactions can be defined via path integrals which by definition contain a sum over all possible lapse functions.

To conclude, wormhole boundary conditions can be stated in the following form: wormhole wave functions must be square integrable in the maximal analytic extension of minisuperspace. This provides us with a wormhole Hilbert space which, as usual, depends on the gauge choice. However, physical results does not, since they can also be written in terms of path integrals which are manifestly gauge invariant.

References

[1] S. W. Hawking, *Phys. Rev.* **D37** (1988) 904.

[2] S. Coleman, *Nucl. Phys.* **B310** (1988b) 643.

[3] L. J. Garay and J. García–Bellido, *Jordan–Brans–Dicke wormholes and Coleman's mechanism*, preprint IEM–FT–59 (1992).

[4] S. W. Hawking and D. N. Page, *Phys. Rev.* **D42** (1990) 2655.

[5] L. J. Garay, *Phys. Rev.* **D44** (1991) 1059.

[6] L. M. Campbell and L. J. Garay, *Phys. Lett.* **B254** (1991) 49.

[7] L. J. Garay and E. Verdaguer, *Non perturbative quantum gravity and quantum cosmology*, Course "Gravitación y cosmología", I.N.T.A., Madrid (1991).

[8] A. Vilenkin, *Phys. Rev.* **D37** (1988) 888.

[9] P. Puig Adam, *Ecuaciones diferenciales*, Biblioteca Matemática, Madrid (1955).

[10] S. W. Hawking and G. F. R. Ellis, *The large scale structure of spacetime*, Cambridge University Press, Cambridge (1973).

[11] L. J. Garay, *Wormholes in quantum cosmology*, Ph. D. Thesis, Universidad Autónoma de Madrid, Madrid (1992).

[12] S. B. Giddings and A. Strominger, *Nucl. Phys.* **B306** (1988a) 890.

STABILITY AND SIGNATURE IN QUANTUM GRAVITY [1 2]

J. Greensite

Physics and Astronomy Dept.
San Francisco State University
San Francisco, CA 94132 USA

Abstract

It is shown how stochastic quantization can be used to overcome the problem that the Euclidean Einstein-Hilbert action is unbounded from below. The possibility that the Lorentzian signature of spacetime is determined dynamically is also discussed.

A "bottomless action" theory is a theory whose action is unbounded from below. Einstein gravity, bosonic string field theory, and certain matrix models are theories of this type, in which the Euclidean functional integral is ill-defined. A method for defining the Euclidean quantum theory of such bottomless theories, which preserves both the classical limit and (if the theory is bounded from below at zeroth order) reproduces the naive perturbative expansion to all orders in any coupling, was proposed in ref. 1. The method can be derived from stochastic quantization plus boundary conditions in the fictitious Langevin evolution time t_5.[2,3] As an example, we consider a scalar field $\phi(x)$. The boundary conditions, at fictitious time $t_5 = \pm\infty$ are $\phi(x,-\infty) = \phi_i(x)$ and $\phi(x,\infty) = \phi_f(x)$ respectively, where $\phi_{i,f}$ are any non-singular field configurations. It is the boundary condition at $t_5 = \infty$ which eliminates the runaway solutions expected in a bottomless theory. Of course, it is important that expectation values so defined do not depend on the choice of $\phi_{i,f}$; this is in fact the case. If $Q[\phi]$ is any given observable, the stochastic quantization prescription is

$$\begin{aligned} < Q > &= \lim_{T\to\infty} \frac{1}{Z_5} \int D\eta \; Q[\phi(x,0)]\delta[\phi(x,-T) - \phi_i(x)] \\ &\quad \times\delta[\phi(x,T) - \phi_f(x)]\exp[-\frac{1}{4\hbar}\int_{-T}^{T} d^5x \; \eta^2] \\ \partial_5\phi &= -\frac{\delta S}{\delta\phi} + \eta \end{aligned} \tag{1}$$

[1]Contribution to the 1st Iberian Meeting on Gravity, Evora, Portugal (Sept. 1992).
[2]Supported by the U.S. Department of Energy under Grant No. DE-FG03-92ER40711.

The Langevin equation can itself be used to change the functional integration variable from η to ϕ; this leads to the "fifth-time action" prescription

$$
\begin{aligned}
<Q> &= \int_{\phi_i}^{\phi_f} D\phi(x,t_5)\, Q(x,0) e^{-S_5} \\
S_5 &= \int d^5x [\frac{1}{2}(\partial_5\phi)^2 + \frac{1}{8\hbar^2}\left(\frac{\delta S}{\delta\phi}\right)^2 - \frac{1}{4\hbar}\frac{\delta^2 S}{\delta\phi^2}]
\end{aligned}
\tag{2}
$$

where the last term in S_5 comes from the Jacobian $det[\delta\eta/\delta\phi]$. S_5 is bounded even if the original action S is not, and, for the usual reasons, expectation values $< Q >$ are insensitive to the boundary conditions $\phi_{i,f}$ placed on $\phi(x,t_5)$ at $t_5 \to \pm\infty$. Eq. 2 defines the stabilized theory corresponding to an action S; it is not hard to see that the classical Euclidean equations of motion are obtained in the $\hbar \to 0$ limit. It can also be readily shown that, providing S is stable at zeroth order, the perturbation expansion based on S_5 reproduces the naive perturbative expansion based on S.[1] Note that stabilization is equivalent to replacing the unstable action S by a four-dimensional, stabilized action S_{eff}, where

$$
e^{-S_{eff}[\varphi(x)]} = \int D\phi\ \delta[\phi(x,0) - \varphi(x)] e^{-S_5}
\tag{3}
$$

Stabilization applied to the Einstein-Hilbert action yields

$$
\begin{aligned}
<Q> &= \frac{1}{Z_5}\int Dg\ \sqrt{G} Q[g_{\mu\nu}(x,0)] e^{-S_5} \\
S_5 &= \frac{1}{4}\int d^5x \left[\frac{1}{\kappa^2} G^{\mu\nu\alpha\beta}\partial_5 g_{\mu\nu}\partial_5 g_{\alpha\beta} + \frac{1}{\kappa^2} g G^{-1}_{\mu\nu\alpha\beta}(R^{\mu\nu} - \frac{1}{2}g^{\mu\nu}R)\right. \\
&\quad \left. \times (R^{\alpha\beta} - \frac{1}{2}g^{\alpha\beta}R) - \hbar\beta\sqrt{g}R\right]
\end{aligned}
\tag{4}
$$

where $G^{\mu\nu\alpha\beta}$ is the supermetric, β is a singular constant, and the DeWitt measure $\sqrt{G}$ (=const. in D=4) is obtained automatically from the Jacobian $det[\delta\eta/\delta g]$. The Einstein-Hilbert action is bottomless already at zeroth order, i.e.

$$
S^0[g_{\mu\nu}] = \int \frac{d^4p}{(2\pi)^4} h_{\mu\nu}(p) p^2 \left[\frac{1}{4}P^{(2)} - \frac{1}{2}P^{(0-s)}\right]_{\mu\nu\alpha\beta} h_{\alpha\beta}(-p)
\tag{5}
$$

where $g_{\mu\nu} = \delta_{\mu\nu} + \kappa h_{\mu\nu}$ and $P^{(2)}$ and $P^{(0-s)}$ are spin-2 and spin-0 projection operators. Using (a generalized version of) Eq. 3, we find for the stabilized action at zeroth order[4]

$$
S^0_{eff}[g_{\mu\nu}] = \int \frac{d^4p}{(2\pi)^4} h_{\mu\nu}(p) p^2 \left[\frac{1}{4}P^{(2)} + \frac{1}{2}P^{(0-s)}\right]_{\mu\nu\alpha\beta} h_{\alpha\beta}(-p)
\tag{6}
$$

Stabilization simply flips the sign of the conformal term, at zeroth order in Newton's constant. Non-perturbatively, S_{eff} is bounded from below, diffeomorphism invariant and yields the usual Einstein equations in the classical limit.

Although S_5 is a higher-derivative theory when based on the Einstein-Hilbert action, this is not the case for S_5 based on the Einstein-Cartan theory, which appears to be reflection-positive in the stabilized form. The stabilized formulation has recently been applied to study the cosmological constant problem,[5,6] and to numerical simulations.[2]

The second topic I would like to discuss in this talk is the dynamics of spacetime signature.[7] In ordinary field theories whose actions are bounded from below, the rotation from Minkowski to Euclidean space is a technique which is needed to define integrals which are otherwise ill-defined, and to obtain the correct $i\epsilon$ prescription for propagators. It is interesting to consider the possibility that Wick rotation, which is essentially a rotation in the spacetime signature, is more than just a technical trick, and actually has physical content. Let $g_{\mu\nu} = e^a_\mu \eta_{ab} e^b_\nu$, and consider the functional integral with generalized signature

$$\begin{aligned} Z &= \int d\mu(e,\phi,\psi,\overline{\psi}) \exp\left[-\int d^D x \sqrt{g}\mathcal{L}\right] \\ \eta_{ab} &= diag\{e^{i\theta},1,...,1\} \end{aligned} \tag{7}$$

We obtain the Euclidean functional integral for $\theta = 0$, and the Feynman integral for $\theta \to \pi^-$ ($\theta \to -\pi^+$ is equivalent to the Feynman integral under a time inversion). The range of theta is taken to be $\theta \in [-\pi,\pi]$, which is the range for which ordinary scalar and vector theories are bounded from below, without the need for stabilization.

Now consider promoting $exp(i\theta)$ from a parameter to a dynamical field. In that case the signature of spacetime is determined dynamically, by the expectation value of this phase field. The expectation value can be determined by calculating the effective potential $V(\theta)$ induced by the quantum fluctuations of all other fields. The one-loop contribution to $V(\theta)$ from a massless scalar field, fluctuating on a flat ($e^a_\mu = \delta^a_\mu$) background is

$$\exp[-\int d^D x V_0(\theta)] = det^{-\frac{1}{2}}[-\sqrt{\eta}\eta^{ab}\partial_a\partial_b] \tag{8}$$

and proper-time regulation of the determinant gives

$$\begin{aligned} V_0(\theta) &= -\frac{1}{2}\int_{1/\Lambda^2}^{\infty} \frac{ds}{s} \int \frac{d^D p}{(2\pi)^D} \exp[-s(e^{-i\theta/2}p_0^2 + e^{i\theta/2}\vec{p}^{\,2})] \\ &= -\frac{\Lambda^D}{D(4\pi)^{D/2}} \exp[-i(D-2)\theta/4] \end{aligned} \tag{9}$$

where Λ is a momentum cutoff which, given the non-renormalizability of gravity, presumably exists at the Planck scale. The p-integration in (9) is only well-defined for θ in the range $[-\pi,\pi]$, which, again, is related to the fact that $Re[\sqrt{g}\mathcal{L}_\phi]$ for a scalar field is only bounded from below for $|\theta| \leq \pi$.

In computing V_0, I have ignored any possible θ-dependence of the measure. Let us assume that for free fields of mass m, the measure is such that contributions to

Z from each (propagating) bosonic degree of freedom are equal, and inverse to the contributions from each fermionic degree of freedom. Thus, e.g., $Z = 1$ at any θ for a supersymmetric combination of free fields. With this assumption, given n_F fermionic and n_B propagating bosonic degrees of freedom, $V(\theta)$ at one loop is given by

$$V(\theta) = (n_F - n_B)\frac{\Lambda^D}{D(4\pi)^{D/2}}\exp[-i(D-2)\theta/4] \tag{10}$$

$V(\theta)$ is complex-valued. To determine $< e^{i\theta} >$ at one-loop level, we look for a value of θ in the range $[-\pi, \pi]$ such that: i) $Re[V]$ is minimized; and ii) $Im[V]$ is stationary. If these two conditions are not satisfied for the same value of θ, then there may be large quantum fluctuations in the signature. From (10), the requirements are seen to be:

$$\left.\begin{array}{r}\cos[(D-2)\theta/4] = 0 \\ \\ \min[Re[V(\theta)]] = 0\end{array}\right\} \quad \theta \in [-\pi, \pi] \tag{11}$$

The first condition is just the requirement that $Im[V]$ is stationary. Then $Re[V] = 0$ at the stationary point of $Im[V]$, and the second condition is the requirement that this is also the minimum of $Re[V]$.

For $n_B > n_F$ there is no solution of the conditions above. For $n_F > n_B$ it is only possible to find a solution in $D = 4$ dimensions, and this solution happens to be $\theta = \pm\pi$. In other words, given some modest assumptions about the measure, $V(\theta)$ singles out both Lorentzian signature and the observed dimension of spacetime. Uniquely in D=4 dimensions, it appears that Lorentzian signature may have a dynamical origin.

References

[1] J. Greensite and M. Halpern, Nucl.Phys. **B242** (1984) 167.

[2] J. Greensite, "Stabilized Quantum Gravity: Stochastic Interpretation and Numerical Simulation", Nucl. Phys. B (in press).

[3] A. Giveon, M. Halpern, E. Kiritsis, and N. Obers, Nucl. Phys. **B357** (1991) 655.

[4] J. Greensite, Nucl. Phys. **B361** (1991) 729;

[5] A. Carlini and M. Martellini, Phys. Lett. **B276** (1992) 36.

[6] J. Greensite, Phys. Lett. **B291** (1992) 405.

[7] J. Greensite, *Dynamical Origin of the Lorentzian Signature of Spacetime*, Copenhagen preprint NBI-HE-92-59, gr-qc@xxx/921008.

Particle Contents of Higher Order Gravity

J. Julve, J.C. Alonso

Instituto de Matemáticas y Física Fundamental
CSIC,Serrano 123, 28006 MADRID, Spain

19 January 1993

Abstract

Theories of gravity quadratic in the curvatures of space-time are shown to be equivalent to Einstein's plus some new fields by a Legendre transform. This makes explicit the actual degrees of freedom of the higher order theory.

Introduction

Theories of gravity containing squared or higher order terms in the curvatures come from QFT in curved spaces and as effective theories from the string.

Theories with squared terms are known to be renormalizable [1] and have interesting renormalization group properties [2]. They may also have important applications in cosmology [3] [4].

Counting the actual degrees of freedom (d.o.f.) of these theories is obscured by the fact that the higher differential order of the equations of motion enlarges the set of initial conditions for the solutions. Thus, while the Einstein's gravity

$$\mathcal{L}_2 = \sqrt{-g}R$$

contains the two d.o.f. of the graviton, the higher order theory

$$\mathcal{L}_3 = \sqrt{-g}\{aR + bR^2 + dR^3 + ...\} \equiv \sqrt{-g}f[R]$$

is known to contain three [4], and the more general case involving the Ricci curvature

$$\mathcal{L}_8 = \sqrt{-g}\{aR + bR^2 + cR_{\mu\nu}R^{\mu\nu} + ...\}$$

describes instead eight d.o.f. [5].

The equivalence of $\mathcal{L}_3$ and Einstein plus a scalar field was explicitely shown [4] at the level of the equations of motion. The theory $\mathcal{L}_8$ in turn generally contains a massles graviton (2 d.o.f.), a massive spin-two field (5 d.o.f.) and a scalar field (1 d.o.f.) [5]. We will concentrate on it in the sequel. In this talk we face the task of finding an Action with independent explicit fields for those 2+5+1 d.o.f. and derivable from $\mathcal{L}_8$ by some recipe.

The conversion of a higher derivative (HD) theory into an equivalent lower derivative (LD) one can be accomplished by a Legendre transform [6] at the obvious price of increasing the number of fundamental variables. However obtaining the final version for the explicit diagonalized fields requires some extra tricks which will be shown here. To ilustrate this we first briefly recall the Legendre and Helmholtz theory. Then we apply it to a simple example of a scalar field. Finally we work out the case of the HD Gravity.

1 Legendre transform

Consider a mechanical system with just one configuration variable q[t]. The Lagrangian function $\mathcal{L}[q, \dot{q}]$ and the stationary action principle yield the Euler's equations of motion which, for second-derivative $\mathcal{L}$, are of second differential order thus requiring two initial conditions.

If the system is non-singular, namely if $\frac{\partial^2\mathcal{L}}{\partial\dot{q}^2} \neq 0$, the Legendre transform can be performed. From the definition of the canonical conjugate momentum

$$p \equiv \frac{\partial \mathcal{L}}{\partial \dot{q}}$$

one can work out $\dot{q}$=Q[q,p] and obtain the Hamilton function

$$\mathcal{H}[q,p] = pQ - \mathcal{L}[q,Q]$$

Now the canonical equations of motion

$$\dot{q} = \frac{\partial \mathcal{H}}{\partial p} \quad , \quad \dot{p} = -\frac{\partial \mathcal{H}}{\partial q}$$

are a system of first order differential equations which is equivalent to Euler's.

However, more in the taste of (covariant) particle physicists, the canonical equations may be derived as the Euler equations (i.e., by a variational principle) from the so called Helmholtz Lagrangian

$$\mathcal{L}_H[q,\dot{q},p] = p\dot{q} - \mathcal{H}[q,p]$$

where p and q are independent variables. This formulation gets from unimportant to useful when HD theories $\mathcal{L}[q,\dot{q},\ddot{q},..]$ are considered. The Legendre transform can be done for any function f of the derivatives of q , with the requisite that it can be worked out of the defining equation $p = \frac{\partial\mathcal{L}}{\partial f}$. Accordingly the canonical equations get modified to

$$f = \frac{\partial \mathcal{H}}{\partial p} \quad , \quad g = \frac{\partial \mathcal{H}}{\partial g}$$

where g is a function of derivatives of q and p which is related to f. In any case the ultimate advantage is that of having an equivalent Lagrangian $\mathcal{L}_H$ of lower differential order than $\mathcal{L}$. The particle interpretation is greatly simplified if one succeeds in getting a 2nd. order $\mathcal{L}_{\mathcal{H}}$ so that the corresponding propagators have 2nd. order poles.

2 Scalar Higher Derivative example

We adopt (+ - - -) as the Minkowsky signature.

Consider the following fourth order theory for a scalar field φ with a source j[x]

$$\begin{aligned}\mathcal{L}[\varphi,\partial\varphi,\partial^2\varphi] &= -\frac{1}{2}\frac{1}{m^2}\varphi\Box(\Box+m^2)\varphi - j\varphi \\ &= \frac{1}{2}[(\partial_\mu\varphi)^2 - \frac{1}{m^2}(\partial_\mu\partial_\nu\varphi)^2] - j\varphi\end{aligned}$$

In this example the terms $(\partial_\mu\varphi)^2$, $(\partial_\mu\partial_\nu\varphi)^2$ and $j\varphi$ are reminiscent of R, R^2 and $R_{\mu\nu}R^{\mu\nu}$, and the couplings to matter respectively in the gravity case.

The equation of motion

$$-\frac{1}{m^2}\Box(\Box+m^2)\varphi = j$$

yields a quartic propagator that may be decomposed in quadratic pieces, namely

$$D[k] = \frac{-m^2}{k^2(k^2-m^2)} = \frac{1}{k^2} - \frac{1}{k^2-m^2}$$

We identify here a massless physical particle and a massive "poltergeist". The problem is to find out a 2nd. order equivalent action for these physical and ghost fields.

As explained before one tries a Legendre transform. The obvious choice here is to define the canonical conjugate variable

$$\Pi = \frac{\partial\mathcal{L}}{\partial\Box\varphi}$$

from which

$$\Box\varphi = -m^2(\Pi + \frac{1}{2}\varphi)$$

Then the "Hamiltonian" is

$$\mathcal{H}[\varphi,\Pi] = -\frac{1}{2}m^2\Pi^2 - \frac{1}{2}m^2\Pi\varphi - \frac{1}{8}m^2\varphi^2 + j\varphi$$

Notice that, as it will be the normal case, it is a purely algebraic expression. The Helmholtz Lagrangian readily writes

$$\begin{aligned}\mathcal{L}_H[\varphi,\partial^2\varphi,\Pi] &= \Pi\Box\Phi - \mathcal{H}[\varphi,\Pi] \\ &= \Pi\Box\varphi + \frac{1}{2}m^2(\Pi^2 + \varphi\Pi + \frac{1}{4}\varphi^2) - j\varphi\end{aligned}$$

The "kinetic" and mass terms come mixed, so a diagonalization is needed. This step will turn out to be crucial and physically meaningful also in the gravity case. Now the suitable field redefinition is

$$\varphi = \tilde{\Pi} + \tilde{\varphi}$$

$$\Pi = \tfrac{1}{2}(\tilde{\Pi} - \tilde{\varphi})$$

which leads to the final transparent expression

$$\mathcal{L}_H = -\frac{1}{2}\tilde{\varphi}\Box\tilde{\varphi} + \frac{1}{2}\tilde{\Pi}(\Box + m^2)\tilde{\Pi} - j(\tilde{\Pi} + \tilde{\varphi})$$

as desired.This tells us that:

1. $\tilde{\varphi}$ is physical and massless with (quadratic) propagator $\frac{1}{k^2}$
2. $\tilde{\Pi}$ is a massive ghost with (quadratic) propagator $-\frac{1}{k^2-m^2}$
3. $\tilde{\varphi} + \tilde{\Pi}$ propagates with an effective (quartic) propagator $\frac{1}{k^2} - \frac{1}{k^2-m^2} = \frac{-m^2}{k^2(k^2-m^2)}$
4. j couples to $\tilde{\varphi} + \tilde{\Pi}$, i.e. the coupling $j\varphi$ unfolds into two interaction vertices, one involving the physical field and one with the poltergeist.

More lessons can be learned if there are gauge symmetries in the game. Work is in progress also on (HD) QED and (HD) YM.

3 Higher Order Gravity

Now we come back to the theory

$$\mathcal{L}_8[g_{\mu\nu}, R_{\mu\nu}] = \sqrt{-g}\{aR + bR^2 + cR_{\mu\nu}R^{\mu\nu}\}$$

If no global topological complications occur, namely if the Gauss-Bonnet identity holds and surface terms can be dropped, it is the most general expression of second order in the curvatures.

Following refs.[6] we now define the canonical conjugate variable

$$\Pi^{\mu\nu} = \frac{\partial \mathcal{L}}{\partial R_{\mu\nu}}$$

which is a tensor density with mass dimension $[\Pi^{\mu\nu}] = 2$ (recall that $[g_{\mu\nu}] = [b] = [c] = 0$ while $[a] = 2$). From it we can work out $R_{\mu\nu}$:

$$\begin{aligned} R_{\mu\nu} &= \mathcal{R}_{\mu\nu}[g, \Pi] \\ &= \frac{1}{2c\sqrt{-g}}[\Pi^{\alpha\beta} g_{\mu\alpha} g_{\nu\beta} - \frac{b}{4b+c}\Pi^{\alpha\beta} g_{\alpha\beta} g_{\mu\nu}] - \frac{1}{2}\frac{a}{4b+c} g_{\mu\nu} \end{aligned}$$

The Hamiltonian function is again an algebraic expression

$$\begin{aligned} \mathcal{H}[g, \Pi] &= \Pi^{\mu\nu}\mathcal{R}_{\mu\nu} - \mathcal{L}_8[g_{\mu\nu}, \mathcal{R}_{\mu\nu}] \\ &= \frac{1}{4c}\frac{1}{\sqrt{-g}}\Pi^{\mu\nu}\Pi^{\alpha\beta} g_{\mu\alpha} g_{\nu\beta} - \frac{1}{4c}\frac{b}{4b+c}\frac{1}{\sqrt{-g}}(\Pi^{\mu\nu} g_{\mu\nu})^2 \\ &\quad - \frac{1}{2}\frac{a}{4b+c}\Pi^{\mu\nu} g_{\mu\nu} + \frac{a^2}{4b+c}\sqrt{-g} \end{aligned}$$

and the Helmholtz LD Lagrangian becomes

$$\mathcal{L}_H[g, R_{\mu\nu}[g], \Pi] = \Pi^{\mu\nu} R_{\mu\nu}[g] - \mathcal{H}[g, \Pi]$$

where the fields $g_{\mu\nu}$ and $\Pi^{\mu\nu}$ occur mixed. This Lagrangian can be diagonalized by new fields $\tilde{g}_{\mu\nu}$ and $\tilde{\Pi}^{\mu\nu}$:

$$\begin{aligned} g_{\mu\nu} &= \tilde{g}_{\mu\nu} + \tilde{\Pi}_{\mu\nu} \\ \Pi^{\mu\nu} &= a\sqrt{-\tilde{g}}\,\tilde{g}^{\mu\nu} \end{aligned}$$

where $[\tilde{g}_{\mu\nu}] = [\tilde{\Pi}^{\mu\nu}] = 0$, $\tilde{g} \equiv det\, \tilde{g}^{\mu\nu}$, and $\tilde{g}^{\mu\nu}$ is the inverse of $\tilde{g}_{\mu\nu}$.

Objets like $g^{\mu\nu}$,$\sqrt{-g}$ and $R_{\mu\nu}[g]$ can be expanded around $\tilde{g}_{\mu\nu}$. After dropping total derivatives, a careful calculation gives:

$$\begin{aligned} \mathcal{L}_8 &= a\sqrt{-\tilde{g}}\{R[\tilde{g}_{\mu\nu}] + \tilde{g}^{\mu\nu}(\Delta^{\alpha}_{\alpha\beta}\,\Delta^{\beta}_{\mu\nu} - \Delta^{\alpha}_{\mu\beta}\,\Delta^{\beta}_{\alpha\nu})\} \\ &\quad - \frac{\sqrt{-\tilde{g}}}{4c(4b+c)}\{a^2(4b+c)(\tilde{\Pi}^{\mu\nu}\tilde{\Pi}_{\mu\nu} - \tilde{\Pi}^2) + a^2(3b+c)\tilde{\Pi}^2 + O(\tilde{\Pi}^3)\} \end{aligned}$$

where

$$\Delta^{\alpha}_{\mu\nu} = \frac{1}{2} g^{\alpha\rho}(\tilde{D}_\mu\tilde{\Pi}_{\nu\rho} + \tilde{D}_\nu\tilde{\Pi}_{\mu\rho} - \tilde{D}_\rho\tilde{\Pi}_{\mu\nu})$$

$\tilde{D}_\mu$ is the covariant derivative for the metric $\tilde{g}_{\mu\nu}$, the indices are rised and lowered by $\tilde{g}_{\mu\nu}$, and $\tilde{\Pi} = \tilde{\Pi}^{\mu\nu} g_{\mu\nu}$. In this final expression one readily sees that $\tilde{g}_{\mu\nu}$ is a graviton, thus accounting for the two d.o.f. of a massless spin-two field. The term $a\sqrt{-\tilde{g}}\tilde{g}^{\mu\nu}\,\Delta^{\alpha}_{[\alpha\beta}\,\Delta^{\beta}_{\mu]\nu}$ contains (by linearizing $\tilde{g}_{\mu\nu} = \eta_{\mu\nu} + h_{\mu\nu}$) a Fierz-Pauli kinetic term for $\tilde{\Pi}_{\mu\nu}$. Therefore if 3b+c=0, $\tilde{\Pi}_{\mu\nu}$ describes a massive spin-two field accounting for five d.o.f. If instead 3b+c $\neq 0$, it contains besides a scalar field, namely $\tilde{\Pi}$, totalling six d.o.f. [7].

In conclusion, we have obtained a diagonalized LD equivalent theory for the eight d.o.f. of the original HD theory. The result shows that if the parameters are such that 3b+c=0, the actual d.o.f. then reduce to seven. Many other points like the conformal limit of the theory, the detailed structure of the propagators, the interaction of these fields with matter and with themselves, the gauge fixation, etc. will be developed elsewhere.

Aknowledgements

One of us (J.J.) wishes to thank J.Pérez Mercader for helpful discussions and for having brought him again to the field of higher derivative theories of gravity.

References

1. K.S.Stelle,Phys.Rev. **D16**(1977)953.
2. I.G.Avramidi,A.O.Barvinsky,Phys.Lett. **B159**(1985)269.
   J.Julve,M.Tonin,N.Cim. **46B**(1978)137.
3. T.Goldman,J.Pérez-Mercader,F.Cooper,M.M.Nieto, Phys.Lett. **B281**(1992)219.
4. B.Whitt,Phys.Lett. **B145**(1984)176.
   J.D.Barrow, S.Cotsakis,Phys.Lett. **B214**(1988)515.
5. N.H.Barth,S.M.Christensen,Phys.Rev. **D28**(1983)1876.
6. G.Magnano,M.Ferraris,M.Francaviglia,Gen.Rel.Grav. **19**(1987)465.
   A.Jakubiec,J.Kijowski,Phys.Rev. **D37**(1988)1406.
   G.Magnano,M.Ferraris,M.Francaviglia,J.Math.Phys. **31**(1990)378.
7. H.Van Dam,M.Veltman,Nucl.Phys. **B22**(1970)397.
   D.G.Boulware,S.Deser,Phys.Rev. **D6**(1972)3368.

EMERGENCE OF CLASSICAL SPACETIME

CLAUS KIEFER[1]
Institut für Theoretische Physik der Universität Zürich,
Schönberggasse 9, CH-8001 Zürich, Switzerland

1. Introduction

Quantum theory is an extremely successful theory. It explains phenomena ranging from the stability of white dwarfs to the decay properties of radioactive nuclei. With the exception of gravity, it is known to describe all known interactions, and attempts are being made to construct a unified theory which also encompasses gravity within a quantum framework. Up to now there is no hint, neither from experiment nor theory, which casts doubt on the *universality of quantum theory.* This, however, leads to a conceptual problem. If quantum theory is assumed to hold universally, the *superposition principle* should always apply, i.e., the superposition of physically allowed quantum states should again be an allowed state – independent of the micro- or macroscopic nature of the system. Only a tiny fraction of possible states corresponds, however, to states which one might call "classical." Nevertheless, only classical states are observed in the macroscopic world. How can this apparent contradiction be resolved?

The key point to its resolution is the observation that physical systems are almost never found in an isolated state – they are coupled to external degrees of freedom, the "environment" of the system[1]. If the coupling of the environment to the system is strong enough, information about phase relations may migrate from the system into its environment and can no longer be observed in the system itself. This process is referred to as *decoherence.* In realistic cases the environment contains a huge number of degrees of freedom in which case the Poincaré recurrence time is so long that the information about phase relations never returns to the system. This leads to an effective *irreversibility* in the system if a special initial condition has been chosen. An illustrative example is the case of a dust grain in interaction with air molecules and light[2]. Even if one started with a superposition of the dust grain being in different positions (such superpositions can be generated by Schrödinger cat-like experiments), it can never be observed since the environment dislocalises the information about interference terms within a fraction of a second and therefore prevents the wave packet from spreading. Formally, the behaviour of the system is described by a density matrix ρ_S which can be found from the density matrix ρ_{tot} of the total system by "tracing out" the environment $\mathcal{E}$,

$$\rho_S = \mathrm{Trace}_{\mathcal{E}}\,\rho_{tot}. \qquad (1)$$

The reduced density matrix ρ_S evolves according to a nonunitary master equation instead of a von Neumann equation. This is the formal reason why superpositions "disappear" from the system.

---

[1]Supported by the Swiss National Science Foundation

It has to be emphasized that the process of decoherence is not equivalent to a "collapse of the wave function." If the total state was initially pure, it will remain pure with all interference terms still present. In realistic cases it is, however, impossible to construct an apparatus which can measure these interference terms[3].

Formally, decoherence occurs if the nondiagonal elements of the reduced density matrix (1) become small in a certain basis. It is important to note that this does not correspond to just diagonalize ρ_S (which can always be done) but to the fact that the coupling to the environment *selects* a preferred basis of states with respect to which nondiagonal terms become small[4]. Superpositions of such states will rapidly decohere. For the harmonic oscillator, for example, it has been shown that these preferred states are the coherent states[4,5].

Since the environment of a system is again coupled to an environment, and so on, one will have eventually to take into account the whole Universe. The dominating force on a cosmological range is gravity, and one is thus naturally led to consider a theory of quantum gravity. How, then, can one understand that space and time behave classically to such a high degree of accuracy? As there is no system external to the Universe, traditional "explanations" which invoke classical observers from the outset do not work. There is thus a strong need to understand the emergence of a classical world intrinsically. How this can be achieved, at least in principle, was first shown in Ref. 6 and then elaborated on from various points of view (see Ref. 7 for a brief review of recent work). It is also the subject of this talk. We first formulate the problem and discuss the formal framework within which it is discussed. We then turn to a similar problem in QED where the conceptual points are well understood and which thus may serve as a guide to an understanding of the analogous problem in quantum gravity. This is discussed in the last section within the context of a simple model.

2. Quantum gravity and the classical world

Although there does not yet exist a consistent quantum theory of gravity, promising attempts are made within the framework of canonically quantising general relativity. The central equation in these approaches reads

$$H\Psi = 0, \tag{2}$$

where H is the total Hamiltonian of all gravitational and non-gravitational fields. In most approaches, Ψ depends – apart from non-gravitational fields – on the metric h_{ab} of a three-manifold Σ (in which case (2) is referred to as the Wheeler-DeWitt equation[8]) or on a complex $SO(3)$ connection which is constructed from both the metric and the extrinsic curvature of Σ[9]. In the following we restrict ourselves to the first, the geometrodynamical, formulation (a discussion of the semiclassical approximation in the connection representation can be found in Ref. 10).

An important property of (2) is the absence of any time parameter. How, then, can one recover from this equation, at least approximately, the familiar notion of time? Formally, this can be achieved by performing in (2) an expansion with respect to the Planck mass[11]. One can then write Ψ approximately as a superposition

$$\Psi \approx \sum_k A_k e^{iS_k[h_{ab}]} \chi_k[h_{ab}, \phi], \tag{3}$$

where S_k depends only on the three-metric and obeys the gravitational Hamilton-Jacobi equation. Each S_k corresponds, in the classical limit, to a family of classical trajectories. The states χ_k depend in addition on the matter field ϕ and satisfy an approximate Schrödinger equation,

$$i\frac{\partial}{\partial t_k}\chi_k = H_M \chi_k, \tag{4}$$

where H_M is the matter Hamiltonian, and

$$\frac{\partial}{\partial t_k} \equiv \nabla S_k \nabla. \tag{5}$$

Thus, t_k ("WKB time") is a parameter labelling classical trajectories in configuration space which run orthogonally to hypersurfaces $S_k = constant$.

There is, however, a problem. Eqs. (4) and (5) hold separately for each component of (3) but *not* for a superposition[12]. One thus has to understand why the various components in (3) can be treated independently and, moreover, why WKB time becomes a classical quantity within one component. Before we discuss how decoherence may solve this problem, we discuss the analogous problem in QED.

3. Decoherence in QED

In the following we show how superpositions between macroscopically different electromagnetic field strenghts can be suppressed by interaction with charged (scalar) matter fields. To make an analogy with the gravitational case (3), we treat the electromagnetic part semiclassically and the matter part fully quantum,

$$\Psi \approx \sum_k C_k[\mathbf{A}] e^{iS_k[\mathbf{A}]} \chi_k[\phi, \phi^+, \mathbf{A}], \tag{6}$$

where S_k obeys the electromagnetic Hamilton-Jacobi equation, and χ_k obeys the approximate Schrödinger equation

$$i\int d^3x \frac{\delta S_k}{\delta \mathbf{A}}\frac{\delta \chi}{\delta \mathbf{A}} \equiv i\frac{\partial \chi_k}{\partial t_k} \approx H_\phi \chi_k. \tag{7}$$

Here, H_ϕ is the total Hamiltonian of QED without the pure electromagnetic part ($\mathbf{A}$ is the vector potential). Eq. (7) can be obtained from the full Schrödinger equation for QED by an expansion with respect to the electric charge[13]. It is important to note that time emerges here "intrinsically" as a parameter in the configuration space of all vector potentials – even if the total state is a stationary one. This is analogous to the simple toy model of Page and Wootters[14]. In QED we have, however, the external Minkowski time available which allows us to *identify* WKB time, in this order of approximation, with Minkowski time (recall the recovery of time from the timeless Jacobi principle in mechanics). In gravity, however, there is no external time and (4), (5) describe the recovery of "everyday's time" as an approximate concept.

Let us be more specific and consider the example where only one component in (6) is present, and the corresponding S is given by

$$S \approx -V\mathbf{EA}, \tag{8}$$

where V is the space volume, and $\mathbf{E}$ is the electric field. We assume that the electromagnetic field is approximately constant in space. Note the analogy of e^{iS} to the momentum eigenstate e^{ipx} in quantum mechanics. For the matter state χ we take a Gaussian functional, using the momentum representation for the scalar fields,

$$\chi = N(t)\exp\left(-\int d^3p\phi^+(\mathbf{p})\Omega(\mathbf{p},t)\phi(\mathbf{p})\right). \tag{9}$$

Inserting this ansatz into (7) one obtains equations for N and Ω which can be explicitly solved. A special application is the derivation of Schwinger's pair creation formula in this framework[13].

If one makes observations on the electromagnetic field only, one has to trace out the scalar field in the total state and to discuss the reduced density matrix[15]

$$\rho(\mathbf{A},\mathbf{A}') = \mathrm{Trace}_{\phi,\phi^+}|\Psi\rangle\langle\Psi| \tag{10}$$

or, alternatively, its Fourier transform $\tilde{\rho}(\mathbf{E},\mathbf{E}')$. For times $T \gg m/eE$ one finds[15]

$$\tilde{\rho} \approx \tilde{\rho}_0 \exp\left(-\frac{8\pi^2 V}{e^3\bar{E}T}(E-E')^2\right), \tag{11}$$

where we have taken the electric field in, say, the z-direction, $\mathbf{E} = E\mathbf{e_z}$, $\bar{E} \equiv (E+E')/2$, and ρ_0 would be the density matrix if there were no scalar field. The electric field can thus be considered as a classical quantity if the width of the Gaussian in (11) is much smaller than the mean value $\bar{E}$ of the field itself. The quantum region is thus the weak field region, a result, which is consistent with the analysis of the measurement of electromagnetic field quantities in Ref. 16. It is also interesting to note that the nondiagonal elements in (11) vanish exactly in the infrared limit $V \to \infty$.

Consider now a superposition of two states

$$\Psi \approx e^{-iVEA}\chi + e^{iVEA}\chi^*, \tag{12}$$

i.e., a (possibly macroscopic) superposition of an electric field pointing "upwards" with one pointing "downwards". We now show that this superposition, in case that it has been generated, becomes suppressed through the correlation with the scalar field state in (12). Assuming that E has a definite value within one component of (12) (this has been discussed above), one can focus on the nondiagonal element $\rho_{\pm}(\mathrm{A})$ of the reduced density matrix along the diagonal in **A**-space which describes interferences between the two components in (12). One obtains[15]

$$\begin{aligned} \rho_{\pm}(A) &= \rho_0 \mathrm{Trace}_{\phi,\phi^+} \chi^2 \\ &= \rho_0 \det \frac{Re(\Omega)}{\Omega} \approx \rho_0 \exp\left(-\frac{Ve^2E^2}{512\pi m}\right). \end{aligned} \tag{13}$$

Again, this vanishes in the infrared limit $V \rightarrow \infty$ where it gives rise to an exact superselection rule. One can estimate from (13) that for, say, $E \approx 10^7 Volts/cm$ interference effects may become noticeable on length scales smaller than about $10^{-4} cm$.

The effect of the "vacuum fluctuations" discussed here should be considered as a lower limit on decoherence. In realistic situations, real charges are present and suppress interference terms even more. In principle, however, the effect of decoherence (13) should be measurable. A superposition like (12) can be generated, for example, by coupling an atomic device to a capacitor. By measuring expectation values of an observable, $\mathcal{O}$, according to $<\mathcal{O}>= \mathrm{Tr}(\rho\mathcal{O})$, the density matrix (13) becomes distinguishable from an ensemble of field up and down configurations.

We conclude this section with two comments.

(1) The above derivation breaks down if the kinetic term for the electromagnetic field (which was neglected here) becomes important. This means that the exact "pointer states" are not **A** or **E** eigenstates but some generalised coherent states in analogy to the quantum mechanical case[4,5].

(2) One can formulate the whole discussion in the notion of *decoherent histories*[3,17], although this is somewhat trivial in the present example. The histories coincide, for each component of the state (6), with the classical trajectories running orthogonally to $S_k = constant$. These histories become consistent histories (i. e., the corresponding interference terms are negligible) through the interaction with the charged scalar fields, as discussed above.

4. Decoherence in quantum gravity

In the above examples the division of the total system into a relevant subsystem and its environment is obvious. This is no longer the case in the realm of quantum cosmology where the whole Universe is described in quantum terms. It has been suggested in Ref. 6 that the relevant degrees of freedom are global degrees like the radius a of the Universe and the homogeneous mode ϕ of a scalar field while the role of the irrelevant degrees of freedom are played by inhomogeneous perturbations describing density perturbations or gravitational waves. Technically, this is discussed through expanding a general three-metric and scalar field into harmonics on the three-sphere (which is the choice for our three-manifold). In the limit where the inhomogeneous degrees of freedom (symbolically denoted by x_n) have small amplitudes, the total state reads, in analogy to (3),

$$\Psi \approx \sum_k C_k(a,\phi)e^{iS_k(a,\phi)} \prod_n \chi_{k,n}(a,\phi,x_n). \tag{14}$$

The only difference to (3) is that the semiclassical part also contains the homogeneous matter mode ϕ while the pure quantum part contains part of the gravitational degrees of freedom. The states $\chi_{k,n}$ (k fixed) obey, in analogy to (4), a Schrödinger equation with respect to the Hamiltonian of each degree of freedom x_n, and the WKB time is defined through S_k. If a and ϕ are the relevant degrees of freedom, all physical information has to be calculated from the reduced density matrix

$$\rho(a,\phi;a',\phi') = \text{Trace}_{\{x_n\}}|\Psi\rangle\langle\Psi|. \tag{15}$$

Taking, for example, for the "environmental" modes the gravitational wave modes only (assumed to be, like (9), in a Gaussian state), one obtains[6]

$$\rho(a,\phi;a',\phi') \approx \rho_0 \exp\left(-\frac{N^3}{6a^2}(a-a')^2 - 3m^2N(a\phi - a'\phi')^2\right), \tag{16}$$

where m is the mass of the scalar field, and N denotes a cutoff in the number of degrees of freedom x_n.

Two comments are in order.

(1) It is recognizable from Eq. (16) that decoherence is stronger for a than for ϕ. In fact, decoherence for a is a necessary requirement for the decoherence of ϕ. This is also the case with other models. The interior degree of freedom in a Kaluza-Klein model, for example, can only decohere if a can already be assumed to be classical[18]. The fact that gravitational degrees of freedom become more classical than matter degrees of freedom is, of course, a direct consequence of the *universal coupling* of gravity to all fields in Nature. This leads to quantum states which strongly correlate gravitational with non-gravitational degrees of freedom.

(2) It is not clear what one should take for the cutoff number N in (16) since this is connected with the ill-understood ultraviolet behaviour of quantum gravity. It has been shown, e. g., that dimensional regularisation is impossible for the decoherence factor[19]. One possible proposal is to consider only modes with wavelength bigger than the Planck scale[20]. Alternatively, one can make a field redefinition[15] which renders the decoherence factor finite. This is of course ad hoc but the motivation is to make all expressions similar to QED where the answer is unambiguous. In this case one obtains for the decoherence of a (neglecting the homogeneous mode of the scalar field for simplicity)

$$\rho(a,a') \approx \rho_0 \exp\left(-\frac{\pi m^3 a}{32}(a-a')^2\right). \tag{17}$$

As the width of this Gaussian decreases with increasing a, the Universe becomes more classical with increasing scales – a result which one would expect in a realistic model. Consider now, in analogy to (12), a superposition of the form

$$\Psi \approx e^{iS}\prod_n \chi_n + e^{-iS}\prod_n \chi_n^*. \tag{18}$$

Integrating out the inhomogeneous modes of the scalar field, one obtains for the nondiagonal element $\rho_\pm(a)$ of the reduced density matrix describing interferences between the two components in (18) (restricting to the diagonal in a space)

$$\rho_\pm(a) \approx \rho_0 \exp\left(-\frac{\pi m H_0^2 a^3}{128}\right), \tag{19}$$

where H_0 denotes the Hubble parameter corresponding to the semiclassical a-evolution. Inserting the values $m \approx 100\ GeV$, $H_0 \approx 55 km/(secMpc)$ and $a \approx H_0^{-1}$, one finds that the exponential factor in (19) is $\exp(-10^{43}) \approx 0$ so that for the present size of the Universe decoherence is very efficient. Decoherence is insufficient if this factor becomes approximately equal to one. This happens if either a or H_0 is small, the first case corresponding to small scales, the second case to the region near the classical turning point. It is also equal to one if the mass of the scalar field vanishes. This is possible since the mass enters the decoherence factor inversely compared to QED (see (13)), a consequence of the different coupling terms in the Hamiltonians.

While one would expect decoherence to be inefficient near the Planck scale, its inefficiency near the region of the classical turning point is somewhat surprising. It would mean that macroscopic, even cosmological, effects of quantum gravity can occur. Whether this effect is only an artifact of this simple model or whether it will survive also in more realistic models, can only be decided by future investigations. That quantum effects might become important near the turning point has been concluded, from a different point of view, through the discussion of wave packets[21] as well as the consequences of a simple initial condition on the quantum state[22,23]. Such quantum effects would be very important for a deeper understanding of the arrow of time[22].

5. Acknowledgements

I would like to thank the organisers of the *First Iberian Meeting on Gravity* for organising such an interesting and stimulating meeting.

6. References

1. H. D. Zeh, Found. Phys. **1**, 69 (1970).
2. E. Joos and H. D. Zeh, Z. Phys. **B59**, 223 (1985).
3. R. Omnès, Rev. Mod. Phys. **64**, 339 (1992).

4. W. H. Zurek, in *Proceedings of the NATO workshop on Physical Origins of Time Asymmetry*, edited by J. Halliwell, J. Perez-Mercader, and W. H. Zurek (Cambridge University Press 1992).

5. O. Kübler and H. D. Zeh, Ann. Phys. (N. Y.) **76**, 405 (1973).

6. H. D. Zeh, Phys. Lett. **A116**, 9 (1986); C. Kiefer, Class. Quantum Grav. **4**, 1369 (1987).

7. C. Kiefer, in *Proceedings of the 10th Seminar on Relativistic Astrophysics and Gravitation*, Potsdam 1991, edited by S. Gottlöber, J. P. Mücket, and V. Müller (World Scientific 1992).

8. B. S. DeWitt, Phys. Rev. **160**, 1113 (1967); J. A. Wheeler, in *Battelle Rencontres*, edited by C. DeWitt and J. A. Wheeler (Benjamin, New York, 1968).

9. A. Ashtekar, *Lectures on non-perturbative canonical gravity* (World Scientific, Singapore, 1991).

10. C. Kiefer, *Topology, Decoherence, and Semiclassical Gravity*, preprint (1993).

11. T. Banks, Nucl. Phys. **B249**, 332 (1985).

12. J. B. Barbour, *A problem with recovery of the time-dependent Schrödinger equation in quantum cosmology*, preprint (1991).

13. C. Kiefer, T. Padmanabhan, and T. P. Singh, Class. Quantum Grav. **8**, L185 (1991); C. Kiefer, Phys. Rev. D **45**, 2044 (1992).

14. D. N. Page and W. K. Wootters, Phy. Rev. D **27**, 2885 (1983).

15. C. Kiefer, Phys. Rev. D **46**, 1658 (1992).

16. N. Bohr and L. Rosenfeld, Det. Kgl. dansk. Vid. Selskab. **XII**, 8 (1933).

17. M. Gell-Mann and J. B. Hartle, in *Complexity, Entropy and the Physics of Information*, ed. by W. H. Zurek (Addison-Wesley 1990).

18. F. Mellor, Nucl. Phys. **B353**, 291 (1991).

19. J. P. Paz and S. Sinha, Phys. Rev. D **45**, 2823 (1992).

20. C. Kiefer, Class. Quantum Grav. **6**, 561 (1989).

21. C. Kiefer, Phys. Rev. D **38**, 1761 (1988).

22. H. D. Zeh, *The physical basis of the direction of time* (Springer, Berlin, 1992).

23. H. D. Conradi and H. D. Zeh, Phys. Lett. **A154**, 321 (1991); H. D. Conradi, Phys. Rev. D **46**, 612 (1992).

Loop spaces and path integral quantization

R. F. Picken

Dep. de Matemática, Instituto Superior Técnico,
Av. Rovisco Pais, 1096 Lisboa Codex, Portugal
and
Centro de Física da Matéria Condensada,
Av. Prof. Gama Pinto 2, 1699 Lisboa Codex, Portugal

Abstract

As a first step towards understanding path integrals in terms of the integration theory and differential geometry of path and loop spaces, a special model, quantum mechanics on a compact Lie group G, is analysed. The path integral for the propagator is shown to equal an integral over ΩG, the space of based loops on G. Geometrical features of ΩG are discussed, and a possible method of calculating the path integral via a formal application of the Duistermaat-Heckman integration formula is described.

1. Introduction

Infinite-dimensional spaces, in particular spaces of functions, are starting to play an increasingly important role in the mathematical description of a variety of physical theories. A prominent class of such spaces are so-called loop spaces, i.e. spaces of continuous maps from the circle to a smooth manifold, possibly satisfying some additional analytic constaints. Loop spaces are an essential ingredient in the non-local descriptions of gauge theory and gravity[1], but also crop up in fields as diverse as string theory (closed strings) and hydrodynamics (loops of vorticity).

In this talk, however, we want to regard loop spaces as a a special case of path spaces in the sense of the Feynman path integral. We recall briefly the fundamental statement concerning the Feynman path integral, restricting ourselves to quantum mechanics for simplicity. Let M be a smooth manifold representing the configuration space of a quantum mechanical system with Hamiltonian H, and let q, q' be two points of M. Then the kernel for the Schrodinger equation, or propagator from an initial position q at time $t = 0$ to a final position q' at time $t = t_1$, denoted $< q, 0|q', t_1 >$ (in the Schrodinger picture) may be represented as a path integral:

$$< q,0|q',t_1 >= \mathcal{N} \int_{\mathcal{P}_{q,q';[0,t_1]}M} d\mu[\gamma] \exp(iS[\gamma]/\hbar) \tag{1}$$

where $\mathcal{N}$ is a normalization constant, $\mathcal{P}_{q,q',[0,t_1]}M$ is a space of continuous paths γ from $[0,t_1] \subset \mathbf{R}$ to M such that $\gamma(0) = q$, $\gamma(t_1) = q'$, $d\mu[\gamma]$ is a measure on this space of paths, and $S[\gamma]$ is the classical action associated with the path γ.

Because of the difficulties in interpreting path integrals mathematically, it seems worthwhile to develop as many perspectives as possible on them. Our philosophy is to regard the right-hand-side of (1) as a geometric integral, to be understood in terms of the infinite-dimensional geometry and geometrical integration theory of path spaces.

In fact, as a first step in this direction, we will mainly concentrate on a special case, namely when the configuration manifold M is a compact Lie group G. The motivation for this choice is that there is a known answer to compare with, which is obtained by canonical methods, under natural assumptions about group invariance. Furthermore the extra group structure helps to simplify the analysis considerably, as will be seen later on, which is advantageous when one is still at the stage of testing ideas.

2. The path integral for the propagator

The model to be considered then is that of a quantum particle constrained to move on a (compact, simple and semi-simple) Lie group G. The central object of interest for this system is the propagator from an initial position g at time $t = 0$ to a final position g' at time $t = t_1$. Under the assmption of invariance of the propagator under the left or right action of G on itself, it is enough to consider propagators of the form

$$K(\phi,t_1) \stackrel{\text{def}}{=} < 1,0|e^{i\phi},t_1 > \tag{2}$$

where 1 is the identity element of G, and $e^{i\phi}$ belongs to some fixed maximal Abelian subgroup, or torus for short, T of G. To avoid the problem that the exponential map $\phi \mapsto e^{i\phi}$ is many-to-one we choose ϕ to be such that the map $\gamma_0 : [0,1] \to T,\ t \mapsto e^{i\phi t}$ is the shortest geodesic between 1 and $e^{i\phi}$ with respect to the natural bi-invariant metric on G (we will exclude the possibility that 1 and $e^{i\phi}$ are conjugate points).

We choose an action which is invariant under the obvious left and right action of G on paths

$$S[\gamma] = (-1/2) \int_0^{t_1} dt < \gamma^{-1}\dot{\gamma}, \gamma^{-1}\dot{\gamma} > \tag{3}$$

for paths $\gamma : [0,t_1] \to G$ starting at 1 and ending at $e^{i\phi}$. Here $<,>$ is the Cartan-Killing bracket on g, the Lie algebra of G. After a rescaling of the time variable the action reads

$$S_{t_1}[\gamma] = (-1/t_1)\int_0^1 dt < \gamma^{-1}\dot{\gamma}, \gamma^{-1}\dot{\gamma} > \tag{4}$$

for paths $\gamma : [0,1] \to G$ such that $\gamma(0) = 1$, $\gamma(1) = e^{i\phi}$.

One could now write down an expression like (1) for the propagator as an integral over all paths γ, but in fact it is possible to use the group structure on G to change variable in the path integral as follows: any path γ may be written as the product $\gamma = \gamma_0 g$, where γ_0 is as before and g satisfies $g(0) = g(1) = 1$. This corresponds to factorising a generic path γ into a fixed reference path γ_0 and a closed loop g on G, starting and finishing at 1. Thus g is an element of the based loop group of G, denoted ΩG ("based" since the loops all start and finish at the same point of the target manifold, and "group" since the space of loops based at the identity of G has the structure of a group under pointwise multiplication of loops).

In terms of the new variable g the action becomes

$$S_{\phi,t_1}[g] = (-1/t_1)\int_0^1 dt < g^{-1}\dot{g} + g^{-1}i\phi g\, ,\, g^{-1}\dot{g} + g^{-1}i\phi g > \tag{5}$$

and the path integral expression (1) for $K(\phi, t_1)$ reads

$$K(\phi, t_1) = \mathcal{N}\int_{\Omega G} d\mu[g] \exp\left(iS_{\phi,t_1}[g]/\hbar\right) \tag{6}$$

where $\mathcal{N}$ is a normalisation constant and $d\mu[g]$ is some measure on ΩG. Note that, as a result of the above manipulations, the two specific features of the problem, ϕ and t_1, have been eliminated from the space of paths being integrated over and transferred inside the integral to become part of the integrand, leaving the "universal" problem of studying the space ΩG and integrals of suitable functions over ΩG.

3. Geometrical features of ΩG

ΩG, the group of smooth loops based at the identity of G, can be equipped with natural geometric structures in such a way as to make it into an infinite-dimensional homogeneous Kähler manifold. In other words ΩG admits a complex structure J and a compatible symplectic structure ω, which are invariant under the action of an infinite-dimensional group (containing ΩG itself) on ΩG.[2,3] Because of the group invariance it is enough to describe these structures at the identity loop **1** of ΩG (satisfying $\mathbf{1}(t) = 1$, $\forall t \in [0,1]$). Identifying the tangent space of ΩG at **1** with $L\mathbf{g}/\mathbf{g}$, the space of smooth loops on $\mathbf{g}$ modulo constant loops, and expressing $X \in L\mathbf{g}/\mathbf{g}$ as

$$X(t) = \sum_{\substack{n=-\infty \\ n\neq 0}}^{\infty} X_n e^{2\pi i n t}, \tag{7}$$

with $X_n \in \mathfrak{g}_c$, the complexified Lie algebra, satisfying $X_n^\dagger = X_{-n}$, the complex structure J is given by

$$(J_1X)(t) = \sum_{n>0} iX_n e^{2\pi int} - \sum_{n<0} iX_n e^{-2\pi int} \tag{8}$$

and the symplectic structure ω by

$$\omega_1(X,Y) = \int_0^1 dt < X(t), \dot{Y}(t) > \tag{9}$$

Notice that the antisymmetry of ω in its arguments follows from partial integration and its non-degeneracy is due to the absence of a constant term in the representation (7).

The fact that ΩG is homogeneous Kähler is not too surprising when one bears in mind that ΩG is an infinite-dimensional analogue of finite-dimensional "flag manifolds". These are homogeneous spaces of the form G/T or $G/C(T_0)$, where T is a torus of the compact, finite-dimensional Lie group G, and $C(T_0)$ is the centraliser of a subtorus T_0 of T in G. Flag manifolds are all homogeneous Kähler, as well as possessing a number of other interesting features.[4] As indicated, ΩG can be shown to be of flag manifold form[3], $\Omega G \cong \mathcal{G}/C(\mathcal{T}_0)$, where $\mathcal{G}$ is the semi-direct product $U(1)\times LG$, with product given by $(e^{ia}, f(.))(e^{ib}, g(.)) = (e^{i(a+b)}, f(.)g(a+.))$, its torus is $U(1) \times T$, where T stands for the constant loops in the torus T of G, and the centraliser of the subtorus $\mathcal{T}_0 = U(1) \times 1$ is $U(1) \times G$.

4. The geometric path integral on ΩG

In view of the analogy between ΩG and flag manifolds, discussed in the previous section, it seems appropriate to look first at integration problems on flag manifolds. There is in fact an important class of such integrals, which can be solved by using a powerful result, going under the name of the Duistermaat-Heckman (DH) integration formula.[5] Here we give a version of their theorem due to Berline and Vergne[6], which is geared towards symplectic geometry:

Theorem Let M be a smooth compact symplectic manifold of dimension $2n$, and let H be a real function on M with only isolated critical points $\{m_k\}_{k\in I}$ for some index set I and whose corresponding Hamiltonian vector field ξ (given by $dH + \iota_\xi\omega = 0$) is associated with a circle action on M. Then

$$(1/2\pi i)^n \int_M e^{iH}\omega^n/n! = \sum_{k\in I} e^{iH(m_k)}/\mathrm{Pf}\, J_\xi(m_k) \tag{10}$$

where $J_\xi(m_k) = -\mathcal{L}_\xi(m_k)$, with $\mathcal{L}$ denoting the Lie derivative, is a linear skew-symmetric automorphism of $T_{m_k}M$ and Pf denotes the Pfaffian.

Some comments are in order (a more detailed discussion can be found in Picken[4,7]):

1. The factor $1/\mathrm{Pf}\, J_\xi(m_k)$ which appears on the right-hand-side of the formula (18) is a kind of winding number of the Hamiltonian vector field of H around its zero at m_k.

2. The DH result is often loosely described as saying that the "semiclassical approximation is exact". However a more useful way of thinking about it is as an example of "localisation" to the critical points (see the fourth remark).

3. The simplest example of the DH formula is provided by the choice $(M,\omega) = (S^2, \sin\theta\, d\theta \wedge d\phi)$, $H = \cos\theta$. H corresponds to the circle action of rotating the angle ϕ and has two critical points at $\theta = 0, \pi$. Absorbing the factors of i, the formula (18) corresponds to

$$\int_{S^2} e^{\cos\theta} \sin\theta d\theta \wedge d\phi = 2\pi(e - e^{-1}) \tag{11}$$

The right-hand-side is given by a sum over the critical points, and the minus sign in front of the second term on the right-hand-side derives from the opposite sense of the winding of the Hamiltonian vector field associated to H around the two critical points.

4. The idea of the Berline-Vergne proof is straightforward: excise from M balls of radius ϵ centred on each critical point of H, and show that the integrand is exact on the remaining manifold with boundary. Then use Stokes' theorem and let the radius ϵ tend to zero.

We return now to the expression (6) for the propagator $K(\phi, t_1)$ as an integral over ΩG. In fact, we know what the result of this integration should be from canonical quantization and spectral theory of the Laplacian on G,[8] namely

$$K(\phi, t_1) = \sum_{\nu \in \check{T}} p(\nu) \exp i S_{\phi, t_1}[g_\nu]/\hbar \tag{12}$$

where $\check{T}$, the coweight lattice, consists of elements ν of the torus Lie algebra satisfying $\exp 2\pi i\nu = 1$, $p(\nu)$ is a complicated group-theoretical prefactor, which we do not give explicitly here, and $g_\nu : [0,1] \to T$, $t \mapsto e^{2\pi i \nu t}$ is the closed geodesic corresponding to ν.

Since g_ν is a critical point of the action S_{ϕ,t_1}, the expression (12) has a similar structure to the right-hand-side of the DH formula (10). Although the DH formula has not yet been extended to infinite dimensions, one can still try to apply it formally to the path integral expression for $K(\phi, t_1)$. After some work to give meaning to the infinite-dimensional symplectic volume element (formally $\omega^\infty/\infty!$) and the factor $\mathrm{Pf}^{-1} J_\xi$, one finds that the DH formula does indeed reproduce the main features of the exact expression (12), modulo some mild regularization of the infinite-dimensional determinants which appear. For further details see Picken[7].

We conclude with a few remarks. The result described is encouraging evidence for the existence of a geometric integration theory in infinite dimensions suitable for calculating path integrals. It may be possible to extend the method based on the DH formula to other special models such as Chern-Simons theory, where there are indications of the exactness of the semiclassical approximation. One might be able to exploit the hint of supersymmetry present in the DH formula (summarized by $\int_{M_{2n}} e^H \omega^n/n! = \int_{M_{2n}} e^{H+\omega}$ and $(d + \iota_\xi)(H + \omega) = 0$, where $d + \iota_\xi$ is interpreted as the supersymmetry operator which annihilates the supersymmetric Hamiltonian $H + \omega$.[9] Although the subject of this talk is fairly far removed from the main theme of this Meeting, there is still a link in that path integrals and "geometry" are important concepts in gravity theory too.

References

[1] A. Ashtekar and C. J. Isham, *Class. Qu. Grav.* **9** (1992) 1069

[2] A. N. Pressley and G. Segal, *Loop Groups*, Oxford University Press 1988

[3] D. S. Freed, in *Infinite Dimensional Groups with Applications (MSRI Publications 4)*, V. Kac (Ed.), Springer, 1985

[4] R. F. Picken, *J. Math. Phys.* **31** (1990) 616

[5] J. J. Duistermaat and G. J. Heckman, *Invent. Math.* **69** (1982) 259, **72** (1983) 539

[6] N. Berline and M. Vergne, *Duke Math. J.* **50** (1983) 539

[7] R. F. Picken, *J. Phys. A: Math. Gen.* **22** (1989) 2285

[8] M. S. Marinov and M. V. Terentiev, *Fortschr. Phys.* **27** (1979) 511

[9] M. Blau, E. Keski-Vakkuri and A. Niemi, *Phys. Lett.* **B246** (1990) 92

ON MINISUPERSPACE GAUGE MODELS WITH FERMIONS

Sergey V. SHABANOV
Laboratory of Theoretical Physics, Joint Institute for Nuclear Research,
Head Post Office, P.O.Box 79
Moscow, Russia

ABSTRACT

Dynamical features inherent to minisuperspace cosmological models with gauge fields and fermions are discussed. A method for calculating a transition amplitude is proposed. It takes into account the true geometric structure of physical phase space of gauge fields and fermionic degrees of freedom. It is argued that the contribution of fermions to the transition amplitude is described by the same minisuperspace model without fermions, but with an extended gauge group. The method proposed can be applied for calculating the Hartle-Hawking wave function for minisuperspace gauge models with fermions.

1. Minisuperspace cosmological gauge models

In the minisuperspace approach to quantum cosmology, one restricts to a finite number of degrees of freedom of gravitational and matter fields to be quantized [1]. To construct such models, one assumes the four-geometries to be spatially homogeneous, isotropic, and closed so that they are characterized by a single scale factor. The metric is given by the Friedman-Robertson-Walken ansatze (see also [2])

$$g_{\mu\nu}dx^{\mu}dx^{\nu} = \frac{2G}{3\pi}\left(-N^2(t)dt^2 + a^2(t)\omega^i\omega^i\right) \tag{1}$$

where $N(t)$ and $a(t)$ are arbitrary non-vanishing functions of time; G is the gravitational constant and ω^i are the left-invariant one-forms ($i = 1,2,3$) on the three-sphere satisfying the condition $d\omega^i = -\varepsilon_{ijk}\omega^j \wedge \omega^k$. To include matter fields into the model, one should take care for a consistency of an ansatze for matter fields with the Einstein equations – the energy-momentum tensor has to give rise to homogeneous spaces in order to obtain the $SO(4)$-invariant metric tensor ansatze (1). While this problem for a comformally invariant scalar field is easily solved just by the requirement that the scalar field is spatially homogeneous [1], it becomes not so obvious in the case of gauge fields [3].

The ansatze, consistent with the geometry (1), for Yang-Mills fields has been proposed in [3] and it leads to the following effective action of the Einstein-Yang-Mills system

$$S=\frac{1}{2}\int_{t_1}^{t_2}dt\frac{N}{a}\left[-\left(\frac{a}{N}\partial_t a\right)^2+a^2-\lambda^2a^4+\left(\frac{a}{N}\partial_t\chi\right)^2+\left(\frac{a}{N}D_t\mathrm{x}\right)^2-2V\right], \quad (2)$$

where the gauge group of the initial field system is assumed to be $SO(n)$, $n>3$; a real scalar $\chi=\chi(t)$, a vector $\mathrm{x}=\mathrm{x}(t)\in\mathrm{R}^l$, $l=n-3$, and a real antisymmetric $l\times l$ matrix $y=y(t)$ entering into the covariant derivative $D_t\mathrm{x}=\dot{\mathrm{x}}+y\mathrm{x}$ describe Yang-Mills field degrees of freedom in the reduced system; $\lambda^2=2G\Lambda/9\pi$ with Λ being the positive cosmological constant, and

$$V=\frac{\alpha}{3\pi}\left[\left(\chi^2+\mathrm{x}^2-\frac{3\pi}{2\alpha}\right)^2+4\chi^2\mathrm{x}^2\right] \quad (3)$$

is the potential of the Yang-Mills fields, $\alpha=g^2/4\pi$ is the fine structure constant.

The action (2) is invariant under two local groups, the reparametrizaton one

$$t\to t'(t), \qquad N(t)\to N(t')\frac{dt'}{dt} \quad (4)$$

and the $SO(l)$ gauge group, under which only variables x and y transform as follows

$$\mathrm{x}\to\Omega\mathrm{x}, \qquad y\to\Omega y\Omega^T+\Omega\partial_t\Omega^T, \quad (5)$$

where $\Omega=\Omega(t)\in SO(l)$. Due to the local symmetries, the variables N and $y=y^\alpha T_\alpha$, T_α are real generators of $SO(l)$, appears to be non-dynamical and play the role of the Lagrangian multipliers. Momenta canonically conjugated to them vanish and form the primary constraints [4].

Carrying out a Legendre transformation with respect to the velocities $\dot{a}$, $\dot{\mathrm{x}}$ and $\dot{\chi}$ we obtain the canonical Hamiltonian

$$H = \frac{N}{a}\left(H_{WD}-y^\alpha\sigma_\alpha\right), \quad (6)$$

$$H_{WD} = \frac{1}{2}\left(-p_a^2-a^2+\lambda^2a^4+p_\chi^2+\mathrm{p}^2\right)+V \quad (7)$$

where $\sigma_\alpha=\mathrm{p}T_\alpha\mathrm{x}$ and p_a, p_χ and p are canonical momenta conjugated to a, χ and x, respectively. Secondary constraints are to be found by calculating the Poisson brackets of the primary constraints with the canonical Hamiltonian and setting the results to zeros [4]. In so doing, we obtain the secondary constraints to be equivalent to the following equalities

$$H_{WD} = 0, \quad (8)$$

$$\sigma_\alpha = 0. \quad (9)$$

As a consequence, $H = 0$. It is easy to be convinced that all constraints in the theory are the first-class ones [4].

Equation (8) is the classical Wheeler-DeWitt equation for the minisuperspace model. The constraints (9) generate simultaneous $SO(l)$-rotations of phase-space variables $\mathbf{x}$ and $\mathbf{p}$. The number of independent constraints among σ_α is $l-1$ since any vector in $\mathbf{R}^l$ has a stationary subgroup $SO(l-1)$ Therefore, a "particle" described by a position vector $\mathbf{x}$ has only one physical degree of freedom. As the quantities σ_α coincide with components of the particle angular momentum in $\mathbf{R}^l$, just radial motion is physical.

2. Phase space structure in minisuperspace gauge models

Physical configuration (phase) space of Yang-Mills fields is known to possess a non-trivial geometry [5]. This geometry should be taken into account when analyzing non-perturbative quantum dynamics of the Yang-Mills fields [6], [7]. The phase space structure of mechanical systems with a gauge group is considered in length in our review [8]. Due to simplicity of the minisuperspace gauge model, one can easily establish the physical phase space geometry and investigate its dynamical manifestation [9].

Total phase space of gauge fields consists of points $(\mathbf{x}, \mathbf{p}) \in \mathbf{R}^{2l}$ and $(\chi, p_\chi) \in \mathbf{R}^2$ (we ignore the pure unphysical degrees of freedom (y_a)). Physical phase space is a surface of constraints $\sigma_\alpha = 0$ on which all points connected by gauge transformations are identified. Variables χ and p_χ are gauge-invariant therefore their phase space is an ordinary plane, $\mathbf{R}^2$. A general solution to Eq.(9) is $\mathbf{p} = \xi\mathbf{x}$ where ξ is an arbitrary function of time (its explicit form is determined by equations of motion). This solution means that only radial excitations are admissible. Further, one can always put a vector $\mathbf{x}$ along one of the coordinate axes with the help a gauge rotation, say, $x_i = \delta_{i1}x$. As a consequence, $p_i = \delta_{i1}p, \;\; p = \xi x$. The continuous gauge arbitrariness is exhausted, while there remain residual discrete gauge transformations forming the Z_2 group. They change the sign of $x \; : \; x \to \pm x$ (the gauge rotations through the angle π). The signs of p and x should change simultaneously due to the equality $p = \xi x$. Hence the points (x,p) and $(-x,-p)$ on the phase plane $\mathbf{R}^2$ are gauge equivalent and should be identified. The phase plane turns into a cone unfoldable into a half-plane that is just the physical phase space because the gauge arbitrariness is exhausted.

In quantum theory, canonical variables becomes operators obeying the Heisenberg algebra. Operators of the constraints (8) and (9) must annihilate physical states [4]. The condition $\hat{\sigma}_\alpha\psi_{ph} = 0$ means that ψ_{ph} being a function of $\mathbf{x}$ has to be invariant under the $SO(l)$-rotations of $\mathbf{x}$, i.e., it is a function of $r = |\mathbf{x}|$. Therefore, to remove unphysical degrees of freedom in the theory, it is sufficient to introduce the spherical coordinates and to neglect a dependence of physical states on the angular variables. Following then the ideology [1], one can calculate the transition amplitude $U_\eta^{ph}(z,z') =< z|\exp(-\eta \hat{H}_{WD}/\hbar)|z' >$ where z implies the set (a,χ,r) and η is the Euclidean conformal time ($d\eta = d\tau/a$, $\tau = -it$). In the limit $\eta \to \infty$, this kernel gives the ground-state wave function of the Universe [1].

The kernel U_η^{ph} satisfies the Schroedinger equation $(\hbar\partial_\eta + \hat{H}_{WD})U_\eta^{ph} = 0$ with the initial condition $U_{\eta=0}^{ph}(z,z') =< z|z' >$ (– the unit operator kernel). Due to the conformal invariance, gravitational and matter field variables separate (see (7)), i.e.,

$$U_\eta^{ph}(z,z') = U_\eta^{grav}(a,a')U_\eta^{mat}(\chi,r,\chi',r') \ . \tag{10}$$

The amplitude U_η^{grav} is found in [1]. The conic structure of phase space of the physical degree of freedom r leads to a modification of the path integral representation for U_η^{mat} [9] (the apmlitude U_η^{mat} is given in [10] where the physical phase space is assumed to be a plane). The modification results that in the path integral for U_η^{mat}, one should take into account not only trajectories going from r' to r, but also the trajectories reflected from the origin $r = 0$ and connecting r' and r [9]. This is valid for the semiclassical approximation too.

Thus, the phase space geometry of gauge fields influences the Hartle-Hawking wave function of the Universe. Another dynamical manifestation of the gauge field phase-space geometry observed in [9] is that the physical phase space structure affects the exponential quantization of wormhole size (non-Abelian wormholes are considered in [11], [3]).

3. Fermions in minisuperspace gauge models

In a gauge theory with fermions, gauge and Dirac fields are simultaneously transformed under gauge transformations. Removing a continuous gauge arbitrariness in the theory by imposing gauge conditions on components of gauge fields we arrive at an ordinary field theory. As we have seen above, there could remain discrete gauge transformations which lead to further reduction of phase space in the theory. These discrete transformations act on all variables in the reduced theory in accordance with the gauge transformation law. Therefore they act on fermionic degrees of freedom giving rise to a reduction of their phase space [6–8].

Thus, if we include fermions into minisuperspace cosmological gauge models, we could expect some dynamical consequences induced by a non-trivial geometry of physical phase space of fermions. Such models can be constructed [12], although their explicit form is not so important to analyze an influence of the phase space geometry of fermions on the transition amplitude U_η^{ph} introduced above. Indeed, if we introduce fermions keeping the conformal invariance of the theory (massless fermions), one could expect that gravitational and matter variables in the Wheeler-DeWitt equation separate *). The transition amplitude will have the form (10)

*) Even if they do not separate, the method proposed below for calculating the transition amplitude can be applied (with a small modification) due to that a number of fermion excitations is finite in any mechanical model.

where U_η^{mat} becomes a function of Grassmann variables described fermionic degrees of freedom. Therefore, the problem is reduced to an investigation of mechanical models with a gauge symmetry and Grassmann variables, which is done in [13], [8].

Consider first a simple example and then discuss possible generalizations. Let the matter quantum Hamiltonian read [13]

$$\hat{H}^{mat} = \frac{1}{2}\hat{\mathrm{p}}^2 + V(\mathrm{x}, \hat{\psi}^+, \hat{\psi}) . \tag{11}$$

Here $\hat{\mathrm{p}} = -i\hbar\partial$, $\hat{\psi}^+ = \psi^*$ and $\hat{\psi} = \hbar\partial/\partial\psi^*$ with x and ψ^* being three-dimensional real vector and three-dimensional complex Grassmann vector, respectively (we use the coordinate representation for boson degrees of freedom and the holomorphic representation for the fermionic ones). The potential V is assumed to be invariant with respect to $SO(3)$-rotations of the vectors x and $\hat{\psi}$

$$\mathrm{x} \to \Omega\mathrm{x} , \quad \hat{\psi} \to \Omega\hat{\psi} , \quad \hat{\psi}^+ \to \hat{\psi}^+\Omega^T \tag{12}$$

where $\Omega \in SO(3)$, $\Omega\Omega^T = 1$.

To convert the model with the global $SO(3)$-symmetry into the corresponding gauge model, we require that operators of generators of the transformations (12) annihilate physical states

$$\hat{\sigma}_\alpha \Phi_{ph}(\mathrm{x}, \psi^*) = \varepsilon_{\alpha\beta\gamma}(\hat{p}_\beta x_\gamma + i\hat{\psi}_\beta^+\hat{\psi}_\gamma)\Phi_{ph}(\mathrm{x}, \psi^*) = 0 , \tag{13}$$

where $\varepsilon_{\alpha\beta\gamma}$ is the unit antisymmetric tensor. Thus, physical states are invariant under $SO(3)$-rotations of vectors x and ψ^*, which means the presence of the gauge symmetry in the theory. The Lagrangian formalism for the system considered is given in [13].

Let us analyze the physical phase space structure in the model. As in the minisuperspace model considered in Sec.2, one can choose the unitary gauge $x_2 = x_3 = 0$ to remove the continuous gauge arbitrariness in the bosonic sector. Indeed, after fixing the gauge, there remain discrete gauge transformations changing the sing of x_1 and $SO(2)$-rotations in the plane (x_2, x_3) (the stationar group of the vector $(x_1, 0, 0)$).

Notice that the residual gauge group acts on the Grassmann variables. The $SO(2)$-rotations in the anticommutative plane (ψ_2^*, ψ_3^*) are generated by $\hat{\sigma}_1^F = i\varepsilon_{1\alpha\beta}\hat{\psi}_\alpha^+ \hat{\psi}_\beta$. In refs. [8] and [13], it is shown that a continuous gauge symmetry in a pure fermionic system (i.e. when generators of gauge transformations are functions just of Grassmann canonical variables) is equivalent to a discrete gauge symmetry. The meaning of a gauge symmetry is that all observables, being functions of canonical variables, should be gauge-invariant. The equivalence of a continuous gauge symmetry to the discrete one implies that the continuous and discrete groups have the same set of invariants on anticommutative phase space. Indeed, the only $SO(2)$-invariant on the anticommutative plane $\psi_{2,3}^*$ is $\psi_2^*\psi_3^*$ that is also the only invariant of the Z_2 group with the elements

$$E = \text{diag}(1,\ 1,\ 1)\ , \qquad S_1 = \text{diag}(1,\ -1,\ -1) \in SO(3)\ . \tag{14}$$

To find the whole set of residual discrete gauge transformations, one should pick out such $SO(3)$-rotations of the first coordinate axis through the angle π that keep the unitary gauge condition. They are

$$S_2 = \text{diag}(-1,\ 1,\ -1)\ , \qquad S_3 = \text{diag}(-1,\ -1,\ 1)\ . \tag{15}$$

Thus, the physical phase (configuration) space is obtained by identification of points on the Euclidean superspace $(p_1, x_1; \psi, \psi^*)$ connected by the group transformations $(E, S_\alpha) \in \text{S}$.

Let us turn now to the quantum theory. To solve Eq.(13), one should introduce the generalized spherical coordinates on the superspace [13] $(\mathbf{x},\ \psi^*)$

$$\mathbf{x} = \Omega\rho\ , \qquad \psi^* = \xi^*\Omega^T\ , \tag{16}$$

where $\rho = (r, 0, 0)$, the matrix $\Omega = \Omega(\theta, \varphi)$ belongs to $SO(3)$ and it is parametrized such that the first equality in (16) determines the ordinary spherical coordinates in $\mathbf{R}^3$ with θ and φ being the spherical angles. Straightforward calculations [13] show us that solutions to Eq.(13) are functions just of ρ and ξ^* which are invariant under transformation from the group S. The Hamiltonian (11) in the physical Hilbert space reads

$$\hat{H}^{mat}_{ph} = \frac{1}{2}\hat{p}_r^2 + \frac{1}{2r^2}\left[(\hat{\sigma}_2^F)^2 + (\hat{\sigma}_3^F)^2\right] + V(\rho, \hat{\xi}^+, \hat{\xi})\ , \tag{17}$$

where $\hat{p}_r = -i\hbar r^{-1}\partial_r \cdot r$ and $\hat{\sigma}_\alpha^F = i\varepsilon_{\alpha\beta\gamma}\hat{\xi}_\beta^+\hat{\xi}_\gamma$.

Having obtained the physical Hamiltonian, one can construct the path integral representation for the transition amplitude $U_t(z^*, z) =< z^*|\exp(-it\hat{H}^{mat}_{ph}|z >$ where z implies the set (r, ξ). We shall not give an explicit path integral form for U_t. It can be obtained by means of a general recipe proposed in [13]. The kernel $U_t(z^*, z)$ appears to be an S-invariant solution to the Schroedinger equation with the Hamiltonian (17). The S-invariance is provided by summing over all trajectories connecting points z^* and $S_\nu z'$ $(S_0 \equiv E)$ in the path integral, which corresponds to the true phase space structure in the model.

Suppose now that the interaction of fermions and gauge fields is of the minimal type $\bar{\psi}A\psi$. Therefore, in the minisuperspace model, the potential V appears to be bilinear in Grassmann variables. If there would not be a gauge symmetry in the theory with the Hamiltonian (11), one could easily do the path integral over Grassmann variables since it is the Gaussian one. The gauge symmetry complicates the problem because an effective action entering into the path integral for U_t contains terms of fourth degrees in Grassmann variables. We shall not do this non-Gaussian integral over Grassmann variables. We propose a simple method for calculating a contribution of fermions to the transition amplitude, which is based on the operator approach.

Notice first of all that the whole physical Hilbert space is split to four orthogonal subspaces. Each subspace is labeled by a number of fermionic excitations, $\mathcal{H}_{ph} = \sum_\nu \oplus \mathcal{H}_{ph}^{(\nu)}$ and $\nu = 0,1,2,3$. Let us neglect a dependence of the potential V on Grassmann variables (i.e., $V = V(\rho^2)$) (a remark about generalizations see bellow) and consider eigenstates of the Hamiltonian (17). Since they should be S-invariant, one can find the following ansatze

$$\Phi_{ph}^{(0)} = \Phi^{(0)}(r^2)\,, \qquad \Phi_{ph}^{(1)} = (\rho,\xi^*)\Phi^{(1)}(r^2)\,, \tag{18}$$

$$\Phi_{ph}^{(2)} = \varepsilon_{\alpha\beta\gamma}\rho_\alpha\xi_\beta^*\xi_\gamma^* r\Phi^{(2)}(r)\,, \qquad \Phi_{ph}^{(3)} = \varepsilon_{\alpha\beta\gamma}\xi_\alpha^*\xi_\beta^*\xi_\gamma^* r^3\Phi^{(3)}(r)\,, \tag{19}$$

where functions $\Phi^{(2)}(r)$ and $\Phi^{(3)}(r)$ are antisymmetric. The states (18-19) differ each other by a number of fermionic excitations, i.e., $\Phi_{ph}^{(\nu)} \in \mathcal{H}_{ph}^{(\nu)}$. It is easy to see that

$$(\hat{\sigma}^F, \hat{\sigma}^F)\Phi_{ph}^{(\nu)} = \hbar^2\nu(\nu+1)\Phi_{ph}^{(\nu)}\,. \tag{20}$$

Substituting the ansatze (18-19) into the Schroedinger equation for eigenstates of the operator (17) we obtain

$$\left(\partial_r^2 + \frac{2(\nu+1)}{r}\partial_r - \hbar^{-2}V(r^2)\right)\Phi^{(\nu)} = -\frac{2E}{\hbar^2}\Phi^{(\nu)}\,. \tag{21}$$

The differential operator entering into (21) is nothing but the radial part of the Laplace operator in $\mathbf{R}^{2\nu+3}$. Thus, we reduce the problem to investigation of a bosonic system. It looks like a boson minisuperspace gauge model with the extended gauge group $SO(2\nu+3)$. The difference however is that one should take antisymmetric solutions to Eq.(21) for $\nu = 3,4$ instead of the symmetrical ones as one could expect in the $SO(2\nu+3)$-gauge model.

Using the spectral representation for the transition amplitude we get

$$U_t(z^*, z') = \sum_E \langle z^*|E\rangle\langle E|z'\rangle \exp(-it\hat{H}_{ph}^{mat}/\hbar) =$$

$$= \sum_{\nu=0}^{3} \mu_\nu(z^*)\mu_\nu(z')U_t^{(\nu)}(r,r')\,; \tag{22}$$

$$U_t^{(\nu)}(r,r') = \sum_E \Phi_E^{(\nu)}(r)\Phi_E^{(\nu)*}(r')\exp(-itE/\hbar)\,, \tag{23}$$

where the factors $\mu_\nu(z^*)$ coincide with the ones of $\Phi^{(\nu)}$ in (18-19) and the functions $\Phi_E^{(\nu)}$ satisfy the Schroedinger equation (21). The path integral for the symmetrical ($\nu = 0,1$) and antisymmetrical ($\nu = 2,3$) transition amplitudes (23) can be found in the standard way. For the symmetrical amplitudes, contributions of reflected trajectories (see discussion at the end of Sec.2) should be taken with the sign plus, while the same contributions to the antisymmetric amplitudes are take with the negative sign.

4. Concluding remarks

Let us list main points of our approach to minisuperspace gauge models with fermions. As gauge and fermionic fields realize a representation of a gauge group, operators of constraints is a sum of generators of gauge transformations of commutative and anticommutative variables. Therefore, *(i)* to solve constraints, one should use curvilinear coordinates on the configuration superspace. *(ii)* The physical Hamiltonian thus constructed depends on the quadratic Casimir operator composed of gauge generators for fermionic fields (see (20) and (17)). Due to the latter property, the boson part of wave functions for excitations with fermions can be found as physical states in a boson model with an extended gauge group. *(iii)* The dependence of wave functions on the Grassmann variables (the factors $\mu_\nu(z^*)$) is determined by basis S-invariant polynomials built of variables spanning the physical configuration superspace.

Let G be a gauge group of a mixed minisuperspace gauge model. Its action splits the bosonic configuration space on gauge group orbits. Motion along the orbits is characterized by the generalized angular momentum (or, group orbit momentum). Its components are generators of gauge transformations in the bosonic sector of the theory. Obviosly, the Casimir operators of G are integrals of such motion. In quantum theory, the orbit motion is determined by eigenvalues of the Casimir operators. Constraints in the mixed gauge models mean that the total group orbit moment (i.e., a sum of the bosonic and fermionic momenta) should vanish. Thus, bosons have the group orbit moment opposite to the fermion one. That is why pp. *(i)* and *(ii)* must be valid in the general case. Notice that if we introduce curvilinear coordinates such that surfaces of constant transverse variables (analogous to r) coincide with gauge group orbits, then the "angular" part of the Laplace-Beltrami operator is proportional to the quadratic Casimir operator [14].

Consider the case when V depends on fermionic fields. The operator $\hat{V}$ does not change a number of fermionic excitation if it is assumed to provide the minimal interaction of fermions and gauge bosons. Due to its gauge invariance, it can be diogonalized simultaneously with the quadratic Casimir operator entering into the Laplace-Beltrami operator. The only modification of Eq. (21) appears to be $V \to V_\nu$.

There remains just to show a regular way of constructing factors μ_ν, i.e., to generalize p. *(iii)*. In ref. [13], a generalized Shevalley theorem is proved. The theorem states that any polynomial on physical configuration superspace invariant with respect to residual gauge transformation can be uniquely continued to a gauge invariant polynomial in the whole configuration space. Thus, the problem of finding the factors μ_ν is reduced to finding a basis of gauge-invariant polynomials built of commutative and anticommutative variables realizing a gauge group representation. The latter is equivalent to listing all irreducible G-invariant symmetrical and antisymmetrical tensors [13].

A few words about the path integral approach. The result (22) can be recovered directly by doing the path integral containing fermionic fields in the fourth degree in the effective action. An important point here is that just the summation over trajectories connecting points z^* and the "reflected" ones $S_\nu z'$ provides unique correspondence to the operator approach developed above. We shall demonstrate it in our more elaborate forthcoming work where we shall also present an application of our method to a concrete minisuperspace cosmological model with gauge and fermionic fields.

Acknowledgments

The author is very grateful to J.M.Mourao for useful discussions on minisuperspace cosmological models.

References

1. J.B.Hartle and S.W.Hawking, *Phys.Rev.D* 28 (1983) 2960.
2. B.DeWitt, *Phys.Rev.D* 160 (1967) 1113;
   C.Isham and J.E.Nelson, *Phys.Rev.D* 10 (1974) 3226;
   W.E.Blyth and C.Isham, *Phys.Rev.D* 11 (1975) 768.
3. O.Bertolami, J.M.Mourao, R.F.Picken and I.P.Volobujev, *Int.J.Mod.Phys.A* 6 (1991) 4149.
4. P.A.M.Dirac, *Lectures on quantum mechanics*, Yeshiva University, New York, 1964.
5. I.M.Singer, *Commun.Math.Phys.* 60 (1978) 7;
   O.Babelon and C.M.Viallet, *Commum.Math.Phys.* 81 (1981) 515;
   L.V.Prokhorov, *Sov.J.Nucl.Phys.* 35 (1982) 192;
   L.V.Prokhorov and S.V.Shabanov, *Phase space structure of the Yang-Mills fields*, JINR preprint E2-90-207, Dubna, 1990;
   L.V.Prokhorov and S.V.Shabanov, *Phys.Lett.B* 216 (1989) 341;
   S.V.Shabanov, *Phase space structure in gauge theories*, JINR lectures P2-89-533, Dubna, 1989 (in Russian).
6. S.V.Shabanov, *Phys.Lett.B* 255 (1991) 398;
   S.V.Shabanov, *Lett.Mod.Phys.A* 6 (1991) 909.
7. S.V.Shabanov, *Lectures on Quantization of Gauge Theories by the Path Integral Method*, IFM preprint, IFM 16/92, Lisbon, 1992.
8. L.V.Prokhorov and S.V.Shabanov, *Sov.Phys.Uspekhi* 34(2) (1991) 108.
9. S.V.Shabanov, *Phys.Lett.B* 272 (1991) 11.
10. O.Bertolami and J.M.Mourao, *Class. Quantum Grav.* 8 (1991) 1271.
11. Y.Verbin and A.Davidson, *Nucl.Phys.B* 339 (1990) 545.

12. I.Volobujev, *A talk given at the first international Sakharov conference on physics*, Moscow, May, 1991;
J.M.Mourao, *private communications*.
13. S.V.Shabanov, *J.Phys.A: Math.Gen.* **24** (1991) 1199.
14. S.Helgason, *Groups and Geometric Analysis*, Academic press, Inc., 1984.

RIEMANN-CARTAN TOPOLOGICAL GRAVITY

S.N.SOLODUKHIN
Department of Theoretical Physics, Moscow University
Moscow, 117234, Russia

ABSTRACT

We show additional (topological) symmetry under arbitrary shifts of vielbeins in the Riemann-Cartan theory of gravity. The action invariant under this symmetry is constructed in dimension $D = 2$ and $D = 4k$. The space of classical gauge inequivalent solutions is found.

1. Introduction

Recently much attention has been payed to the investigation of topological gravity[1-4]. The metrical degrees of freedom don't have dynamics in these theories. In known models it is reached by one of following ways[3].

1. The action doesn't depend on metric, i.e. the variation of action with respect to metric is zero:

$$\delta_g S = 0.$$

This class of models includes the Chern-Simons theory (together with 3D gauge gravity of the Chern-Simons type) and B-F systems[2].

2. The variation of action with respect to metric takes the form of BRST - commutator:

$$\delta_g S = \{\hat{Q}, X\}$$

The prototype for this class of topological theories is four-dimensional Donaldson theory[1]. The known models of 2D topological gravity are of this type[4].

The main idea which we exploit here is to consider the theory with too large group of symmetry that all local freedoms of metric (or vielbeins) are the gauge ones. So it is only topological degrees of freedom that are essential. This way to construct topological theory can be called the gauge one in the sense that the model is invariant under local shifts of metric:

$$g_{\mu\nu} \rightarrow g_{\mu\nu} + w_{\mu\nu} \tag{1}$$

or under local shifts of vielbeins:

$$h^a_\mu \rightarrow h^a_\mu + w^a_\mu \tag{2}$$

The invariance with respect to these transformation is reached by introduction additional field ω (ω will shown to be the Lorentz connection one-form) the transformation law of which compensates the shifts given by eqs.1-2 :

$$\delta_g S[g,\omega] + \delta_\omega S[g,\omega] = 0$$

Thus, though the invariant action can depend on metric (for instance via the Hod ge operator $*$), this dependence is fictive since the action is invariant under eqs.1-2 and consequently the local metrical degrees of freedom are the gauge ones. Notice that the prototype for such a theory is model considered in details by Horowitz[2]:

$$S = \int_M F \wedge B - \frac{1}{2} B \wedge B \tag{3}$$

where differential forms $F = dA$ and B have rank $2m$. This action is invariant under transformations

$$A \to A + w, \quad B \to B - dw \tag{4}$$

One can see that this symmetry is of that type as given by eqs.1-2, i.e.

$$\delta_A S[A,B] + \delta_B S[A,B] = 0.$$

2. Two-Dimensional Model

In two dimensions the metric has only one dynamical freedom - the conformal m ode (and zweinbeins h^a_μ correspondingly two: conformal and Lorentz modes), so constructing the model of last type in two dimensions it is sufficien tly to consider the invariance under local conformal-Lorentz transformations:

$$h^a_\mu \to e^\sigma(\cos\phi\,\delta^a{}_b + \sin\phi\,\varepsilon^a{}_b) h^a_\mu, \tag{5}$$

where $\varepsilon_{ab} = -\varepsilon_{ba}$. So one should consider the Riemann-Cartan-Weyl geometry[5,6] where the tran sformation law for conformal-Lorentz connection one-form $\omega^a{}_b = W\delta^a{}_b + B\varepsilon^a{}_b$ takes the form:

$$\omega^a{}_b \to \omega^a{}_b - \delta^a{}_b d\sigma - \varepsilon^a{}_b d\phi \tag{6}$$

Then the action invariant under local conformal-Lorentz transformations is following:

$$S = \frac{1}{2}\int_\Sigma *Q^a \wedge Q_a \tag{7}$$

where $Q^a = dh^a + \omega^a{}_b \wedge h^b$ is torsion one-form.

One can show that the equations of motion are equivalent to the zero-torsion condition: $Q^a = 0$. We denote the space of the gauge inequivalent classical solutions as follows:

$$\Phi = \Phi_\omega \oplus \Phi_h \equiv \frac{\{\omega^a{}_b, h^a:\ Q^a=0\}}{Diff \oplus Lor \oplus Conf}$$

One can see[6] that Φ_ω is moduli space of gauge inequivalent f lat connections on Σ. It is characterized by their holonomies around non-contractible loops. This is equivalent to specifying homomorphism χ from $\pi_1(\Sigma)$ into group G (up to conjugation by G). In our case the group $G = R_+ \oplus SO(2)$ (R_+ is the group of positive numbers with respect to multiplication). On the other hand[6] the space Φ_h is shown to be the factor space $\Phi_h = Ker\nabla_0^{(1)}/Im\nabla_0^{(0)}$, where ∇_0 is the external differential operator of the elliptic complex $\{\Lambda^k_\chi,\ \nabla_0^{(k)}\}$, where Λ^k_χ is the space of k-forms with the values in the fibr es of the flat bundle with base Σ and group G, correspondingly to the homomorphism χ from $\pi_1(\Sigma)$ into the group G.

3. Model in Higher Dimensions

Let h^a, $a = 1, ..., D$ be an orthonormal basis of one-forms (veilbeins) on D-dimensional manifold M^D and $\omega^a{}_b$ is the Lorentz connection two-form ($\omega_{ab} = -\omega_{ba}$). Then the torsion one-form $Q^a = dh^a + \omega^a{}_b \wedge h^b$ has standard transformation under local Lorentz rotations $h^a \to S^a{}_b h^b$, $\omega \to S\omega S^{-1} + SdS^{-1}$: $Q^a \to S^a{}_b Q^b$. However there is a more wide class of transformations under which the Q^a is invariant. Let $det(h^a_\mu)$ is not zero. Then we have $Q^a = dh^a + T^a$, where we denote $T^a = \omega^a{}_b \wedge h^b$. Then torsion Q^a is obviously invariant unde r transformations:

$$\begin{aligned} h^a &\to h^a + u^a \\ T^a &\to T^a - du^a \end{aligned} \tag{8}$$

The transformation law for $T^a = \frac{1}{2}T^a_{\mu\nu}dx^\mu \wedge dx^\nu$ induces the corresponding law for $\omega^a{}_b = \omega^a{}_{b\mu}dx^\mu$ and infinitesimally we get

$$\begin{aligned} \omega^a{}_b &\to \omega^a{}_b + \Delta\omega^a{}_b \\ \Delta\omega^a{}_b \wedge h^b &= -\nabla u^a \\ \Delta\omega_{\mu\nu,\alpha} &= \frac{1}{2}[(\nabla u^a)_{\mu\nu}h_{a\alpha} - (\nabla u^a)_{\alpha\nu}h_{a\mu} - (\nabla u^a)_{\mu\alpha}h_{a\nu}] \end{aligned} \tag{9}$$

where $(\nabla u^a)_{\mu\nu}) = (\nabla_\mu u^a_\nu - \nabla_\nu u^a_\mu)$, $\nabla_\mu u^a_\nu = \partial_\mu u^a_\nu + \omega^a{}_{b,\mu}u^b_\nu$ and $\omega_{\mu\nu,\alpha} = \omega_{ab,\alpha}h^a_\mu h^b_\nu$. Thus there is here an additional topological symmetry under arbitrary shifts of veilbeins and the Lorentz connection one-form plays the role of a gauge field corresponding to this symmetry.

One can suggest the following action in dimension $D = 4k$, $k = 1, 2, ...$

$$S = \int_{M^{4k}} \frac{1}{2^k}(Q^{a_1} \wedge Q^{a_1})...(Q^{a_k} \wedge Q^{a_k}) \tag{10}$$

which is obviously invariant under both Lorentz rotations and transformations eq.8-9. In four dimensions we get

$$S = \int_{M^4} \frac{1}{2} Q^a \wedge Q^a \tag{11}$$

This action takes the form

$$S = \int \frac{1}{2}(dh^a \wedge dh^b) + dh^a \wedge T^a + \frac{1}{2}(T^a \wedge T^a) \tag{12}$$

if one denotes $T^a = \omega^a{}_b \wedge h^b$. It is worth noting that omitting the first term (which is total derivative) this expression has the form of action considered by Horowitz[2] for topological abelian gauge fields (see eq.3).

One obtains the following equations of motion from the action eq.11 :

$$\begin{aligned} &Q^a \wedge h^b - Q^b \wedge h^a = 0 \\ &\nabla Q^a \equiv R^a_b \wedge h^b = 0 \end{aligned} \tag{13}$$

The first equation is equivalent to zero-torsion condition $Q^a = 0$ if orthonormal basis h^a is non-degenerate (i.e. $deth^a_\mu \neq 0$). This condition can be rewritten in the form $dh^a + T^a = 0$. Under topological transformations (see eqs.8-9) T^a transforms as $T^a \to T^a - du^a$. The eq.13 means that $dT^a = 0$. So the space of gauge inequivalent solutions for the field T^a is $H^2(M) \oplus ... \oplus H^2(M)$. For every T^a one can obtain the corresponding expression for $\omega^a{}_b$. If manifold has non-trivial cohomology space $H^2(M)$ then the basis h^a is not globally determined. More precisely h^a can be considered as a section of more general bundle as ordinary.

Acknowledgments. I would like to thank Yu.N.Obukhov for very useful discussions. I also thank R.Myers and C.Burgess for kind hospitality and support at McGill University where part of this work was finished.

References

1. E. Witten, *Comm.Math.Phys.* 117 (1988), 353.
2. G. T. Horowitz, *Comm.Math.Phys.* 125 (1989), 417.
3. D. Birmingham, M. Blau, M. Rakowski and G. Thompson, *Phys.Rep.* 209 (1991), 290.
4. E. Verlinde and H.Verlinde, *Nucl.Phys.* B348 (1991), 435.
5. M. Katanaev and I.Volovich, *Ann.Phys.(N.Y.)* 197 (1990), 1.
6. S. N. Solodukhin, *Topological 2D Riemann-Cartan-Weyl Gravity*, preprint, 1991.

Particle Creation in the Collision of Plane Waves

MIQUEL DORCA and ENRIC VERDAGUER
Grup de Física Teòrica
Universitat Autònoma de Barcelona
08193 Bellaterra (Barcelona), Spain

Abstract

The creation of particles in a colliding wave background is computed. The background represents the head on collision of two gravitationl shock waves followed by gravitational radiation. Such collision leads to the formation of a caustic. The "in" and "out" vacua states are defined for a massless scalar field. They are, respetively, related to the flat "in" vacuum in the spacetime region before the collision and to a physically reasonable vacuum at the caustic. It is found that particles are spontaneously created with an spectrum inverse to the frequency, the spectrum is compatible in the long wavelenght limit with a thermal spectrum with a temperature which is inversely proportional to the focusing time of the waves.

1. Introduction

Colliding wave spacetimes are exact solutions of Einstein equations which represent the head on collision of two exact gravitational plane waves and as such they are some of the simplest exact highly dynamical gravitational fields. The nonlinearity of General Relativity leads to some conspicuous features here[1]. When such gravitational fields interact with quantum matter one expects copious production of particles, it is the purpose of this work to compute such production. For this we quantize a massless scalar field in a colliding wave background and identify two physically meaningfull vacua: the "in" and "out" vacua. We show that the "in" vacua contains "out" particles and therefore particles are spontaneously created as a consequence of the interaction of the quantum field with gravity. First we describe the physics of colliding waves.

2. Colliding Plane Waves

We start with single gravitational plane wave. These are very simple solutions of Einstein's field equations with a group G_5 of isometries and an Abelian subgroup G_3 acting on null hypersurfaces[2]. Yet as it was first pointed out by Penrose[3] they have interesting global properties like the absence of spacelike Cauchy hypersurfaces as a consequence of their focusing properties. Exact plane waves focus null geodesics which collide with the wave. They are suposed to represent some physical situations in which strong gravitational waves may be produced such as in black hole collisions[4] or in some particular cases of travelling waves on cosmic strings[5] (in particular for strings with a deficit angle of π). The simplest way to represent plane waves is with the use of the so called *harmonic coordinates* (X, Y, U, V). If we restrict ourselves to one polarization,

$$ds^2 = dUdV - dX^2 - dY^2 + h(U)\left(X^2 - Y^2\right)dU^2, \tag{1}$$

where $h(U)$ is an arbitrary positive bounded function such that when it is different from zero only in the interval $U \in (a, b)$, $(a, b$ real parameters) leads to a sandwich wave. Note that for $U \notin (a, b)$ the spacetime is flat. The above coordinates do not display some of the symmetries of the spacetime. These are better seen with the use of the so called *group coordinates*, (x, y, u, v), which put (1) in the form: $ds^2 = dUdV - F^2(u)\,dX^2 - G^2(u)\,dY^2$. Here $U = u$, $V = v + x^2FF' + y^2GG'$, $X = xF(u)$, $Y = yG(u)$, where F and G are solutions to the equations $F''/F = h = -G''/G$. These coordinates are adapted to the Killing fields ∂_x and ∂_y. Using the weak energy condition one can see[6,7] that there is some $u = u_f$ (focusing surface) where the determinant of the transversal two dimensional metric vanishes, i.e. $F(u_f)G(u_f) = 0$. The group coordinates become singular at u_f although the spacetime is not. The focusing properties can be seen as follows. Assume, for instance, that $F(u_f) = 0$ and consider two particles following geodesics at some fixed coordinates (x_0, y_0) and $(x_0 + \Delta x_0, y_0)$, then at $u = u_f$ the particles will focus at the same time (we use u as an affine parameter).

The quantization of a field in this background is rather trivial[6]: no particle creation or vacuum polarization take place. For this quantization one takes a null hypersurface $u = const.$ (with $u < a$) as substitute Cauchy surface although null geodesics with fixed u do not register on that surface.

Let us now consider the head on collision of two one polarization plane waves. Due to mutual focusing one expects rather dramatic consequences when two of such waves frontaly collide. In fact, the first solution representing the head on collision of two shock waves was obtained by Kahn and Penrose[8] and at the focusing surfaces of the waves true singularities of the spacetime are formed. Field quantization in that spacetime was considered by Yurtsever[9] but particle creation could be only computed approximately.

The collision of two plane waves does not always lead to a singularity, we shall consider a spacetime in which a Cauchy horizon, which may be interpreted as a *caustic* is formed at the focusing hypersurfaces of the plane waves. The spacetime may, in group coordinates, be separated into four regions, region IV ($u < 0$, $v < 0$) is flat (spacetime before the collision) i.e. ${ds^2}_{IV} = 4L_1L_2\,dudv - dx^2 - dy^2$, the single plane wave regions: the u-wave (region II), ($0 \leq u < \pi/2$, $v \leq 0$), with metric,

$${ds^2}_{II} = 4L_1L_2\left[1 + \sin(u)\right]^2 dudv - \frac{1 - \sin(u)}{1 + \sin(u)}dx^2 - \left[1 + \sin(u)\right]^2 \cos^2(u)dy^2,$$

and the v-wave (region III), ($0 \leq v < \pi/2$, $u \leq 0$), with the same metric as region II but changing u for v. And finally the interaction region (region I), is given by[10]

$$ds_I^2 = 4L_1L_2\left[1 + \sin(u + v)\right]^2 dudv - \frac{1 - \sin(u + v)}{1 + \sin(u + v)}dx^2 - \left[1 + \sin(u + v)\right]^2 \cos^2(u - v)dy^2.$$

The parameters L_1 and L_2 have dimension of length and give the focusing times of the u-wave and v-wave respectively. Note that in the interaction region the metric has a group G_2 of isometries with ∂_x and ∂_y Killing fields along the transversal directions of the waves. The spacetime is regular everywhere but the coordinates become singular at the focusing surfaces $u = \pi/2$ and $v = \pi/2$ of the u and v waves respetively, and at the caustic $u + v = \pi/2$. The metric is continuous across the boundaries of the four regions $u = 0$ and $v = 0$, the first tangential derivatives are also continuous but the normal derivatives are not. Thus it satisfies the O'Brian-Synge conditions[11], they guarantee that the Ricci tensor is zero everywhere (i.e. it is a vacuum solution) but the Weyl tensor acquires distributional values at the boundaries. The solution may be interpreted[12,13] as representing two shock gravitational waves followed by gravitational radiation.

To understand the global properties of that spacetime one must use appropriate coordinates in each region. In the flat region IV, the group coordinates are well adapted Lorentzian coordinates. But in the plane wave regions II and III we must use harmonic coordinates (X, Y, U, V). In these coordinates the metric, in region II, has the form (1) with $h(U) = 3/16L_1L_2(1+\sin u)^5$. It is useful to analyze geodesics in that region and one can see that only null geodesics with exact perpendicular incidence i.e. with momenta $p_x = p_y = 0$ will hit the surface $u = \pi/2$ all others reach the interacion region. One sees in these harmonic coordinates that the boundary $v = 0$ is folded back in such a way that $u = \pi/2$ is the closing line of the boundary. This indicates that one may identify the "line" $u = \pi/2$ with the "point" $u = \pi/2$, $v = 0$. That point is a topological singularity called a *folding singularity* and is an accumulation point for the geodesics.

The interaction region is locally isometric to a region of the interior of the Schwarzschild spacetime with mass $M = \sqrt{L_1L_2}$. In that isometry the Schwarzshild coordinates are defined as: $t = x$, $r = M[1 + \sin(u + v)]$, $\phi = 1 + y/M$, $\theta = \pi/2 - (u - v)$. The caustic is mapped to the black hole horizon. Since appropriate coordinates to describe the horizon are the Kruskal-Szekeres coordinates, it is useful to introduce in region I a similar set of Kruskal-Szekeres like coordinates difined as follows. We first introduce time and space coordinates (ξ, η) by $\xi = u + v$, $\eta = v - u$ and define $\xi^* = 2M\ln\left[(1+\sin\xi)/2\cos^2\xi\right] - M(\sin\xi - 1)$. Then introduce null coordinates $\tilde{U} = \xi^* - x, \tilde{V} = \xi^* + x$ and finally $U' = -2M\exp\left(-\tilde{U}/4M\right) \leq 0$, $V' = -2M\exp\left(-\tilde{V}/4M\right) \leq 0$. The metric (2), in region I, reads

$$ds_I^2 = \frac{2\exp\left[(1-\sin\xi)/2\right]}{(1+\sin\xi)}dU'dV' - M^2(1+\sin\xi)^2d\eta^2 - (1+\sin\xi)^2\cos^2\eta dy^2, \qquad (2)$$

with $U'V' = 8M^2\cos^2\xi/(1+\sin\xi)\exp\left[(\sin\xi - 1)/2\right]$, $U'/V' = \exp(x/2M)$. Notice that the metric looks manisfestly flat in these coordinates at the caustic $\xi = \pi/2$. This hypersurface is defined by $U' = 0$ and $V' = 0$ in the new coordinates and the vectors $\partial_{U'}$ and $\partial_{V'}$ become null Killing fields there.

4. Particle Creation

Let us now consider the quantization of a massless scalar field ϕ in the colliding wave spacetime. We have to solve the Klein-Gordon equation,

$$\Box_g \phi = 0, \qquad (3)$$

in the different spacetime regions. Before the collision in region IV the spacetime is flat and the solutions of (3) may be expressed in terms of the ordinary flat space modes, which we define as "in" modes,

$$u_k^{in(IV)}(u,v,x,y) = \frac{1}{\sqrt{2k_-(2\pi)^3}} \mathrm{e}^{-i2L_2k_-v-i2L_1k_+u+ik_xx+ik_yy}, \tag{4}$$

where k_-, k_x and k_y are independent separation constants and $k_+k_- = (k_x^2 + k_y^2)/4$. These constitute a set of positive definite frequency modes which are well normalized on the boundary of region IV with regions II and III. The field operator ϕ can be expressed, as usual, in terms of the positive and negative frequency modes with creation and annihilation operators. These operators may be used to defined the "in" vacuum $|0\,\mathrm{in}>$ and physically meaningful states for "in" particles. Although the boundary of region IV is not a Cauchy surface for the spacetime, since all modes propagate towards the interaction region they are well normalized on the boundary of the interaction region with the plane waves. That null surface is a Cauchy surface for the interaction region. The modes on such surface uniquely define the solutions in the interaction region. Since we will define the "out" modes in the caustic we will be interested in propagating the "in" modes up to the caustic. Now the Klein-Gordon equation cannot be solved exactly in region I however the above task is greatly simplified because one can show that the "in" modes become blueshifted near the boundary between region I and II when $u \to \pi/2$ and between region I and III when $v \to \pi/2$. Thus the relevant part of the modes in the interaction region have high frequencies near the points $u = \pi/2,\ v = 0$ and $v = \pi/2,\ u = 0$ and consequently one may use the geometrical optics approximation to propagate them up to the caustic. The problem is thus reduced to compute the geodesics in region I near the above points and match them to the geodesics in region II (and III). It is worth to note that the geometrical optics approximation is exact in regions II and III. The final expression for the "in" modes near the caustic is[14]

$$u_k^{in(I)}\Big|_{\xi \lesssim \pi/2} = \frac{1}{\sqrt{(2\pi)2|k_x|}} \sum_l C_l^{(1)}\, Y_l^m \begin{cases} (-U'/2M)^{i4M|k_x|} \ ; \ k_x \geq 0 \ \ (V'=0). \\ (-V'/2M)^{i4M|k_x|} \ ; \ k_x \leq 0 \ \ (U'=0). \end{cases} \tag{5}$$

where $C_\alpha^{(1)}$ are coefficients and $Y_l^m\,(y/M, \pi/2 - \eta)$ are spherical harmonics. We have assumed for simplicity that the transversal coordinate y is a cyclic coodinate with associated momentum m. The coefficients $C_\alpha^{(1)}$ are restricted by mode normalization on the caustic.

Now we shall define the "out" vacuum $|0\,\mathrm{out}>$. This can be easily done if we note that a natural set of positive frequency modes can be defined at the caustic using the null Killing fields $\partial_{U'}$ and $\partial_{V'}$. These modes are

$$u_k^{out} = \frac{1}{\sqrt{(2\pi)2\omega_\pm}} \begin{cases} \mathrm{e}^{-i\omega_+U'} \sum_l R_l\, Y_l^m\, \mathrm{e}^{-i\left(l(l+1)/16M^2\omega_+\right)V'} \ , \ V' \simeq 0 \\ \mathrm{e}^{-i\omega_-V'} \sum_l R_l\, Y_l^m\, \mathrm{e}^{-i\left(l(l+1)/16M^2\omega_-\right)U'} \ , \ U' \simeq 0. \end{cases} \tag{6}$$

with the coefficients R_l satisfying certain normalization conditions and with ω_- and $\hat{\omega}_+\hat{\omega}_-$ two independent separation constants. It is easy to show that modes (6) are normalized at the caustic. The field operator ϕ can be writen in terms of these modes with the corresponding creation and annihilation "out" operators. Using such operators one may construct the "out" vacuum and the corresponding "out" particle states. This "out" vacuum is physically reasonable: a particle detector in free fall near the horizon will detect no particles if the quantum state is the "out" vacuum.

We can now calculate spontaneous particle creation by evaluating the "out" number operator N^{out} in the in vacuum, $< 0\,\mathrm{in}|N^{out}|0\,\mathrm{in} >$. This is obtained as usual[15] by means of the Bogoliubov coefficients $\beta_{k\tilde{k}} = -(u_k^{in}, u_{\tilde{k}}^{out*})$, i.e. $N_{\tilde{k}}^{out} = \int d^3\vec{k}|\beta_{k\tilde{k}}|^2$ where $\int d^3\vec{k} = (2\sqrt{L_1L_2})^{-1}\sum_m \int dk_- \int dk_x$. The final result is[14]

$$N_{\tilde{k}}^{out} = \frac{\mathcal{G}}{8\pi\sqrt{L_1L_2}\,\omega_\pm}, \tag{7}$$

where $\mathcal{G}$ is a geometrical factor independent of the frequency $\omega_\pm$. Equation (7) indicates that particles are produced with an spectrum which is inverse to the frequency. Note that in the long wavelength limit the spectrum is compatible with a thermal spectrum with a temperature $T = 1/8\pi\sqrt{L_1L_2}$, i.e. inversely proportional to the focusing time of the waves. This result is not unexpected since $1/4\sqrt{L_1L_2}$ is the surface gravity of the caustic.

6. References

1. J. B. Griffiths, *Colliding waves in general relativity* Clarendon Press, Oxford, (1991)
2. D. Kramer, H. Stephani, M. MacCallum and E. Herlt, *Exact solutions of Einstein's field equations* (Cambridge University Press, Cambridge, England, 1980).
3. R. Penrose, *Rev. Mod. Phys.* **37**, 215 (1965).
4. P. D. D'Eath, in *Sources of gravitational radiation*, edited by L. Smarr (Cambridge University Press, Cambridge, England, 1979).
5. D. Garfinkle, *Phys. Rev.* **D41**, 1112 (1989) (1987).
6. G. W. Gibbons, *Commun. Math. Phys.* **45** 191 (1975).
7. J. Garriga and E. Verdaguer, *Phys. Rev.* **D43**, 391 (1991).
8. K. Kahn and R. Penrose, *Nature (London)* **229**, 185 (1971).
9. U. Yurtsever, *Phys. Rev.* **D40**, 360 (1989).
10. S. Chandrasekhar and B. Xanthopoulos, *Proc. R. Soc. (London)* **A408**, 175 (1986).
11. S. O'Brian and J. L. Synge, *Commun. Dubl. Inst. Adv. Stud.* **A9**, (1952).
12. S. A. Hayward, *Class. Quantum Grav.* **6**, 1021 (1989).
13. U. Yurtsever, *Phys. Rev.* **D38**, 1706 (1988).
14. M. Dorca and E. Verdaguer. In preparation.
15. N. D. Birrell and P. C. W. Davies, *Quantum fields in curved space* (Cambridge University Press, Cambridge, England, 1982).

PARTICIPANTS

Abreu, João P.
Departamento de Física, Universidade de Lisboa
Portugal

Alonso, Juan Carlos
Instituto de Matemáticas y Fisica Fundamental, Madrid
Spain

Ashtekar, Abhay
Physics Department, Syracuse University
U.S.A.

Bastero Gil, Mar
LAEFF, C.S.I.C., Madrid
Spain

García-Bellido, Juan
Instituto Estructura de la Materia, C.S.I.C., Madrid
Spain

Bento, Luís
Departamento de Física, Universidade de Lisboa
Portugal

Bento, Maria C.
Departamento de Física, IST, Lisbon
Portugal

Bertolami, Orfeu
Departamento de Física, IST, Lisbon
Portugal

Bettencourt, Luís M.A.
Departamento de Física, IST, Lisbon
Portugal

Bleyer, Ulrich
KAI Gravitationsprojekt bei der Universitat Potsdam
(ehem Einstein Laboratorium)
Germany

Bowick, Mark J.
Physics Department, Syracuse University
U.S.A.

Branco, Gisela
Departamento de Física, IST, Lisbon
Portugal

Branco, Gustavo C.
Departamento de Física, IST, Lisbon
Portugal

Brito, Ana
Physics Department, Moscow State University
Russia

Brotas, António
Departamento de Física, IST, Lisbon
Portugal

Caetano, Alcides J.
Departamento de Matemática, Universidade de Lisboa
Portugal

Campos Lorenzo, Antonio
Grup de Física Teórica, Universitat Autònoma de Barcelona
Spain

Carlini, Alberto
SISSA/ISAS, Trieste
Italy

Donets, Evgueni E.
Laboratory of Theoretical Physics, JINR, Dubna
Russia

Dorca Callao, Miquel
Grup de Física Teórica, Universitat Autònoma de Barcelona
Spain

Ferreira, Luís
Departamento de Física, IST, Lisbon
Portugal

Frolov, Valery
Copenhagen Univ. Observatory
Denmark

Furtado, João
Departamento de Física, IST, Lisbon
Portugal

Garay, Luis J.
Instituto de Optica, C.S.I.C., Madrid
Spain

Greensite, Jeff
Department of Physics, San Francisco State University
U.S.A.

Henriques, Alfredo B.
Departamento de Física, IST, Lisbon
Portugal

Julve, Jaime
IMAFF, C.S.I.C., Madrid
Spain

Kapetanakis, Dimitris
Physics Dept., Technical University München
Germany

Kiefer, Claus
Institute for Theoretical Physics, University of Zurich
Switzerland

Knutsen, Henning
Hogskolesenteret i Rogaland, Stavanger
Norway

Lavrelashvili, George
Max-Planck-Institut fur Physik, München
Germany

Lemos, José S.
Departamento de Física, IST, Lisbon
Portugal

Levinas, Marcelo
Instituto de Astronomia e Fisica del Espacio, Buenos Aires
Argentina

Llorente Ibañez, Alvaro
LAEFF, C.S.I.C., Madrid
Spain

Lobo Fernández, Faustino
C.S.I.C., Madrid
Spain

Macedo, Paulo Gali
Centro de Astrofísica da Universidade do Porto
Portugal

Marcos, Joaquim
Centro de Física da Matéria Condensada, Lisbon
Portugal

Mendes, Luís
Departamento de Física, IST, Lisbon
Portugal

Pérez-Mercader, Juan
LAEFF, C.S.I.C., Madrid
Spain

Mimoso, José P.
Departamento de Física, Universidade de Lisboa
Portugal

Moniz, Paulo V.
Departamento de Física, Universidade de Lisboa
Portugal

Mottola, Emil
Theoretical Division, Los Alamos National Laboratory
U.S.A.

Mourão, Ana M.
Departamento de Física, IST, Lisbon
Portugal

Mourão, José M.
Departamento de Física, IST, Lisbon
Portugal

Nogueira, Isabel
Departamento de Física, IST, Lisbon
Portugal

Patrício, Anita
Departamento de Física, IST, Lisbon
Portugal

Pessoa-Lopez, Isabel
Dept. Física, Univ. de Lisboa
Portugal

Picken, Roger
Departamento de Matemática, IST, Lisbon
Portugal

Pires, Maria Margarida
Departamento de Física, IST, Lisbon
Portugal

Rodrigo Blanco, Carlos
LAEFF, C.S.I.C., Madrid
Spain

Ruffini, Giulio
Physics Department, UC Davis
U.S.A.

Sá, Nuno
Dept. de Física, Univ. de Lisboa
Portugal

Sá, Paulo
C.F.M.C., Departamento de Física, IST, Lisbon
Portugal

Sahni, Varun
Inter University Centre for Astronomy and Astrophysics, Pune
India

Seixas, João Pedro
Departamento de Física, IST, Lisbon
Portugal

Shabanov, Sergei
Laboratory of Theoretical Physics, JINR, Dubna
Russia

Silva, Amaro Rica da
Departamento de Física, IST, Lisbon
Portugal

Silva, Caroline Santos da
Departamento de Física, Universidade do Porto
Portugal

Silva, João
Departamento de Física, IST, Lisbon
Portugal

Sisterna, Pablo
Departamento de Física, Univ. Nacional de Mar del Plata
Argentina

Solà, Joan
Grup de Física Teórica, Universitat Autònoma de Barcelona
Spain

Solodukhin, Sergei
Theoretical Physics Dept., Physics Faculty, Moscow State University
Russia

Tavares, João N.
Departamento de Matemática Pura, Universidade do Porto
Portugal

Tinebra, Fabrizio
I.N.F.N., Sezione di Bologna
Italy

Umerez, Norberto
Instituto de Astronomia e Fisica del Espacio, Buenos Aires
Argentina

Vaz, Estelita
Departamento de Matemática, Universidade do Minho, Braga
Portugal

Veneziano, Gabriele
CERN, Theory Division, Geneve
Switzerland

Verdaguer, Enric
Grup de Física Teórica, Universitat Autònoma de Barcelona
Spain

Vilar, Luis
Centro Brasileiro de Pesquisas Físicas, Rio de Janeiro
Brasil

Wald, Robert
Enrico Fermi Institute, University of Chicago
U.S.A.

Will, Clifford
Department of Physics, Washington University
U.S.A.

Zakharov, Alexander
Institute of Theoretical and Experimental Physics, Moscow
Russia

www.ingramcontent.com/pod-product-compliance
Lightning Source LLC
LaVergne TN
LVHW082255241026
850861LV00008B/52
* 9 7 8 9 8 1 0 2 1 3 6 9 5 *